# Concise Textbook of Equine Clinical Practice Book 5

This concise, practical text covers the essential information veterinary students need to succeed in equine medicine and surgery, focusing on the nervous system, eyes, cardiovascular disorders and haemolymphatic systems. Written for an international readership, the book conveys the core information in an easily digestible, precise form with extensive use of bullet points, tables, flow charts, diagrams, lists, protocols and extensive illustrations.

Part of a five-book series that extracts and updates key information from Munroe's *Equine Surgery, Reproduction and Medicine*, Second Edition, the book distils best practice in a logical straightforward clinical-based approach. It details clinical anatomy, physical clinical examination techniques, diagnostic techniques and normal parameters, emphasising the things regularly available to general practitioners with minimal information of advanced techniques.

- The nervous system section discusses in detail the neurological exam, mentions relevant diagnostic tests and divides the diseases of the system into congenital, infectious, traumatic, toxin-induced and miscellaneous causes.
- In the eye section, the relevant anatomy and the examination of the eye are followed by discussion of useful diagnostic tests and procedures. Diseases of the eye are discussed under the headings of congenital/neonatal, neoplastic, infectious and inflammatory disorders, neurological based and parasitic problems.
- The cardiological examination and relevant diagnostic tests are followed by sections on arrhythmias, congenital and acquired cardiac diseases, pericardial disease and miscellaneous cardiovascular disease.
- Finally, the haemolymphatic section discusses in detail relevant diagnostic approaches and laboratory aids, followed by sections on anaemia, haemostasis disorders, infections and neoplasia.

Ideal for veterinary students and nurses on clinical placements with horses, as well as practitioners needing a quick reference 'on the ground'.

# Concise Textbook of Equine Clinical Practice Book 5

## Nervous System, Eyes, Cardiovascular Disorders and Haemolymphatic System


Erin Beasley

François-René Bertin

Edited By

Graham Munroe

CRC Press is an imprint of the Taylor & Francis Group, an **informa** business

First edition published 2024
by CRC Press
2385 NW Executive Center Drive, Suite 320, Boca Raton, FL 33431

and by CRC Press
2385 NW Executive Center Drive, Suite 320, Boca Raton, FL 33431



CRC Press is an imprint of Taylor & Francis Group, LLC



ISBN: 9781032588766 (hbk)
ISBN: 9781032586007 (pbk)
ISBN: 9781003451921 (ebk)

DOI: 10.1201/9781003451921

Typeset in Sabon
by Evolution Design & Digital Ltd (Kent)

Printed in the UK by Severn, Gloucester on responsibly sourced paper

# Table of Contents

# Preface

A vast array of clinical equine veterinary information is available for the under- and postgraduate veterinarian and veterinary nurse to peruse. This is contained in textbooks, both general and specialised, and increasingly online at websites of varying quality and trustworthiness. It is easy for the veterinary student or nurse, recent graduate and busy general or equine practitioner to become overwhelmed and confused by this diverse range of information. Often what is required, particularly in the clinical situation, is a distillation of the essential knowledge and best practice required to treat a horse in the most suitable way. This concise, practical text is designed to provide the essential information needed to understand and treat clinical cases in equine practice.

This book focuses on medical and surgical conditions of the nervous system, the eye, the cardiovascular system and the haemolymphatic system. It is part of a five-book series, which between them will cover all the areas of equine clinical practice. The information is extracted and updated from *Equine Clinical Medicine, Surgery and Reproduction* (Second Edition), which was published in 2020. It is written for an international readership and is designed to convey the core, best-practice information in an easily digested, quick reference form using bullet points, lists, tables, flow charts, diagrams, protocols and extensive illustrations and photographs.

The nervous system section discusses in detail the neurological exam, mentions relevant diagnostic tests, and divides the diseases of the system into congenital, infectious, traumatic, toxin-induced and miscellaneous causes. In the eye section, the relevant anatomy and the examination of the eye are followed by discussion of useful diagnostic tests and procedures. Diseases of the eye are discussed under the headings of congenital/neonatal, neoplastic, infectious and inflammatory disorders, neurological based and parasitic problems. The cardiological examination and relevant diagnostic tests are followed by sections on arrhythmias, congenital and acquired cardiac diseases, pericardial disease and miscellaneous cardiovascular disease. Finally, the haemolymphatic section discusses in detail relevant diagnostic approaches and laboratory aids, followed by sections on anaemia, haemostasis disorders, infections and neoplasia. All the material is approached in the same logical, straightforward, clinical-based way. There are details of relevant clinical anatomy, physical clinical examination techniques, normal parameters, aetiology/pathophysiology, clinical examination findings, differential diagnosis, diagnostic techniques, management and treatment and prognosis. The emphasis is on information tailored to general equine clinicians with just enough on advanced techniques to make the practitioner aware of what is available elsewhere.

The intention of this series of books is for them to be used on a day-to-day basis in clinical practice by student and graduate veterinarians and nurses. The spiral binding format allows them to lie open on a surface near to the patient, readily available to the veterinary student or practitioner whilst looking at, or treating, a clinical case.

# About the Authors

**Erin Beasley** is a Clinical Associate Professor in Large Animal Internal Medicine at the University of Georgia in Athens, GA, USA. Her doctoral research was in cardiovascular function in horses with gastrointestinal disease. Equine cardiology remains a special interest, although all things related to internal medicine make each day in this profession exciting and challenging.

**François-René Bertin** graduated with a DVM from the National Veterinary School of Nantes (France) and completed an internship at the National Veterinary School of Alfort (France). He trained in Equine Internal Medicine at Purdue University (USA) and became a diplomate of the American College of Veterinary Internal Medicine (ACVIM). Dr Bertin completed his PhD at McGill University (Canada). He joined The University of Queensland (UQ) in 2016 and has authored several research articles, book chapters and the first textbook on the diagnosis and management of equine endocrinopathies. He leads the Equine Endocrinology research group at UQ and is a member of international expert panels to elaborate guidelines for the management of insulin dysregulation and PPID.

**Graham Munroe** qualified from the University of Bristol with honours in 1979. He spent 9 years in equine practice in Wendover, Newmarket, Arundel and Oxfordshire, and a stud season in New Zealand. He gained a certificate in equine orthopaedics and a diploma in equine stud medicine from the RCVS whilst in practice. He joined Glasgow University Veterinary School in 1988 as a lecturer and then moved to Edinburgh Veterinary School as a Senior Lecturer in Large Animal Surgery from 1994 to 1997. He obtained by examination a Fellowship of the RCVS in 1994 and Diplomate of the ECVS in 1997. He was awarded a PhD in 1994 for a study in neonatal ophthalmology. He has been visiting equine surgeon at the University of Cambridge Veterinary School, University of Bristol Veterinary School and Helsingborg Hospital, Sweden. He was Team Veterinary Surgeon for British Driving Teams, 1994–2001; British Dressage Teams, 2001–2002; and the British Vaulting Team in 2002. He was FEI Veterinary Delegate at the Athens 2004 Olympics. He currently works in private referral surgical practice, mainly in orthopaedics. He has published over 60 papers and book chapters.

# Abbreviations

| | |
|---|---|
| 5-FU | 5-fluorouracil |
| A | Aortic valve |
| AA | Aplastic anaemia |
| ACE | Angiotensin-converting enzyme |
| AF | Atrial fibrillation |
| AID | Anaemia of inflammatory or chronic disease |
| AIVR | Accelerated idioventricular rhythm |
| ALL | Acute lymphocytic leukaemia |
| AO | Atlanto-occipital |
| APCs | Atrial premature contractions |
| APTT | Activated partial thromboplastin time |
| ASD | Atrial septal defect |
| AV | Atrioventricular |
| B | Basophils |
| BAB | Blood–aqueous barrier |
| BCG | Bacillus Calmette–Guérin |
| BOB | Blood-ocular barrier |
| BoNTs | Botulinum neurotoxins |
| CLL | Chronic lymphocytic leukaemia |
| CN | Cranial nerve examination |
| CNS | Central nervous system |
| CSF | Cerebrospinal fluid |
| CSNB | Congenital stationary night blindness |
| CT | Computed tomography |
| CVSM | Cervical vertebral stenotic myelopathy |
| DIC | Disseminated intravascular coagulation |
| DMSO | Dimethyl sulphoxide |
| E | Eosinophils |
| EDM | Equine degenerative myeloencephalopathy |
| EDTA | Ethylenediaminetetraacetic acid |
| EEE | Eastern equine encephalitis |
| EEG | Electroencephalography |
| EHM | Equine herpesvirus myeloencephalopathy |
| EIA | Equine infectious anaemia |
| EIAV | Equine infectious anaemia virus |
| ELISA | Enzyme-linked immunosorbent assay |
| EMG | Electromyography |
| EMND | Equine motor neuron disease |
| EPM | Equine protozoal myeloencephalitis |
| ERG | Electroretinography |
| ERU | Equine recurrent uveitis |
| EVA | Equine viral arteritis |
| F | Facial artery |
| FNA | Fine-needle aspirate |
| GA | General anaesthesia |

| | |
|---|---|
| GABA | Gamma-aminobutyric acid |
| GI | Gastrointestinal |
| ICA | Iridocorneal angle |
| IgG | Immunoglobulin G |
| IgM | Immunoglobulin M |
| IMHA | Immune-mediated haemolytic anaemia |
| IMMK | Immune-mediated keratitis |
| IMTP | Immune-mediated thrombocytopenia |
| IOP | Intraocular pressure |
| KCS | Keratoconjunctivitis sicca |
| L | Lymphocytes |
| LA | Local anaesthetic |
| LMN | Lower motor neuron |
| LS | Lumbosacral |
| LSA | Lymposarcoma |
| M | Monocytes |
| MCH | Mean corpuscular haemoglobin |
| MCHC | Mean corpuscular haemoglobin concentration |
| MR | Mitral regurgitation |
| MRI | Magnetic resonance imaging |
| MRLS | Mare reproductive loss syndrome |
| MV | Mitral valve |
| NEE | Near Eastern encephalitis |
| NLS | Nasolacrimal system |
| NSAIDs | Non-steroidal anti-inflammatory drugs |
| OA | Osteoarthritis |
| OAAM | Occipitoatlantoaxial malformation |
| ONH | Optic nerve head |
| PCR | Polymerase chain reaction |
| PCV | Packed cell volume |
| PDA | Patent ductus arteriosus |
| PIFM | Pre-iridial fibrovascular membrane |
| PLR | Pupillary light response |
| PMI | Point of maximum intensity |
| PMNs | Polymorphonuclear cells |
| PN | Peripheral nerves |
| PRCA | Pure red cell aplasia |
| PSSM | Polysaccharide storage myopathy |
| PT | Prothrombin time |
| RBC | Red blood cell |
| REM | Rapid eye movement |
| RPE | Retinal pigment epithelium |
| RT-PCR | Reverse transcription polymerase chain reaction |
| SCCs | Squamous cell carcinomas |
| SPL | Subpalpebral lavage system |
| STT | Schirmer tear test |
| TeNT | Tetanospasmin |
| THO | Temporohyoid osteoarthropathy |
| TIBC | Total iron-binding capacity |
| TMH | Trigeminal-mediated headshaking |
| TR | Tricuspid regurgitation |
| TV | Tricuspid valve |
| TVEC | Transvenous electrical cardioversion |
| UMN | Upper motor neuron |

| | |
|---|---|
| UV | Ultraviolet |
| VEE | Venezuelan equine encephalitis |
| VPCs | Premature ventricular complexes |
| VSD | Ventricular septal defect |
| VT | Ventricular tachycardia |
| WBC | White blood cell |
| WEE | Western equine encephalitis |
| WNV | West Nile virus |

CHAPTER 1

# Nervous System

## NEUROLOGICAL EXAMINATION

### Introduction

- thorough neurological examination is critical to reaching a diagnosis in equine neurological disease.
- **quickly carried out in its entirety in the field without any specialised equipment.**
- logical routine developed so that no phase of the examination is omitted.
- clinical presentation of neurological disease in the horse is often determined by the anatomical site of the problem rather than a specific cause:
  - neurological assessment designed to identify which sections of the nervous system have any dysfunction.
- neurological examination helps to answer two questions:
  - are clinical findings consistent with a neurological deficit?
  - can the signs be localised to a single or multiple neuroanatomical location/s?
- once those questions are answered:
  - list of differential diagnoses and additional diagnostic tests can be considered.

### History

- age: congenital *vs* degenerative lesions.
- breed: genetic predisposition (Table 1.1).
- sex.
- dam/sire history:
  - hereditary conditions.
- diet: exposure to toxins, including at pasture.
- environment and management:
  - transport.
  - new animals on the premises.
- recent illness and drug administration:
  - immunisation records.
- description of reported signs:
  - some disorders not likely to be apparent at examination.
  - video recordings often useful.

### Observation of the patient

- **Distant examination:**
  - horse in its own environment.
  - preliminary step to assess cerebrum, cerebellum and need for safety measures.
- **Mental state and behaviour:**
  - level of consciousness and awareness:
    - **decreased:**
      - depression.
      - lethargy.
      - lack of response to environmental stimuli.
      - somnolence.
      - stupor or semi-comatose or comatose states.
    - **increased:**
      - aggression.
      - restlessness.
      - compulsive behaviours.
      - usually, result of damage to the cerebral cortex and ascending reticular activating system (ARAS) in the brainstem.
- **Head posture and coordination:**
  - **Head turn:**
    - lateral deviation of the whole head with flexion of the neck.
    - ipsilateral cerebral lesion.
  - **Head tilt:**
    - lateral deviation of the poll (Fig. 1.1).
    - described by direction of poll deviation.
    - ipsilateral vestibular lesion.

DOI: 10.1201/9781003451921-1

**TABLE 1.1** Common breed predispositions for neurological diseases in horses

| BREED(S) | DISEASE | PRESENTING SIGNS | AETIOLOGY AND DIAGNOSTIC TESTING |
|---|---|---|---|
| Arabian | Juvenile idiopathic epilepsy | Self-limiting idiopathic epilepsy of Arabian foals (median 2 months of age) with generalised tonic–clonic seizures | Unknown, but possibly autosomal dominant mode of inheritance |
| Arabian | Lavender foal syndrome | Congenital episodic tetany where recumbent foals develop extreme rigidity and opisthotonos when stimulated or trying to right themselves. Never capable of standing unassisted | Autosomal recessive gene (MYO5A) linked to coat colour (lavender or silver hue to coat). Genetic testing available commercially |
| Arabian (also seen in Welsh, Trakehner and Bashkir Curly horses) | Cerebellar abiotrophy | Signs appear from birth or within a few months and include head and neck sway, loss of menace response, spasticity, rearing with extended forelimbs when handled and an intention tremor of the head | Post-natal degeneration of Purkinje cells due to an autosomal recessive genetic mutation of the TOE1/MUTYH gene. Genetic testing available commercially |
| Arabian (Appaloosa, Friesian, Miniature horse, Quarter horse are also overrepresented) | Occipitoatlantoaxial malformation (OAAM) | Familial OAAM of the Arabian horse is associated with occipitalisation of the atlas and atlantalisation of the axis, which are fused together with connective tissue. Signs reflect compression of the cervical spinal cord with severe tetraparesis, ataxia and extended head and neck posture | Deletion of a focus near HOXD3 might be responsible for Arabian OAAM |
| Friesian (Thoroughbred, Miniature horse and Standardbred also overrepresented) | Hydrocephalus | Stillborn or weak foals that die shortly after birth with gross enlargement of, or dome-shape to, the head. Dystocia might have been present | Distension of the ventricular system in the brain due to jugular foramen narrowing. Autosomal recessive inheritance associated with a nonsense mutation in B3GALNT2 |
| Miniature horse | Narcolepsy | Idiopathic narcolepsy | Presumed inherited characteristic amongst miniature horses but unknown genetic basis at this time |

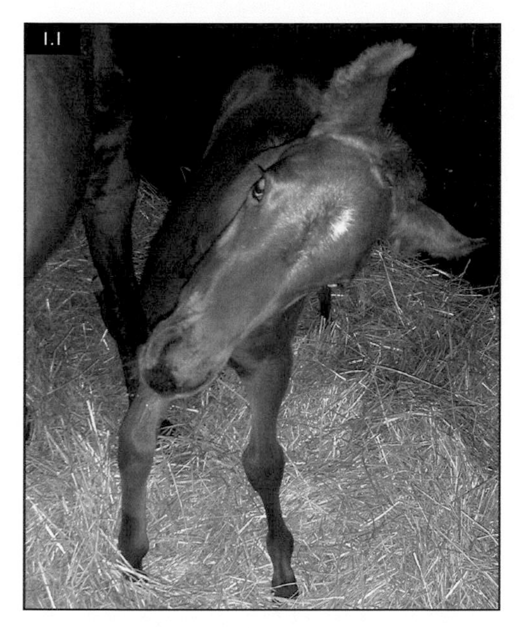

FIG. 1.1 Three-week-old foal with a left-sided head tilt secondary to trauma. The foal was diagnosed with vestibular disease secondary to haemorrhage within the petrous temporal bone.

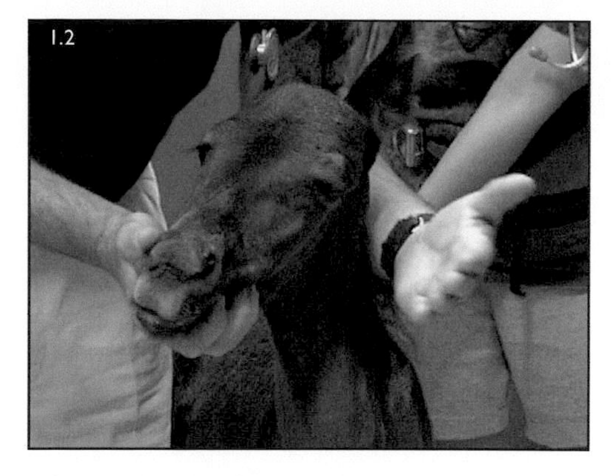

FIG. 1.2 Two-week-old foal with dementia and normal menace response in the left eye. (Photo courtesy FT Bain.)

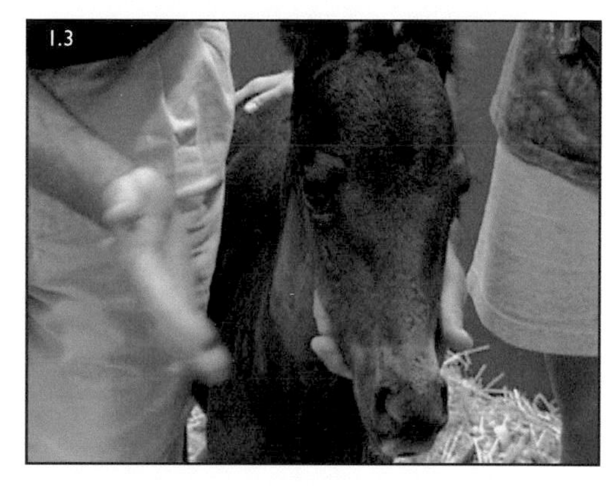

FIG. 1.3 Same foal as in Fig. 1.2 with no menace response in the right eye. The foal was also noted to circle continually to the left. The foal was diagnosed with a left-sided cortical lesion. (Photo courtesy FT Bain.)

## Physical examination

- **Complete examination of the body:**
  - identify disease of other body systems that may account for neurological signs.
  - identify signs possibly associated with nervous system abnormality:
    - patchy sweating pattern.
    - muscular asymmetry or fasciculations.
  - stance and postural abnormalities.
  - identify need for other urgent medical treatment for other abnormalities.

## Cranial nerve (CN) examination

- detailed assessment of each of 12 pairs of cranial nerves.
- abnormalities suggest disease near or within the brainstem:
  - anywhere along the nerve, its nucleus or the body part innervated.

### Olfactory nerve (CN I)

- ability to smell is almost impossible to assess in horses:
  - damage to the nerve is rare.

### Optic nerve (CN II)

- part of the visual pathway.
- nerve and vision can be assessed in several ways:
  - ability to navigate safely in a usual environment or a constructed obstacle course.
  - **Menace response** (Table 1.2):
    - hand flicking towards each eye in turn causing a consistent blink +/– head jerking away (Figs. 1.2, 1.3).
    - false positives if the hand movement:
      - too vigorous and creates an air current detected by the cornea.
      - touches the whiskers or eyelashes.

- false negatives can occur:
  - neonates have no response.
  - horse is depressed or stoic.
  - test repeated too many times in quick succession.
  - horse has a facial nerve (CN VII) lesion preventing blinking.

## Oculomotor nerve (CN III)

- ability to move the eye and constrict the pupil:
  - constrictor muscles of iris innervated by parasympathetic fibres of CN III:
    - pupil constricts in response to light:
    - **pupillary light response** (PLR).
  - dilator muscles controlled by sympathetic fibres from cranial cervical ganglion:
    - pupil dilates in response to fear or excitement.
- PLR:
  - check no physical obstructions to inhibit iris movement, such as synechiae.
  - shine a bright light into each eye in turn in a dark area:
    - observe changes in the pupil size.
    - immediate direct (ipsilateral) and consensual (contralateral) constriction of the pupil is normal.
    - widely dilated (mydriatic) pupil not responding to direct light; in an eye with normal vision, indicates CN III lesion.
    - normal PLR with no menace response indicates:
      - normal CN II and CN III.
      - thalamic or cerebral lesion (central blindness).
    - no PLR in either eye with no menace response is consistent with:
      - retrobulbar lesion affecting CN II and CN III.

## Oculomotor (CN III), Trochlear (CN IV) and Abducens (CN VI) nerves

- responsible for the position and movement of the eye:
  - dysfunction results in persistent deviation of the globe (strabismus).
- close proximity within the skull often means damaged simultaneously.

**TABLE 1.2** Causes of a negative menace response

| CAUSE OF NEGATIVE MENACE RESPONSE | ADDITIONAL DIAGNOSTICS | INTERPRETATION OF FURTHER TESTS |
|---|---|---|
| True visual deficit | • Obstacle course<br>• Ophthalmological examination | • Unilateral central blindness is usually in the contralateral eye to the lesion because of the almost complete cross-over at the optic chiasm in the horse<br>• Central blindness is usually bilateral, due to cerebral swelling<br>• Eye and optic nerve disease will cause ipsilateral blindness and abnormalities of the pupillary light response<br>• Lesions of the optic tracts and lateral geniculate nucleus within the brain will cause contralateral blindness |
| Facial nerve paralysis (inability to blink) | • Rest of cranial nerve examination<br>• Will usually have head jerk even if blink is absent | Usually there is additional evidence of facial paralysis, such as facial drooping on the same side and muzzle pulled to the contralateral side |
| Cerebellar disease | • Normal vision but no menace response<br>• Other signs of cerebellar disease (ataxia, intention tremor) | Unknown mechanism possibly involving the upper motor neuron to the eye |

- assess function by moving the horse's head up and down and from side to side:
  - physiological nystagmus with a slow phase of eyeball rotation away from the direction of head movement, and then a rapid 'catch-up' in the direction of the head movement is normal.
  - lateral and slightly ventral strabismus is consistent with CN III lesions.
  - dorsomedial strabismus is consistent with CN IV lesions.
  - medial strabismus is consistent with CN VI lesions.
  - fixed strabismus cannot be moved out of this position
  - rarely seen alone.

## Trigeminal nerve (CN V)

- responsible for facial sensation and movement of the muscles involved in mastication.
- sensory function assessed by:
  - stimulate the skin at the ears, medial/lateral canthi of the eyes, nostrils and lips:
    - watch for a corresponding reflex in that part of the head.
    - normal reflex requires an intact CN V and facial nerve (CN VII).
- motor function is assessed by:
  - palpate the jaw and cheeks bilaterally.
  - lesion of mandibular branch CN V:
    - dropped jaw with a protruding tongue that retracts when stimulated.
    - weak jaw, difficulty chewing and muscle atrophy.

## Facial nerve (CN VII)

- motor nerve for the muscles of facial expression.
- paralysis:
  - drooping of eyelid and ear on the side of the lesion.
  - retraction of the muzzle away from the side of the lesion (Fig. 1.4).
- other clinical presentations include:
  - corneal ulceration due to an inability to close the eyelids.
  - inspiratory stertor at exercise due to nostril collapse.
  - difficulty in prehension of food due to lip paralysis.

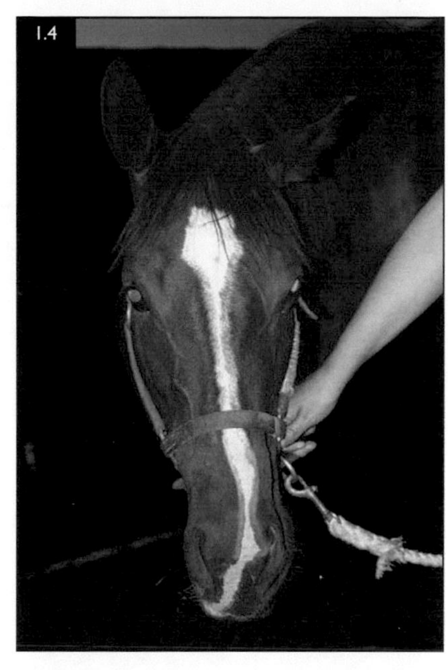


**FIG. 1.4** Left-sided facial nerve injury. Note how the muzzle is deviated to the unaffected side (right).

## Vestibulocochlear nerve (CN VIII)

- **Cochlear** portion is responsible for hearing:
  - bilateral middle ear disease causes deafness:
    - picked up by assessing response to a handclap.
    - unilateral disease is hard to detect.
- **Vestibular** system comprises the sensory structures in the inner ear, the vestibular branch of CN VIII and central components in the cerebellum and medulla:
  - responsible for maintaining orientation of eyes, head, neck and limbs.
  - loss of function results in loss of ipsilateral antigravitational tone:
    - staggering
    - circling.
    - falling over.
    - head tilt (Fig. 1.5).
    - spontaneous nystagmus.
  - CN VIII damage usually results in horizontal nystagmus:
    - fast phase away from lesion and direction of head tilt.

- visual cues usually compensate over time reducing severity of vestibular signs:
  - blindfolding (with care) exacerbates the signs (Fig. 1.6).

## Glossopharyngeal nerve (CN IX)

- responsible for afferent innervation to the pharynx.
- motor lesion results in dysphagia (Table 1.3) – partial if unilateral problem.
- sensory innervation of pharynx assessed by:
  - presence of 'gag reflex' when nasogastric tube is passed, or by touching the wall of the pharynx with an endoscope.

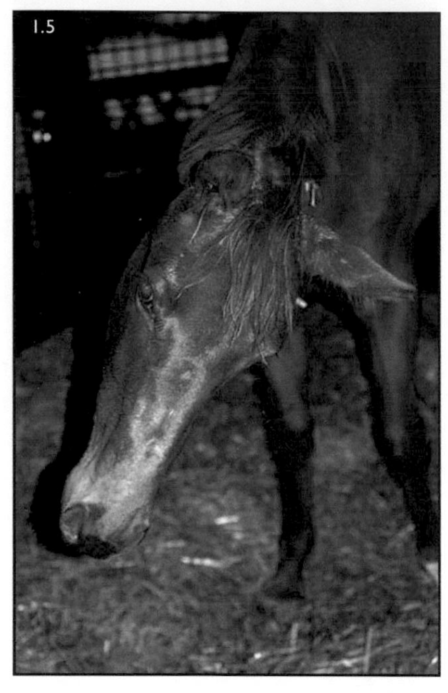

**FIG. 1.5** Horse with head tilt to the left secondary to trauma on the left aspect of the calvarium.

**FIG. 1.6** Blindfolding of a horse with exaggeration of vestibular disease. Note the leaning to the right.

**TABLE 1.3** Neurological causes of dysphagia

| SITE OF ABNORMALITY | ASPECT OF EATING DISRUPTED | TEST TO DIFFERENTIATE |
|---|---|---|
| Trigeminal nerve (CN V) | Muscles of mastication | Facial reflexes |
| Facial nerve (CN VII) | Lip muscle function | Other evidence of facial paralysis |
| Glossopharyngeal nerve (CN IX) | Pharyngeal sensation | 'Gag' reflex |
| Vagus nerve (X) | Pharyngeal musculature | Endoscopy of upper respiratory tract |
| Hypoglossal nerve (CN XII) | Tongue musculature | Check the tongue for atrophy, pull tongue and check retraction is sufficient |
| Supranuclear palsy (cerebral) | Severe lesions might depress or prevent the voluntary effort required for swallowing | Cranial nerve tests are normal |
| Neuromuscular paralysis | Lower motor neuron (LMN) (botulism) | Accompanying additional signs of LMN disease (trembling, weakness, ptosis) |

## Vagus nerve (CN X)

- responsible for afferent and efferent innervation to larynx:
  - plus, efferent supply to pharyngeal musculature.
- most recognised dysfunction is peripheral neuropathy of left recurrent laryngeal nerve:
  - laryngeal hemiplegia and 'roaring'.
  - best detected with endoscopy of the upper respiratory tract.

## Accessory nerve (CN XI)

- innervation of the trapezius and sternocephalicus muscles.
- injuries are rare:
  - paralysis +/– muscle atrophy of the above muscles.

## Hypoglossal nerve (CN XII)

- innervates the tongue and damage leads to paralysis.
- unilateral dysfunction results in tongue muscle atrophy:
  - weak retraction of the tongue when pulled out of side of the mouth.
- bilateral lesions result in tongue protrusion and dysphagia:
  - inability to withdraw the tongue back into the mouth when protruded.

## Syndromes of cranial nerve disease

- cranial nerve damage can result from brainstem nuclei or peripheral nerve injury:
  - brainstem lesions result in weakness or depression and lesions in multiple CNs.
  - peripheral nerve lesions might affect several nerves owing to their juxtaposition:
    - CNs III, IV and VI by fractures at the base of the skull.
    - CNs VII and VIII with inner ear disease or temporohyoid osteopathy.
    - CNs IX and X with guttural pouch disease.
    - CNs II and V with sphenopalatine sinus disease.
  - Horner's syndrome is the most well-known multiple CN syndrome:
    - damage to sympathetic supply to:
      - ocular structures.
      - other structures in the head and neck.
    - clinical signs are ipsilateral to the lesion and include:
      - excessive sweating of the head and neck.
      - drooping of the eyelid and constriction of the pupil.
      - protrusion of the nictitating membrane and enophthalmos.
    - common causes include:
      - damage to the brachial plexus.
      - space-occupying lesions of the neck.
      - perivascular injection near the jugular.
      - guttural pouch disease.
      - basilar skull fractures.
      - retrobulbar masses.
    - rare causes include:
      - damage to the cervical and cranial thoracic spinal cord.

## Neck and trunk examination

- examine neck in detail:
  - symmetry, swellings, localised sweating, and normal range of motion (Fig. 1.7).
- using a blunt instrument placed on the skin, check for:
  - skin sensation over the whole horse (Fig. 1.8).
  - spinal reflexes tested in the standing horse:
    - generally, conscious perception and a skin twitch.

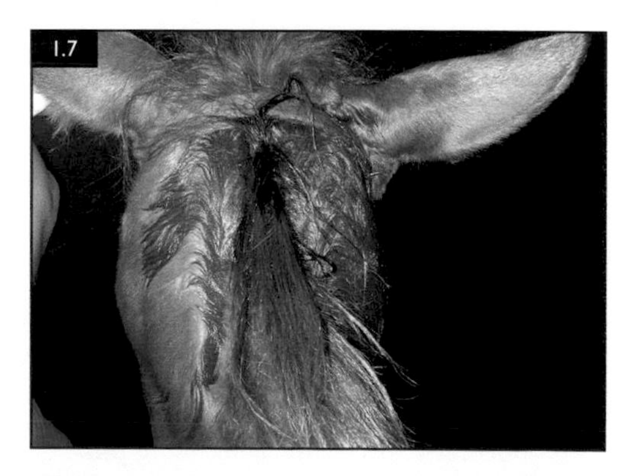


**FIG. 1.7** A foal with a fracture of the atlas. Note the localised sweating at the region of the injury.

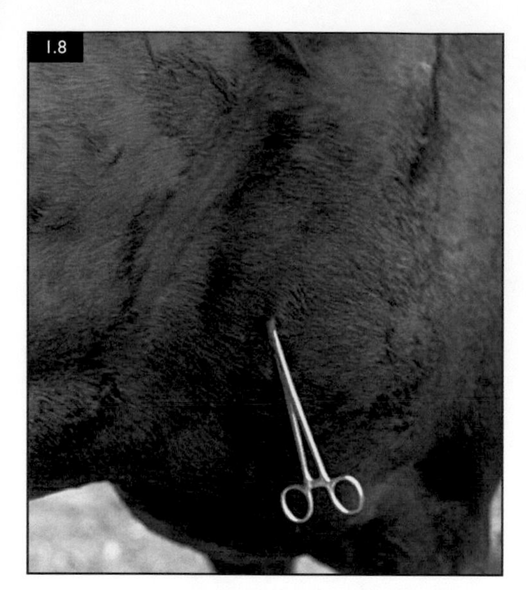


**FIG. 1.8** Fracture of C7 and T1, with hypoalgesia.

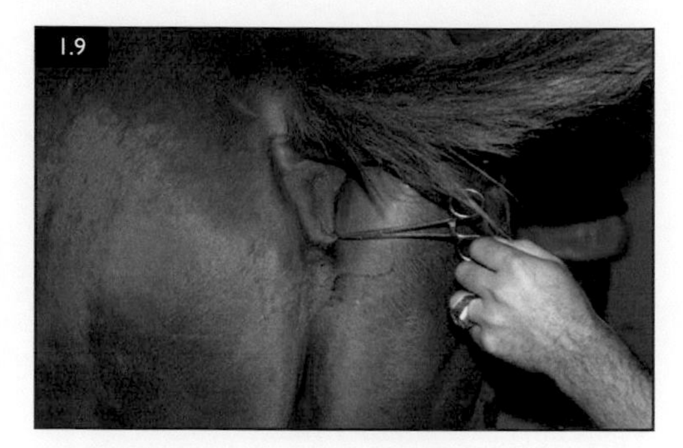


**FIG. 1.9** A lack of anal tone is noted in a horse suffering from equine herpesvirus-1 (EHV-1) myelitis.

        - performed on the neck, there should be a cervicofacial reflex:
            - involves an ipsilateral ear twitch, eyelid closure and contracture of the lip commissure.
        - over the thorax, should be a response in the cutaneous muscle.
        - extension and flexion of the thoracolumbar spinal cord should also be tested.
    - tests aim to check the function of segments of the spinal cord grey matter and peripheral nerve function.
    - **most spinal reflexes are impossible to test in the adult standing horse.**
    - some reflexes can be assessed in foals or recumbent horses:
        - patellar, gastrocnemius and triceps.
        - upper motor neuron (UMN) – hyperresponsiveness.
        - lower motor neuron (LMN) – hyporesponsiveness.
- tail tone can be assessed for paralysis by manipulation:
    - varies in strength depending on horse's disposition.
- perineal reflex tested by prodding or pinching the perineum:
    - normally results in clamping of tail and contraction of anal sphincter (Fig. 1.9).

## Gait and posture

- systematic assessment of the horse standing, walking and trotting:
    - determine whether the horse's gait is normal or abnormal:
        - any apparent musculoskeletal disease.
        - evidence of weakness, pain or irregular gait (ataxia).
- **Weakness or paresis:**
    - decreased or absent voluntary movements:
        - horse drags its limbs
        - wears the dorsal aspect of the hooves.
        - low arc of the swing phase of their stride.
    - severe weakness:
        - all four limbs, without evidence of ataxia or spasticity, suggestive of neuromuscular disease.
        - one limb suggestive of a peripheral neuropathy or muscle lesion.
- **Ataxia:**
    - **Proprioception** – control of normal body movement and spatial orientation:
        - position of the body and spontaneous correction of abnormal positions.
    - **deficiency of this system leads to ataxia:**
        - poor coordination of movement:
            - **inconsistent and unpredictable.**
        - lameness causes consistent and predictable pattern of movement.

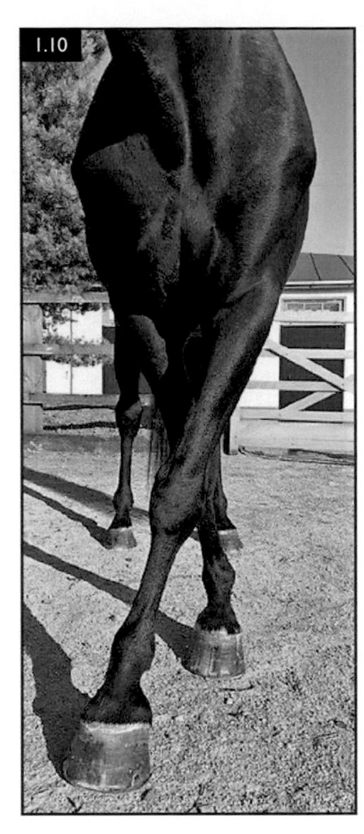

FIG. 1.10 Ataxia secondary to proprioceptive deficits in both forelimbs. This animal was diagnosed with cervical vertebral compressive myelopathy.

FIG. 1.11 Tail pull to assess weakness in the hindlimbs.

- ♦ inappropriate placement of limbs, trunk, head and neck at rest and in motion (Fig. 1.10).
- Assessment of gait and posture:
  - begin with assessment of posture:
    - ♦ severe proprioceptive deficits wide-based stance and inappropriate placement of the limbs.
    - ♦ tested by placing a foot in abduction or adduction and evaluating the time to correction by the horse.
    - ♦ false positives possible in stoic, dull or well-behaved horses.
    - ♦ false negatives are possible with excessive abduction.
  - range of movement of the neck tested at this point with 'carrot stretches':
    - ♦ reduced range in osteoarthritis (OA) of articular facet joints of cervical vertebrae.
  - observe all muscles of horse for evidence of tremors:
    - ♦ at rest – sign of muscle weakness (UMN disease).
  - check for resistance while the horse is standing square and being pushed:
    - ♦ normal horses:
      - – react and push back against the clinician's steady, firm pressure on the shoulder or hip.
      - – pull back against the standing tail pull test.
    - ♦ easily pushed or pulled over:
      - – suggestive of weakness.
- Assessment of horse in motion:
  - exacerbate neurological deficits by conducting increasingly challenging tasks.
  - obvious deficits should curtail the tests due to danger to horse and handler.
  - observe the horse walking +/– trotting in a straight line:
    - ♦ stumbling, missteps, knuckling, toe-dragging or low arc of cranial phase of stride.
  - Hypermetria:
    - ♦ exaggerated movement or range of motion (over-reaching).
  - observe at walk in a straight line with the head elevated:
    - ♦ possible exacerbation of proprioceptive deficits and hypermetria.
  - pull on the tail while the horse is moving to assess weakness (Fig. 1.11).
  - turn the horse in small circles with the horse's head on a relatively free rein:
    - ♦ important that horse keeps moving forward and does not turn too tightly.
    - ♦ exacerbates proprioceptive deficits:
      - – stumbling.
      - – wide circumduction of the outside hindlimb.

- forelimb deficits may pivot on forelimbs rather than crossing over.
- leave limbs in abnormal positions when asked to stop suddenly.
- observe the horse walk in a tight serpentine:
  - increased speed may make the test more sensitive.
  - stumbling, pelvic swaying, interference, missteps and circumduction.
- observe horse walk backwards:
  - normal horses walk backward from one diagonal to the other:
    - ataxic horses break the pattern.
  - severely ataxic horses may resist or fall.
- walk the horse up and down kerbstones:
  - stumbling and missteps.
- blindfolding can exacerbate low-grade neurological deficits at walk:
  - removes visual clues from horse.
  - **can lead to marked increase in ataxia and possible danger to horse/handler.**

- Grading gait abnormalities:
  - several grading systems exist but the most used is the Mayhew scale intended for assessment of cervical vertebral stenotic myelopathy.
  - objective measurement of the severity of the neurological deficits.
  - useful for professional communication and interpretation of follow-up examinations (Table 1.4).

**TABLE 1.4** Grading neurological gait deficits

| GRADE | DESCRIPTION |
|---|---|
| 0 | No gait deficits, normal strength and coordination |
| 1 | Deficit not detectable during walk and inconsistent with special tests |
| 2 | Deficit inconsistent during walk and exaggerated by special tests |
| 3 | Deficit consistent during walk and horse might fall when asked to perform special tests |
| 4 | Spontaneous stumbling, tripping or falling while standing or during walk or trot |
| 5 | Recumbent horse that is unable to rise |

## Neurolocalisation of spinal lesions

- spinal cord divided into 4 regions based on the clinical signs that are exhibited when they are damaged:
  - **Lumbosacral region** (L4–S3):
    - normal forelimbs.
    - weak and ataxic hindlimbs including short strides, dragging of the hindlimbs and, in severe cases, paraplegia (dog-sitting).
    - decreased tail, anal and hindlimb tone, and atrophy of the pelvic muscles.
    - +/– urinary incontinence and hypoalgesia/analgesia of tail, anus and hindlimbs.
  - **Thoracolumbar region** (T3–L3):
    - normal forelimbs.
    - hindlimb ataxia and/or spasticity.
    - normal tail and anal tone, with no neurogenic atrophy.
    - possible hypoalgesia or analgesia caudal to the cranial border of the lesion.
    - +/– focal sweating and urinary incontinence.
  - **Caudocervical region** (C7–T2):
    - forelimb weakness.
    - hindlimb spasticity.
    - ataxia and paresis all four limbs.
    - normal to increased hindlimb reflexes and tone.
    - decreased to absent forelimb reflexes and tone.
  - **Craniocervical region** (C1–C6):
    - ataxia and paresis all four limbs but one grade less severe in the forelimb.
    - normal to increased reflexes in all limbs.
    - possible hypoalgesia or analgesia caudal to the cranial border of the lesion.

## Diagnostic tests

### Cerebrospinal fluid (CSF) collection

- analysed for cellular and biochemical constituents, antibodies and infectious agents.

#### Atlanto-occipital (AO) CSF collection

- cerebellomedullary cistern at the AO site (Fig. 1.12).
- under general anaesthesia with the horse in lateral recumbency:
  - recovery from general anaesthesia can be a challenge for ataxic horses.
  - recently, standing collection under sedation and local anaesthetic (LA) has been reported.
- not performed if increased intracranial pressure (e.g. a space-occupying lesion) suspected:
  - protocol for collection:
    - sterile preparation of the skin.
    - insert 18–20 gauge 9-cm (3.5-inch) spinal needle at intersection of lines:
      - caudally in the dorsal plane from external occipital protuberance.
      - transversely across the cranial border of the wings of the atlas.
    - needle should be parallel with the ramus of the mandible:
      - inserted steadily with stylet removed to check that CSF appears at the hub of the needle.
    - once CSF appears at the needle hub, a sample is obtained by flow.
    - extension with a sterile syringe can be used so no inadvertent movement of the spinal needle during sample collection.
    - loss of resistance or 'pop' is not reliably detected when the subarachnoid space is penetrated.
    - blood contamination is possible (many large veins at the AO site):
      - amount of blood decreases as more CSF flows.
      - use several sample vials rather than one large one.
      - submit the least contaminated for analysis.
      - haemorrhage part of the disease process:
        - amount of blood visible will not decrease.
      - sample of CSF is centrifuged, and supernatant is clear:
        - blood is contamination from the collection procedure.
      - supernatant is xanthochromic (yellow):
        - suggests that haemorrhage is part of the disease process.

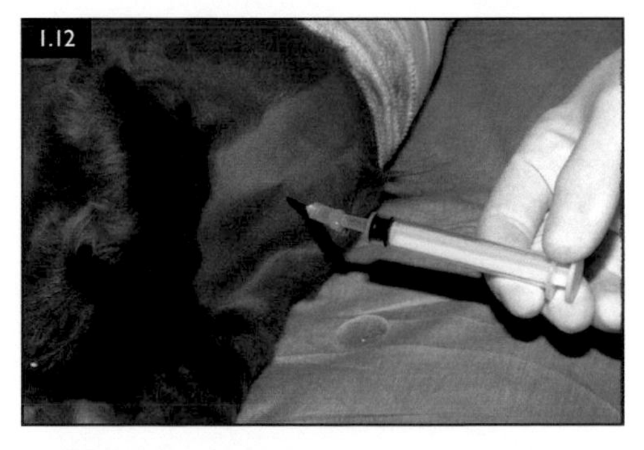

**FIG. 1.12** Location for atlanto-occipital CSF collection.

#### Lumbosacral (LS) CSF collection

- obtained from the subarachnoid space at the LS articulation in the horse (Fig. 1.13).
- horse restrained in stocks plus LA skin infiltration:
  - sedation may cause the horse to shift its weight onto one hindlimb, making the procedure more challenging.
  - sedation often necessary in excitable or unpredictable horses.

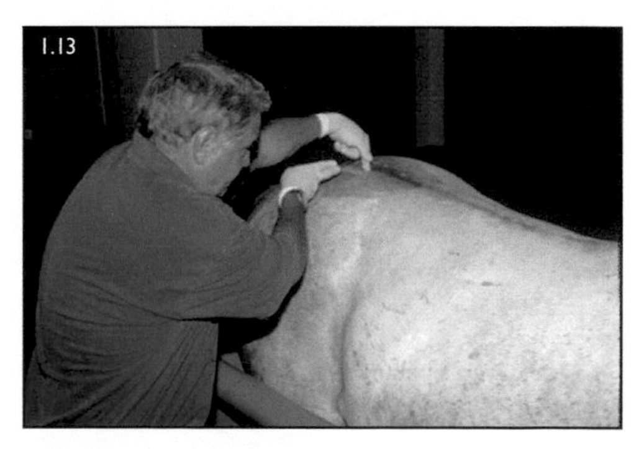

**FIG. 1.13** Lumbosacral CSF collection. The clinician's left hand is placed at the site for needle puncture and the fingers of the right hand are at the right tuber sacrale.

- collection technique:
  - plait and tie up the tail.
  - palpate depression cranial to the cranial edges of each tuber sacrale and caudal to the caudal edge of the spine of L6.
  - sterile preparation of the skin.
  - 2–5 ml of LA placed in the subcutis.
  - skin incision made before insertion of the needle.
  - 18-gauge 15–18-cm (6–7-inch) spinal needle (adults).
  - 18- or 20-gauge 9-cm (3.5-inch) spinal needle (foals).
  - right-handed person should stand on the right of the horse, resting their right wrist on the right dorsal midline of the horse.
  - needle inserted perpendicular to the dorsum of the horse.
  - depth of needle penetration to reach the LS space in the horse depends on their size and obesity:
    - approximately 12.5 cm (5 inches) in a 450-kg horse.
    - useful to have an assistant standing back from the horse to advise on needle direction.
  - prepared for a reaction in the horse as dura–arachnoid is penetrated by the needle:
    - tail flick, hindlimb flexion and muscle contraction.
    - reaction can be violent, and the horse handler should be prepared.
  - advance needle slowly, removing the stylet to check for CSF at 1 mm intervals.
  - aim is to remove CSF from the dorsal subarachnoid space:
    - may fail to flow, and the needle reaches the ventral subarachnoid space on the bony floor of the vertebral canal.
    - caudal sacral segments will have been penetrated in adult horse:
      - needle should be retracted slowly back to dorsal subarachnoid space.
  - CSF collected in serial small syringe aliquots:
    - one with the least blood contamination submitted for analysis.

## Imaging

### Radiography

- **Plain radiography** of the skull and vertebral column of the horse:
  - identify fractures, malformations and luxations:
    - standard views are laterolateral and dorsoventral (neck only).
  - latero-oblique cervical radiography techniques may increase:
    - sensitivity, specificity, positive and negative predictive values for detecting cervical vertebral lesions.
- **Myelography** is indicated in horses with neurological signs that can be attributed to a cervical spinal cord compressive lesion on neurological examination.

### Computed tomography (CT)

- wide-bore gantry openings (up to 80 cm) allow the entire cervical vertebral column and cranial thoracic vertebral column (T1–T2) to be imaged in the adult horse under general anaesthesia:
  - where available provides superior results.

## Neuroelectrodiagnostic testing

**Electromyography** (EMG)

- assessment of the electrical activity of muscle.
- used to confirm which spinal cord segment is affected, or to give more precise differentiation of nerve plexus or peripheral nerve disorders.
- standing sedated patient or under general anaesthesia.
- findings are not specific or pathognomonic for a particular aetiology or neuromuscular disorder.

**Electroencephalography** (EEG)

- graphic recording of rhythmic electrical activity arising from the cerebral cortex.
- normal patterns of activity and abnormal activity have been reported in horses.
- abnormalities associated with intracranial lesions such as:
  - hydrocephalus, encephalitis, meningitis and abscessation (*Streptococcus equi* subsp. *equi*).
- may help assess the functional significance of suspected intracranial disorders.

# NEUROLOGICAL DISEASES

## *Congenital disorders*

## Hydrocephalus

### Definition/overview

- accumulation of excessive volumes of CSF in the brain or cranial cavity.
- rare and most often seen in neonatal foals as a congenital malformation.

### Aetiology/pathophysiology

- inherited defect proposed in some horse breeds:
  - nonsense genetic mutation linked to hydrocephalus in Friesians.
- acquired following meningitis or haemorrhage.
- **normotensive hydrocephalus:**
  - passive CSF expansion subsequent to loss of brain parenchyma after destructive pre- or post-natal infection or injury.
- **hypertensive hydrocephalus:**
  - obstruction of the CSF conduit between the sites of production and absorption by the arachnoid villi in the subarachnoid space.
  - subsequent to hypoplasia/aplasia of a part of this system or may be acquired.

### Clinical presentation

- stillbirth or death shortly after birth.
- usually associated with dystocia in the mare.
- varied presentation in live foals due to compression of the cerebral cortex:
  - poor suckle reflex, blindness, lack of affinity for the mare, depression, ill-thrift and growth retardation (Fig. 1.14).
  - inconsistent and sometimes inapparent – grossly enlarged dome-shaped head (Fig. 1.15).

### Differential diagnosis

- hypoxic–ischaemic encephalopathy.
- electrolyte imbalances
- meningitis.
- hypoglycaemia
- premature weak foal
- septicaemia.

### Diagnosis

- clinical signs of impaired mental function and a domed skull.
- CT and magnetic resonance imaging (MRI) in neonates.
- necropsy.

### Management

- treatment is not recommended, and euthanasia should be considered.

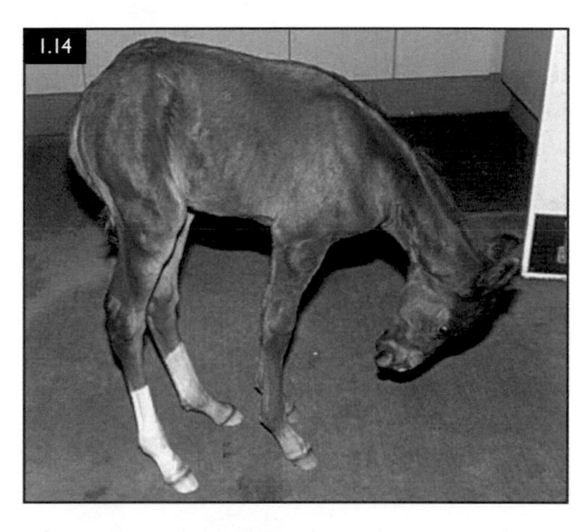

**FIG. 1.14** Hydrocephalus in a 6-week-old male foal. Foal was unable to lift his head and pull his tongue into the mouth plus suffered from severe perinatal asphyxia and developed necrosis of the cerebrum. CSF took the place of the lost parenchyma (compensatory hydrocephalus). (Photo courtesy FT Bain.)

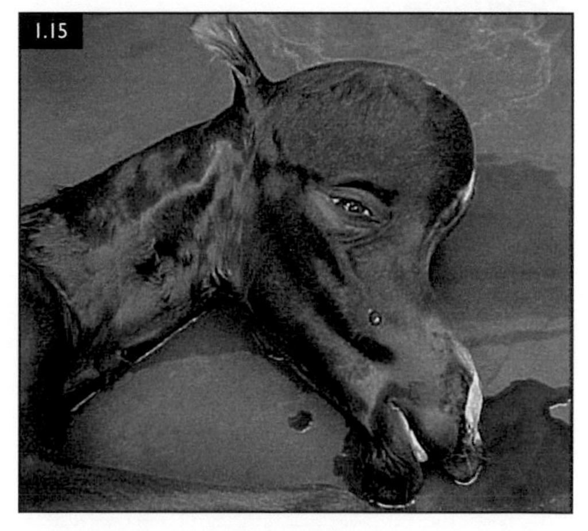

**FIG. 1.15** Congenital hydrocephalus in an aborted foal with domed appearance of the skull.

### Prognosis

- grave as any surviving animals have severe residual neurological deficits.

## Cerebellar abiotrophy

### Definition/overview

- inherited neurodegenerative disease causing post-natal destruction of the Purkinje fibres of the cerebellum.
- greatest frequency in part-bred/purebred Arabian horses but other breeds affected.

### Aetiology/pathophysiology

- autosomal recessive inheritance.

### Clinical presentation

- signs from birth to the first few months of life.
- intention tremor and loss of menace response (inconsistent).
- signs may appear suddenly and remain static, or gradually progress and plateau.
- spasticity or hypermetria in the gait (Fig. 1.16) with rearing and falls.

### Differential diagnosis

- meningitis
- equine protozoal myeloencephalitis (EPM)
- cerebellar abscess.
- hypoxic–ischaemic encephalopathy
- parasitic migrans in the cerebellum.
- West Nile virus (WNV) encephalomyelitis.

### Diagnosis

- history and clinical signs.
- genetic testing for the TOE1 genotype is available commercially.
- MRI.

FIG. 1.16 Cerebellar abiotrophy in an Arabian. Note the hypermetria of the forelimb. (Photo courtesy FT Bain.)

### Management and prognosis

- no effective treatment and animals are unsafe.
- discourage future breeding of the parents of the affected animal.

## Occipitoatlantoaxial malformation (OAAM) (see Book 1, page 201)

### Definition/overview

- rare congenital malformation which is not one single defect.

### Aetiology/pathophysiology

- deletional mutation at the HOX gene cluster in Arabian foals.
- occipitalisation of the atlas (resembles occiput).
- atlantalisation to the axis (resembles atlas).
- symmetrical or asymmetrical lesions.
- variations of OAAM have been reported in other breeds.

### Clinical presentation

- clinical signs reflect the degree of spinal cord compression in the atlas foramen:
  - worst cases – recumbency from birth and spastic tetraparesis:

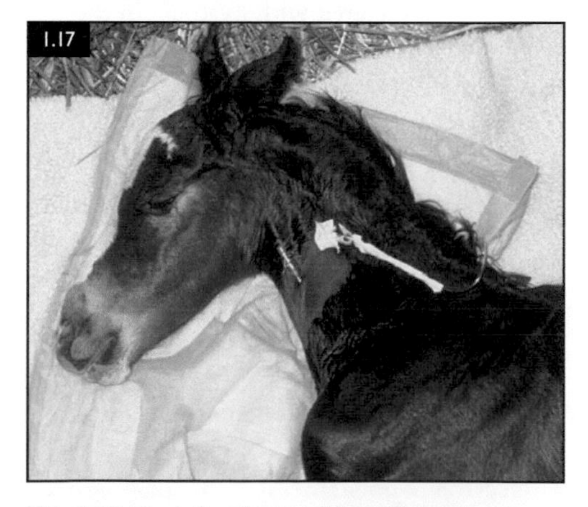


FIG. 1.17 Occipitoatlantoaxial malformation in an Arabian foal. Note the exaggerated dorsal flexion of the proximal cervical vertebrae. (Photo courtesy FT Bain.)

- especially when neck is extended during suckling movements (Fig. 1.17).
  - mild cases have no neurological signs.
- movement of the head occasionally is associated with an audible click.
- some foals – signs not apparent for weeks or months after birth (occasionally few years) and are steadily progressive.
- abnormal head and neck posture (both in extension) and appear stiff.

### Differential diagnosis

- AO luxation
- cervical trauma.

### Diagnosis

- neurological examination localises abnormalities to the cervical spine.
- neck palpation may detect lack of movement of AO joint and physical abnormality.
- radiography +/– myelography or CT confirms the abnormality.

### Management

- laminectomy to alleviate spinal compression.
- surgical fusion of the atlantoaxial joints, with or without laminectomy.

### Prognosis

- guarded to grave, with many cases being euthanised.

## *Infectious neurological diseases*

## Borna disease (Near Eastern encephalitis [NEE])

### Definition/overview

- sporadically occurring encephalitic disease caused by Borna disease virus.
- reported in horses, humans, ruminants and rabbits.
- reported in Central Europe and the Middle East.

### Aetiology/pathophysiology

- severe, usually fatal, encephalitis caused by RNA virus of the *Flaviviridae* family.
- reservoir host in Europe thought to be the shrew (*Crocidura leucodon*):
  - sheds high levels of virus in body fluids.
- incubation period from weeks to several months after exposure.

### Clinical presentation

- initially non-specific clinical signs including:
  - dullness, depression, anorexia and pyrexia.
- neurological signs consistent with an encephalitis follow:
  - head-pressing, ataxia, circling, compulsive behaviours.
- eventually become paretic, develop central blindness, coma and die (80%).

### Differential diagnosis

- other causes of viral encephalitis, depending on the geographical area.

### Diagnosis

- identification of antibodies in serum and CSF ante-mortem.
- histopathology post-mortem.

### Management

- supportive care.

### Prognosis

- >80% mortality rate:
  - most cases are euthanised due to severity of disease and potential for latent and persistent infection.

## Alphavirus encephalitis of horses

### Definition/overview

- seasonal, geographically important causes of potentially severe encephalitis:
  - Western equine encephalitis (WEE).
  - Eastern equine encephalitis (EEE).
  - Venezuelan equine encephalitis (VEE).

### Aetiology/pathophysiology

- *Alphavirus* of the *Togaviridae* family.

**Eastern equine encephalitis**

- one virus with two antigenic variants: North and South American.
- primarily in the USA east of the Mississippi River and occasionally SE Canada.

- mosquito vector, *Culiseta melanura*, and passerine birds as amplifying hosts.
- seasonal variation in disease, with peak incidence in late summer or early autumn.
- infected horses and humans are dead-end hosts.

**Western equine encephalitis**
- seven virus species.
- throughout most of the Americas, with extensive epizootics in Argentina.
- principal enzootic vector, *Culex tarsalis*, and epizootic mosquito vector, *Aedes* species.
- humans and horses are dead-end hosts.

**Venezuelan equine encephalitis**
- one virus with six antigenic subtypes (I–VI).
- primarily in South America, with extension into Central America and Texas.
- enzootic cycles maintained via *Culex* spp. and small vertebrate hosts.
- *Aedes* and *Psorophora* mosquito species transmit epizootic viruses in association with many different vertebrates.
- high mortality in both humans and horses.
- horses can act as an amplifier of the disease.

## Clinical presentation

- approximately 7-day initial incubation period associated with biphasic viraemia:
  - biphasic fever in EEE and WEE.
  - consistently elevated temperature in VEE.
- non-specific signs during the viraemic phase include lethargy and stiffness.
- neurological deficits more evident as disease progresses:
  - EEE and WEE:
    - ataxia, somnolence, conscious proprioceptive deficits.
    - stiff neck and compulsive behaviours.
  - more severe progression with EEE:
    - blindness, circling, excitement and aggressive behaviour.
    - laryngeal, pharyngeal, tongue paralysis and dysphagia (Fig. 1.18).
    - head tilt, nystagmus, strabismus and pupil dilation (Fig. 1.19).
  - VEE:
    - may cause inapparent infections.
    - signs similar to other encephalitis viruses.
    - signs such as epistaxis, pulmonary haemorrhage, oral ulcers and diarrhoea.
- seizures and sudden death can occur with any of the alphavirus infections.

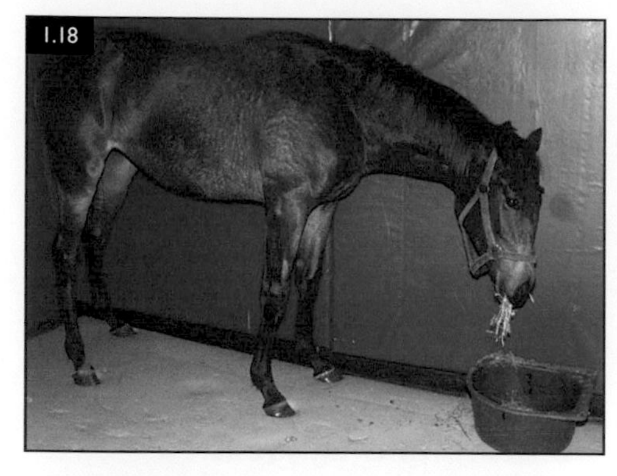

**FIG. 1.18** Dysphagia and dementia associated with EEE. (Photo courtesy FT Bain.)

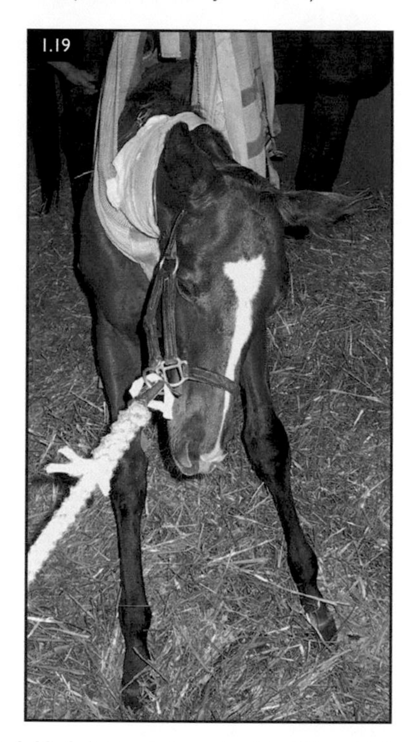

**FIG. 1.19** A horse infected with EEE which presented with severe ataxia and recumbency and required the use of a sling. The horse eventually died.

## Differential diagnosis

- other causes of viral encephalitis.

## Diagnosis

- clinical signs are non-specific.
- serological testing with a fourfold rise in antibody titres:
  - delay in taking the acute sample may lead to sampling during the peak titre.
  - rapid death in EEE may remove the chance of a paired sample.
- high immunoglobin M (IgM) titres suggest recent exposure to EEE virus – detected via enzyme-linked immunosorbent assay (ELISA).
- viral isolation or identification of viral nucleic acid by reverse transcription polymerase chain reaction (RT-PCR) in CNS tissue or CSF.
- post-mortem virus isolation or immunohistochemistry testing from/on brain tissue.
- virus isolation from serum is usually unsuccessful.

## Management

- notifiable diseases in many regions.
- no specific treatment is available.
- supportive care.
- anti-inflammatory drugs, including flunixin or p/o dimethyl sulphoxide (DMSO).
- good nursing care, including hydration and protection from self-induced trauma.
- prevention is based on vaccination and limiting exposure to vectors.
- monovalent, bivalent and trivalent killed vaccines are available:
  - recommendations vary depending on the region, vaccine and disease.
- eliminate mosquitoes or their habitats using repellents and stabling at dawn and dusk.

## Prognosis

- varies greatly among the different diseases:
  - grave for EEE, with mortality rates of 75–100%.
  - guarded to poor for VEE, with mortality rates of 40–80%.
  - guarded for WEE, with mortality rates of 25–50%.
- residual neurological deficits might be present in survivors.

# West Nile virus (WNV) encephalomyelitis

## Definition/overview

- geographically important cause of encephalitis:
  - first reported in Africa and the Middle East.
  - now important in Europe and North America.

## Aetiology/pathophysiology

- *Flavivirus* within the family *Flaviviridae*:
  - two lineages (1 and 2) reported to affect horses.
  - several epidemics of neuroinvasive equine cases in Europe (lineage 2) in the last decade.
- mosquito-borne disease affecting broad range of animals including birds, cats, dogs, horses and humans.
- cycles between bird reservoir hosts and mosquitoes.
- horses become infected via the bites of infected mosquitoes, particularly *Culex* genus.
- humans, horses and most other mammals are considered incidental dead-end hosts:
  - no equine to human or vice versa transmission occurs.
- equine infections most commonly detected in late summer to early autumn.
- no age or breed predisposition.

## Clinical presentation

- majority of infections in horses are subclinical.
- vague signs including pyrexia, anorexia and depression within 9–11 days of infection.
- onset of neurological signs is rapid and include:
  - changes in mentation, possibly progressing into aggression, somnolence and coma.
  - hyperaesthesia.
  - muscle fasciculations, especially of the muzzle.
  - CN abnormalities include tongue weakness, muzzle deviation, head tilt and swallowing difficulties.
  - spinal cord pathology presents as:

- ataxia (sometimes asymmetrical and may involve multiple limbs).
- stiff gait confusable with a lameness.

### Differential diagnosis

- other causes of viral encephalitis.
- hepatic encephalopathy.
- tremorgenic toxicity (ryegrass staggers).
- hypocalcaemia
- head trauma.
- equine herpes myeloencephalopathy.
- botulism.
- verminous meningomyeloencephalitis.

### Diagnosis

- clinical signs in an area and at a season in which WNV activity has been documented.
- specific IgM antibodies in serum and/or CSF.
- fourfold increase in competitive ELISA immunoglobin G (IgG) titres, ideally 10 days apart.
- Plaque reduction neutralisation test-positive serum samples taken during the acute and convalescent phases.
- histopathology of post-mortem samples of the mid- and hindbrain.

### Management

- no antiviral therapies licensed for use in neuroinvasive flavivirus infections.
- supportive care:
  - anti-inflammatory treatment.
  - anti-oedema therapy with i/v mannitol.
- commercial hyperimmune serum or plasma products available but unproven.
- prevention is based on limiting exposure to mosquito vectors and on vaccination:
  - protection against WNV lineage 2 is most important.

### Prognosis

- guarded overall:
  - 30% of horses with neurological disease die or are euthanised 3–4 days after the onset of clinical signs.
  - fair for animals that remain standing.
  - poor for horses:
    - unable to stand because of hindlimb paralysis.
    - signs of cerebral lesions.
  - complete resolution of clinical signs in weeks to months is possible:
    - persistent neurological deficits have been reported.

## Other arboviral diseases

### Equine encephalosis

- acute arthropod-borne viral disease caused by an *Orbivirus* of the family *Reoviridae*.
- endemic in most parts of South Africa:
  - *Culicoides* spp. is the presumed vector of the disease.
- most infections are subclinical.
- clinical cases develop ataxia, stiffness and facial swelling:
  - similar to African horse sickness (AHS).
- infected horses are viraemic for 4–7 days and infective for vectors during this time.
- no evidence that horses can become carriers of the disease.

### Equine infectious anaemia (EIA)

- member of the *Retroviridae* family.
- most common neurological abnormality is symmetrical ataxia of trunk and limbs:
  - other signs include circling, gait alterations and behavioural changes.
- neurological signs rarely present alone:
  - usually with signs of haemolymphatic dysfunction.

### Other diseases

- Louping ill, Japanese B, St. Louis, Murray Valley, Semliki Forest, Russian spring–summer, Powassan and Ross River encephalitis viruses, which are all members of the family *Flaviviridae*.
- members of the family *Bunyaviridae* such as Main Drain viruses and California group viruses may also cause neurological disease in horses.

## Rabies

### Definition/overview

- fatal neurological disease of mammals.
- rare in horses but of concern due to risk to humans.
- widespread around the world.

## Aetiology/pathophysiology

- neurotropic *Lyssavirus* (*Rhabdoviridae*).
- two classic cycles: canine (urban) rabies and wildlife (sylvatic) rabies.
- domestic animals generally regarded as dead-end hosts:
  - usually infected following contact with wildlife vectors.
- most commonly transmitted by salivary contamination of a bite wound:
  - infection by inhalation, oral or transplacental routes have been demonstrated.
- incubation period varies from 2 weeks to several months.
- predilection for replication in the cell bodies (grey matter) of the CNS:
  - dysfunction of neurons leads to behavioural changes and abnormalities of the cranial and peripheral nerves.
- death by respiratory paralysis after infection of the medulla.
- shedding of the virus in nasal and salivary secretions predates onset of clinical signs.

## Clinical presentation

- highly variable and should be considered in any neurological case in endemic areas.
- signs include:
  - anorexia ○ depression ○ blindness.
  - mania ○ hyperaesthesia.
  - muscle twitching ○ lameness.
  - paresis ○ ataxia ○ recumbency.
  - colic ○ urinary incontinence.
  - sudden death.
- early on may mimic other diseases.
- signs of an animal bite are rarely present.
- rapidly progressive once signs are seen with death in 3–10 days.

## Differential diagnosis

- many differential diagnoses including other conditions with signs of grey matter disease such as polyneuritis equi, herpesvirus myeloencephalitis, EPM and Sorghum–Sudan grass poisoning.
- other cerebral diseases, such as hepatoencephalopathy, leucoencephalomalacia, alphavirus encephalitides, space-occupying masses and meningitis.

## Diagnosis

- difficult ante-mortem as clinical signs are not diagnostic.
- indirect fluorescent antibody test on brain tissue is 98% accurate.
- microscopic examination of haematoxylin and eosin-stained brain sections reveals non-suppurative encephalitis and diagnostic Negri bodies.

## Management

- invariably fatal.
- rarely, where an ante-mortem diagnosis is reached:
  - animal should be euthanised to avoid further human contact.
- horse to human transmission has never been reported but all necessary precautions should be taken for animals showing neurological signs in endemic areas.
- government/public health authorities should be contacted, to coordinate management of exposed humans.
- inactivated annual vaccines for protection of horses in endemic areas.

## Prognosis

- always fatal within 3–10 days following development of neurological signs.

# Equine herpesvirus myeloencephalopathy (EHM)

## Definition/overview

- common cause of CNS disease with worldwide distribution.
- up to 80% of horses carry a latent infection.
- neurological disease is an uncommon sequela to EHV-1 infection.
- adults are more commonly affected, and outbreaks have been reported.

## Aetiology/pathophysiology

- only EHV-1 has been associated with neurological disease:
  - $D_{752}$ genotype more strongly associated with EHM than $G_{752}$ genotype.

## Pathogenesis

- infection through respiratory epithelium with virus replication in local lymph nodes and infected leucocytes released into the circulation.
- viraemia for 4–10 days post-infection.
- secondary infection of the vascular endothelium, including in spinal cord and brain:
  - leads to vasculitis, thrombosis, ischaemia and myelomalacia.
- also responsible for abortion storms, sporadic abortions in mares, neonatal death and respiratory disease in young horses.

## Clinical presentation

- multiple horses in a herd are usually affected but sporadic individual cases do occur.
- history may include contact with horses demonstrating signs of herpesvirus respiratory disease.
- outbreaks may occur following introduction of new horses.
- neurological signs occur 6–10 days after initial infection:
  - often preceded by fever, lethargy and inappetence.
  - acute onset and rarely progress after 2–3 days.
  - symmetrical hindlimb ataxia and paresis.
  - urinary incontinence.
  - hypotonia and hypoalgesia of the perineum, tail and anus.
  - spastic tetraparesis and paralysis if the cervical spinal cord is involved:
    - recumbency and rarely able to stand again.
  - occasional cases develop cortical, brainstem or vestibular lesions:
    - related signs include depression, head tilt, CN deficits and coma.

## Differential diagnosis

- Equine degenerative myelitis
- EPM.
- WNV encephalomyelitis
- trauma.
- Equine motor neuron disease (EMND).
- verminous meningomyeloencephalitis (*Halicephalobus deletrix*).
- Cervical vertebral stenotic myelopathy (CVSM).
- hyperammonaemia.

## Diagnosis

- characteristic neurological signs and history.
- xanthochromia and increased total protein in CSF.
- virus isolation or positive PCR from nasopharyngeal swabs or blood samples.
- fourfold rise in virus-neutralising or complement-fixing antibody titres (preferable) between acute and convalescent samples taken 7–10 days apart.
- post-mortem diagnosis is based on histology, immunohistochemistry or PCR to detect viral antigen in CNS tissues.

## Management

- supportive care:
  - evacuation of the bladder and rectum.
  - sling support for paretic horses.
  - maintenance of adequate nutrition and hydration.
  - anti-inflammatory drugs.
  - antiherpetic drugs and corticosteroids are frequently used but definitive evidence is lacking:
    - valacyclovir has been reported for antiherpetic use.
- vaccines currently used to prevent EHV-1 respiratory disease and abortion do not offer protection against myeloencephalopathy:
  - may reduce the incidence of other EHV-1 diseases.
- prevent introduction and subsequent dissemination of EHV-1 infection:
  - isolation of new animals for 3 weeks.
  - maintain distinct herd groups according to age, sex and occupation.
  - isolation of pregnant mares and minimise stress.
  - ability of EHV-1 to remain latent for prolonged periods of time and then re-emerge complicates infection-control measures.

## Prognosis

- fair provided affected horses remain standing:
  - full recovery is possible, but weeks to months may be required.

- poor if recumbent, clinical deterioration or no improvement over the first week of treatment.

## Listeriosis

- rare bacterial infection caused by *Listeria monocytogenes*.
- associated with septicaemia, abortion, keratoconjunctivitis, diarrhoea and meningoencephalitis.
- ubiquitous organism and sites of infection include:
  - wounds, umbilical remnants, oral or nasal mucosa and conjunctiva.
- immunocompromise may have a role in the pathogenesis.
- multifocal microabscessation in the brainstem, medulla and pons.
- antimicrobials routinely used in horses, with the exception of ceftiofur, are effective against *L. monocytogenes*.

## Bacterial meningoencephalomyelitis

### Definition/overview

- highly fatal disease that is rare in adults, but more common in neonatal foals.

### Aetiology/pathophysiology

- meningitis is usually bacterial and occurs in one of three ways:
  - haematogenous spread from other sites.
  - direct extension of a suppurative process in close proximity to the calvarium.
  - secondary to penetrating wounds or fractures of the calvarium.
- neonate foals usually subsequent to generalised sepsis:
  - predominance of Gram-negative organisms:
  - *Escherichia coli*, *Actinobacillus equuli* and *Klebsiella* species.
- among older animals, Gram-positive organisms are more frequent:
  - such as *Staphylococcus* spp. and haemolytic *Streptococcus* spp.
- septic meningitis is the most common and results in profound neurological signs:
  - secondary CNS oedema/obstructive hydrocephalus worsens clinical picture.

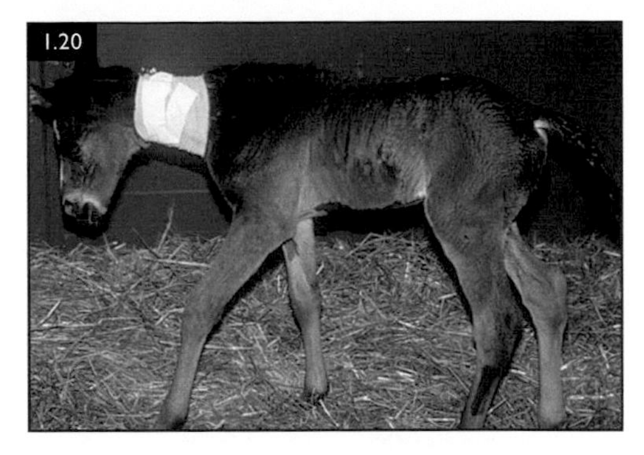


**FIG. 1.20** A foal with meningitis wandering aimlessly.

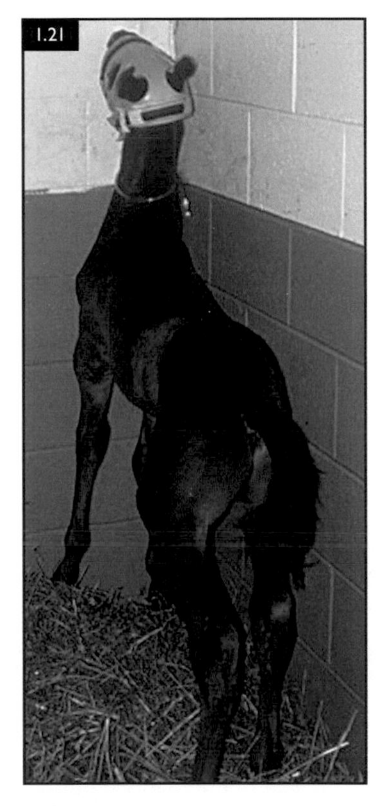


**FIG. 1.21** Seizure activity and head-pressing in a foal with meningitis.

### Clinical presentation (Figs. 1.20, 1.21)

- neonate in initial stages:
  - aimless wandering, depression, loss of affinity for dam and abnormal vocalisation.

- fever of unknown origin.
- later stages:
  - hyperaesthesia, muscular rigidity, blindness, CN deficits, ataxia/paresis all 4 limbs.
- without treatment:
  - recumbency, coma, seizures and death.

### Differential diagnosis

- EPM hypoxic–ischaemic encephalopathy.
- viral encephalitis • trauma.
- Leucoencephalomalacia.
- cholesterol granuloma of the choroid plexus.
- verminous meningoencephalomyelitis (*Halicephalobus deletrix*).
- hepatic or intestinal hyperammonaemia.

### Diagnosis

- **early diagnosis is essential.**
- evidence of bacteria, increased inflammatory cells, elevated protein concentration and low glucose concentration in CSF.
- culture/cytology of CSF.
- blood culture or culture of other available septic sites.
- PCR for identification of certain pathogens in the CSF.

### Management

- broad-spectrum or combination antimicrobials – culture/sensitivity patterns unknown.
  - commonly used drugs are:
    - aminoglycoside and penicillin combinations.
    - potentiated sulphonamides.
    - third-generation cephalosporins.
    - chloramphenicol.
  - meningeal inflammation considerably improves penetration of drugs into the CSF.
  - minimum of 14 days, continuing for 7 days after the resolution of clinical signs.
- aggressive supportive care, including nutritional support.

### Prognosis

- poor, even with aggressive supportive care, with >50% of foals dying.

## Borreliosis (Lyme disease)

### Definition/overview

- immune-mediated disorder caused by *Borrelia burgdorferi*.
- occasionally causes neurological disease (neuroborreliosis).
- significant controversy regarding disease in equine practice.

### Aetiology/pathophysiology

- transmission to horses occurs during feeding activity of infected adult female ticks in the summer, autumn or late winter.
- high seroprevalence in horses in certain regions.

### Clinical presentation

- variable, but include:
  - low-grade fever, hyperaesthesia, muscle fasciculations, lethargy and behavioural changes:
    - less common – spinal muscle atrophy, facial nerve dysfunction, ataxia and paresis.
  - systemic signs may accompany neuroborreliosis including uveitis, cardiac arrhythmias and polyarthritis (joint swelling and stiffness).

### Differential diagnosis

- bacterial meningitis
- WNV encephalomyelitis.

### Diagnosis

- challenging and the disease maybe overdiagnosed, especially in high seroprevalence areas:
  - based on geographical area, clinical signs and ruling out other causes.
  - positive Western blot test.
- **serological tests are unable to distinguish between active infection and subclinical exposure – interpret with caution:**
  - magnitude of antibody titres no relationship to the likelihood of clinical disease.
  - horses may stay seropositive for months or even years.
- PCR test is available for use on tissue (including skin and muscle), ticks, synovial fluid or whole blood.

## Management

- Oxytetracycline: 6.6 mg/kg i/v q12 h or q24 h for 7 days (up to 28 days), or doxycycline: 10 mg/kg p/o q12 h for 7 days (up to 28 days).
- clinical signs may return after treatment is stopped.
- supportive treatment includes:
  - chondroprotective agents.
  - non-steroidal anti-inflammatory drugs (NSAIDs).
- prevention of tick exposure and prolonged tick attachment.

## Prognosis

- good if appropriate treatment is provided.

# Equine protozoal myeloencephalitis (EPM)

## Definition/overview

- common and variable neurological disease in North and South America:
  - cases have occurred all over the world in horses imported from the USA.
  - months to years after arrival.

## Aetiology/pathophysiology

Agents

- *Sarcocystis neurona* in most cases and *Neospora hughesi* in some others.

Host

- Virginia opossum (*Didelphis virginiana*) host *S. neurona* in the USA.
- related opossum (*Didelphis albiventris*) carries *S. neurona* in South America.
- armadillos, raccoons and skunks are natural intermediate hosts.
- no definitive host identified for *N. hughesi*.
- horses considered accidental aberrant intermediate hosts for *S. neurona*:
  - non-contagious and dead-end hosts.

Disease

- seroprevalence at 26–60%.
- incidence of new disease up to 0.51%, so development of disease is uncommon:
  - risks factors may include:
    - spring, summer and autumn.
    - animals between 1 and 5 years of age.
    - animals living in wooded terrain.
    - previous EPM on site.
- infection by ingestion of sporocysts from food/water contaminated by definitive host:
  - sporocysts excyst in the small intestine.
  - release of sporozoites into the bloodstream as a cell-associated parasitaemia.
  - haematogenous spread to the CNS and localisation anywhere except peripheral nerves (PN).
  - some types of immunosuppression and stress may alter likelihood of CNS invasion and development of disease.
  - location of infection within the CNS determines the clinical presentation.

## Clinical presentation

- diversity of clinical signs:
  - lesions of varying size, locations and severity in any part of CNS.
  - onset insidious or acute.
- spinal cord involvement occurs in most cases (95%).
- highly variable disease course:
  - progression over hours or years.
  - waxing and waning of signs over extended periods.
- neurological signs in decreasing order are:
  - spinal cord lesions:
    - ataxia, paresis, proprioceptive deficits and/or spasticity (usually asymmetrical).
    - muscle atrophy.
    - CN deficits from brainstem lesions.
  - central cerebral lesions, e.g. seizures.
  - regions of hypalgesia, hyperaesthesia and increased sweating.
  - urinary incontinence and cauda equina syndrome symptoms.

## Differential diagnosis

- EPM can mimic almost any other neurological disease.

## Diagnosis

- **Challenging:**
  - no definitive ante-mortem diagnosis can be achieved by clinical pathology.
  - clinical diagnosis reached on accumulation of data:
    - relevant history and compatible clinical signs.

- clinical progression.
- laboratory and other diagnostic aids.
- exclusion of other possible causes.
- response to treatment.

- laboratory and other diagnostic aids have been divided into three categories:
  - positive titres for *S. neurona* antibodies, which supports the diagnosis.
  - negative titres, which tends to exclude the diagnosis.
  - tests that support alternative diagnoses and indirectly exclude a diagnosis of EPM.
- CSF antibody detection can result from inadvertent contamination of the sample with serum antibodies and produce false-positive results:
  - CSF and serum for antibodies ratio using surface antigen ELISA.
- for *S. neurona*:
  - PCR for surface antigen.
  - indirect fluorescent antibody test using whole tachyzoites.
- post-mortem findings.

### Management

- anticoccidial medication:
  - combinations of pyrimethamine (1 mg/kg p/o q24 h for 3–6 months) and sulphonamides (20–30 mg/kg p/o q12 h for 3–6 months).
  - recommended that the horse is not fed for 1 hour before and after administration.
  - adverse effects include bone marrow suppression and anorexia:
    - **do not administer to pregnant mares (teratogenic).**
- Benzeneacetonitriles:
  - diclazuril and ponazuril are licensed for use against EPM in the USA.
- antioxidant therapy as adjunctive therapy against oxidative stress damage:
  - vitamin E (10,000 IU p/o q24 h for 30 days).
- anti-inflammatory therapy using NSAIDs or corticosteroids in the first few days of anticoccidial therapy to minimise damage due to the death of the parasite.
- preventive measures include:
  - not feeding from the ground.
  - avoid ground-water consumption by offering clean fresh-water sources.
  - prevent access of wildlife to stables and pasture.
- vaccination is under investigation.

### Prognosis

- good if ataxia of grade 2 or less.
- poor if animal becomes recumbent.
- some animals have sustained permanent nerve injury following recovery.

## Trypanosomiasis

- blood-borne protozoa.
- *Trypanosoma evansi* transmitted by haematophagous flies and vampire bats:
  - Surra important in Asia and South America.
- *T. equiperdum* sexually transmitted and arthropod-borne disease:
  - Dourine in Asia, South Africa, India and Russia:
    - eradicated from North America.
- variable clinical signs and there may be subclinical infections:
  - pyrexia, anaemia, weight loss and lymphadenopathy.
  - often fatal.
  - meningoencephalomyelitis may be seen with signs of:
    - muscle atrophy.
    - facial nerve paralysis.
    - limb ataxia.
    - weakness, worse in the hindlimbs.
- treatment with trypanosomides, but resistance is encountered:
  - treatment of Dourine is not recommended.

## Verminous meningoencephalomyelitis

### Definition/overview

- rare and sporadic disorder caused by aberrant migration of parasites through the CNS.

### Aetiology/pathophysiology

- several possible agents including:
  - *Strongylus*, *Hypoderma*, *Habronema*, *Draschia*, *Halicephalobus gingivalis* and *Setaria* species.

## Clinical presentation

- variable and depends on the area of the CNS involved.
- *Halicephalobus* spp.:
  - intracranial lesions with acute onset of rapidly progressing head-pressing, loss of proprioception, recumbency, coma and death.

## Diagnosis

- clinical signs and post-mortem findings.
- inconsistent CSF changes with some haemorrhage and increased numbers of inflammatory cells, such as eosinophils and neutrophils.

## Management

- anthelmintics (fenbendazole, 50 mg/kg p/o q24 h for 3 days).
- anti-inflammatories (flunixin meglumine, vitamin E and/or dexamethasone).

## Prognosis

- guarded to grave.

# Equine ehrlichiosis

- tick-borne pathogen *Anaplasma phagocytophila* (formerly *Ehrlichia equi*).
- transient truncal and limb ataxia, and weakness:
  - may be severe, leading to falls and injuries.
- inflammatory, vascular or interstitial lesions in the brains of affected animals.
- rapid improvement with oxytetracycline (6.6–7.5 mg/kg i/v q12–24 h for 3–7 days).

# Tetanus

## Definition/overview

- fatal infectious disease caused by exotoxins produced by *Clostridium tetani*.
- characterised by muscular rigidity and hyperaesthesia in horses of all ages.
- common in areas of the world where vaccination not widely used.

## Aetiology/pathophysiology

- *C. tetani*, Gram-positive, spore-forming, rod-shaped anaerobic bacterium:
  - found in animals, birds, soil and organic matter.
  - spores are ubiquitous and resistant.
- anaerobic environment (e.g. devitalised tissue) must be present for infection to occur:
  - deep puncture wounds.
  - subsolar abscesses.
  - surgical incisions.
  - post-foaling damage in the female reproductive tract.
  - site of entry may have healed by the start of clinical signs.
- anaerobic environment germination of *C. tetani* spores and exotoxin production:
  - one of the toxins (tetanospasmin) binds to the motor neuron nerve terminal.
  - transported in retrograde fashion up the motor neuron axon to the spinal cord:
    - accumulates in the ventral horn of the grey matter.
  - migrates trans-synaptically into the inhibitory interneurons:
    - cleaves synaptobrevin which decreases the release of inhibitory neurotransmitters and the suppression of muscular activity.
  - tetanolysin causes tissue necrosis.

## Clinical presentation

- usually unvaccinated animals.
- clinical signs within 2–21 days of a wound being infected (Table 1.5):
  - muscle hypertonia and superimposed clonic paroxysmal muscular spasms.
- advanced and severe cases of tetanus:
  - are unable to stand or eat.
  - die from respiratory failure (spasm of the diaphragmatic and intercostal muscles).
- complications of severely affected animals include:
  - decubital ulceration from increased recumbency.
  - ileus, gastrointestinal (GI) impactions, pneumonia.
  - myopathies, urine retention and cystitis.
  - dysphagia may lead to aspiration pneumonia.
  - muscle hyperactivity and sweating cause dehydration and electrolyte abnormalities.

**TABLE 1.5** Clinical signs of tetanus in the horse

| SEVERITY OF DISEASE | CLINICAL SIGNS |
|---|---|
| **Mild** | • Mild hyperaesthesia that might be exacerbated by performing a menacing gesture/slapping the neck and assessing response (muscular spasm or a flashing nictitans membrane) (Fig. 1.22)<br>• Rigid posture: 'sawhorse' stance and stiff gait, but still able to walk (Fig. 1.23)<br>• Elevated tail head<br>• Altered facial expression and dysphagia: retracted lips, flared nostrils, caudal pointing/erect ears, flicking/prolapsing nictitans membrane |
| **Moderate** | • Hyperaesthesia<br>• Limbs stiff, with generalised clonic muscular spasms<br>• Walks with difficulty<br>• Trismus present ('lockjaw'), but can still eat and drink slowly |
| **Severe** | • Severe hyperaesthesia, whereby mild external stimuli might lead to extreme extensor muscle rigidity/clonic muscle spasm. In some cases, this clonic paroxysmal muscular spasm might cause a horse to fall, or to fracture long bones/pelvis<br>• Unable to walk, might be capable of standing<br>• Opisthotonos might be seen in recumbent cases (or standing foals)<br>• Severe trismus and reduced ability to prehend, masticate and swallow, or unable to ingest water or food |

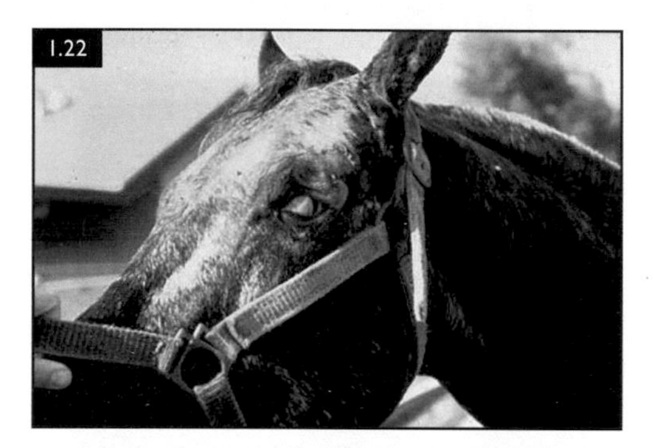

**FIG. 1.22** Prolapse of the nictitating membrane in a horse with tetanus.

**FIG. 1.23** 'Sawhorse' muscle rigidity noted in a horse with tetanus.

- hyperaesthesia combined with extensor muscle rigidity:
  - may cause horse to fall in response to external stimulation.
  - fractures of the pelvis, skull, femur, tibia and radius have been reported.
- some severe cases may develop apparent seizures.

- survivors usually begin to show signs of improvement within 10–14 days:
  - clinical signs may persist for weeks to months.

## Diagnosis

- presumptive is based on history, clinical signs and response to treatment.
- definitive is difficult:
  - microbial culture is difficult.
  - no characteristic post-mortem lesions.

## Differential diagnosis

Table 1.6

**TABLE 1.6** Differential diagnoses for tetanus for horses in Europe

| DIFFERENTIAL DIAGNOSIS | DIAGNOSIS |
|---|---|
| Laminitis (stiffness) | Pounding digital pulses; lateral distal phalanx radiographs |
| Meningitis (stiffness) | CSF cytology |
| Myopathies (stiffness) (e.g. polysaccharide storage myopathy (PSSM), immune-mediated myositis, nutritional myopathy, exertional myopathy) | Serum muscle enzyme concentrations, plasma fibrinogen and serum amyloid A |
| Cervical vertebral fracture/pain | Cervical radiographs |
| EMND (muscle fasciculations, reluctance to move, dysphagia) | Retinal lesions, muscle biopsy (sacrocaudalis dorsalis medialis) |
| Hyperkalaemic periodic paralysis (stiffness, muscle fasciculations) | Episodic, often associated with hyperkalaemia, Appaloosa/Quarter horse, 'Impressive' genetic lineage; DNA testing available |

## Management

- aims of treatment are:
  - inhibition of tetanospasmin (TeNT) production:
    - local and parenteral antimicrobial therapy immediately:
      - tetracyclines and metronidazole now preferred to penicillin.
    - wound debridement and lavage.
  - neutralise unbound TeNT:
    - passive immunisation – administration of tetanus antiserum:
      - no significant difference in survival.
      - no clear clinical advantage.
  - establish immunity:
    - active immunisation:
      - vaccination with tetanus toxoid in the face of clinical disease.
      - more robust antibody response than natural challenge.
  - supportive care:
    - quiet and darkened environment.
    - good footing, such as rubber matting and shavings:
      - deep and clean to help reduce decubital ulcers if recumbent.
    - monitor hydration and respond with intravenous fluids and electrolytes.
    - nutritional support including parenteral administration.
  - muscular relaxation:
    - quiet environment.
    - drugs such as:
      - acepromazine.
      - barbiturates.
      - alpha-2 agonists.
      - benzodiazepines.

## Prognosis

- survival rate varies from 25% to 41%.
- poor prognosis with recumbency and respiratory distress.
- survivors show improvement over a few days, and if live longer than 7 days.

## Prevention

- easily preventable with toxoid vaccination:
  - **should be recommended to all owners.**
- tetanus antitoxin and tetanus toxoid administration recommended in:
  - situations at risk of infection and not effectively vaccinated.
  - wound, surgery, unclear history, etc.

# Temporohyoid osteoarthropathy (THO) (see Book 3, page 67)

## Definition/overview

- well-recognised clinical syndrome, mainly in middle-aged horses.

## Aetiology/pathophysiology

- osseous proliferation of the proximal styloid bone and juxtaposed petrous portion of the temporal bone.

- unknown aetiology.
- pain, chronic inflammation and reduced motion of the temporohyoid joint.
- ankylosis of the joint.
- acute neurological signs associated with fracture of the petrous temporal bone and damage to overlying CNS.

## Clinical presentation

- damage to the vestibular system with vestibular ataxia and head tilt (Fig. 1.24):
  - and/or facial nerve damage:
    - facial paralysis.
    - exposure keratitis/corneal ulceration.
    - reduced tear production.
    - facial hyperaesthesia.
- clinical signs are usually unilateral, but pathology may be bilateral.
- less common neurological signs include:
  - cerebral sequelae due to opportunistic infection:
    - seizures, coma and death.
  - rarely, damage to CN IX and X, resulting in mastication difficulties/ dysphagia.
- headshaking, ear-rubbing and resentment of bridle.

**FIG. 1.24** Right-sided head tilt and facial and vestibulocochlear nerve deficits with right-sided otitis interna/media and THO. Note the muzzle deviation to the left.

## Differential diagnosis

- EPM
- WNV encephalomyelitis.
- trauma
- ear ticks.

## Diagnosis

- Guttural pouch endoscopy (Fig. 1.25).
- CT, preferably under standing sedation:
  - Stylohyoid bone enlargement (Fig. 1.26).

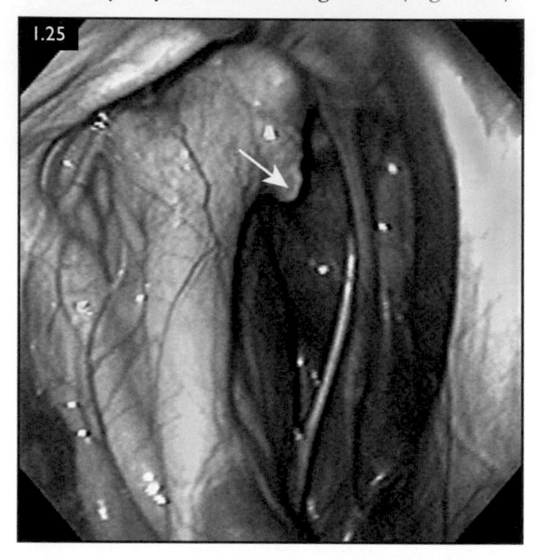


**FIG. 1.25** Endoscopy of the guttural pouch showing right-sided THO. Note the osseous proliferation of the temporohyoid joint (arrow).

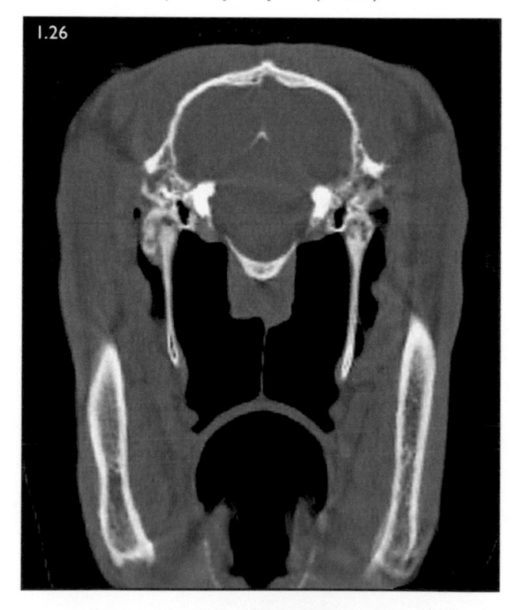


**FIG. 1.26** Reconstructed transverse CT image. Left is to the right of the image. The right temporohyoid joint is enlarged as a result of callus formation. There are less severe degenerative changes affecting the left temporohyoid joint.

- false-negatives with plain radiography are common.

### Management

- antimicrobial therapy is justified in all cases because of the risk of CNS infection.
- anti-inflammatory drugs to reduce pain and inflammation.
- oral gabapentin.
- monitor and treat any exposure keratitis with:
  - subpalpebral lavage (SPL) catheter for cornea lubrication and ulcer treatment.
  - tarsorrhaphy.
  - avoid mouth gags and nasogastric intubation.
- surgery usually under general anaesthesia/standing sedation – ceratohyoidectomy.

### Prognosis

- fair to good:
  - majority of horses return to previous level of competition following treatment.
- vestibular and facial nerve deficits may take up to a year to improve.

## *Central nervous system trauma/accidents*

## Cerebral trauma

### Definition/overview

- **most frequently occurs after trauma to the frontal or parietal regions of the head:**
  - compound and/or displaced fractures.
  - more concerning when the trauma is caused by a narrow and irregular object:
    - hoof or narrow post.
    - broader trauma often causes upper cervical vertebral fractures and spinal cord injury.

### Aetiology/pathophysiology

- abnormal neurological signs due to:
  - mechanical injury to brain.
  - cerebral oedema.
  - brain haemorrhage/ischaemia caused by swelling and intravascular clotting.
- violent trauma to the brain injures the intracranial vessels, cell bodies, glia and axons.
- skull trauma initiates marked neuronal depolarisation and a transient coma.
- optic nerves may be stretched, leading to optic nerve avulsion or secondary ischaemic/reperfusion injury.
- secondary bacterial colonisation of the CNS is possible.

### Clinical presentation

- unconsciousness of variable length in severe cases.
- central depression.
- lateral recumbency with minimal reflexes for up to several hours:
  - may remain in sternal recumbency for a further period before rising.
- wander towards the side of the lesion if remain standing or rise quickly.
- head and neck turn towards the direction of circling.
- variable PLR with possible asymmetry of the pupils and miosis (Figs. 1.27, 1.28):

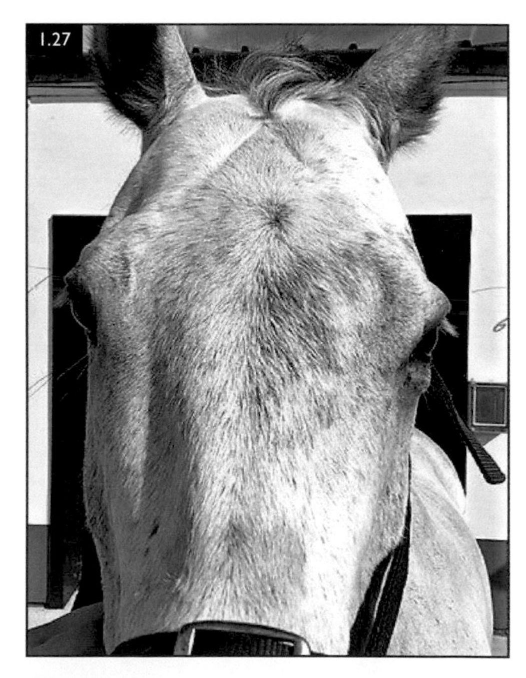


**FIG. 1.27** Evidence of ptosis and abnormal left eye position associated with an unobserved head trauma in an adult horse. In the left eye, there was an absent menace response, mydriasis and absent direct and indirect PLR, suggestive of a lesion affecting CNs II and III (and possibly CNs IV and VI), probably due to pathology in the retrobulbar space.

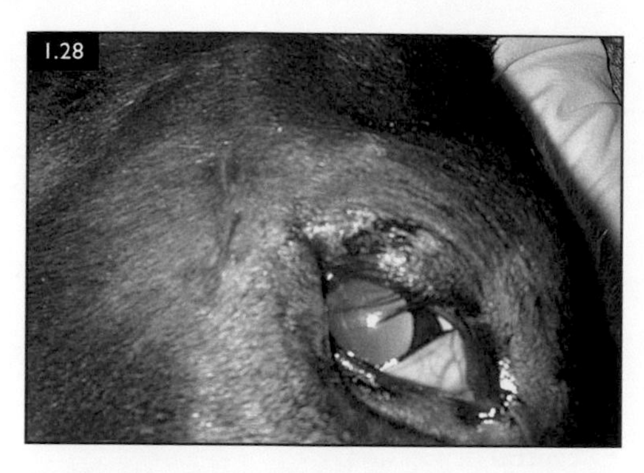

**FIG. 1.28** Foal that was found down in the pasture. The foal had non-responsive pupillary light reflexes and was obtunded. It made a full recovery from the cerebral trauma after 5 days of treatment.

- optic nerve avulsion causes reduced/ absent PLRs and dilated pupils rapidly after injury.
- swelling of the cerebral hemispheres may lead to midbrain herniation:
  - dilated pupils and tetraparesis.

- depressed menace response and central blindness.
- non-neurological signs include:
  - epistaxis from sinuses, ethmoids or nasal turbinates.
  - neurogenic pulmonary oedema and respiratory distress.
  - myocardial damage and sympathetic arterial hypotension.

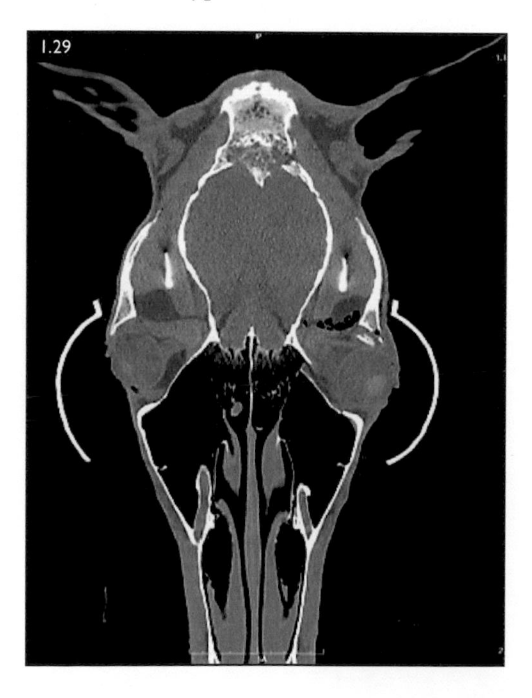

**FIG. 1.29** A reconstructed coronal CT image of the head of the horse in Fig. 1.27. Left is to the right. There is evidence of a comminuted fracture of the left orbit, and the supraorbital process of the frontal bone is displaced axially. The zygomatic process of the malar bone, orbital part of the frontal bone and orbital surface of the lacrimal bone are affected and displaced into the caudal maxillary sinus. There is marked soft-tissue swelling within the retrobulbar region, soft-tissue attenuating material extending into the underlying sinus and gas attenuating material within the soft tissues at the caudal aspect of the globe (air). The orbital nerve is poorly visible owing to the soft tissue and gas in the surrounding area. These findings are consistent with the suspected neuroanatomical localisation of the lesion mentioned earlier.

## Diagnosis

- history of head trauma or clinical evidence of trauma (abrasions, wounds).
- clinical picture:
  - localise lesion neurologically and establish baseline to monitor progression.
  - rule out other conditions such as primary seizure and secondary head trauma.
- diagnostic imaging including CT (Fig. 1.29).

## Management

- early surgery may be required in depressed skull fractures.
- otherwise, early management is supportive and medical:
  - attend to other life-threatening conditions first, especially respiratory function.
- address seizures or excessive thrashing:
  - sedation or short-term anaesthesia:
    - diazepam (5 mg [foal] or 50–100 mg [adult] i/v or i/m).
    - phenobarbital (12 mg/kg i/v loading dose, then 6 mg/kg i/v q12 h).
    - pentobarbital:
      - 150–1000 mg i/v for foals.
      - adults: slow increments of 500–1000 mg i/v to effect.
  - repeated seizures may require long-term therapy and multiple drugs:

- begin at low dose and increase daily until seizures controlled.
  - **alpha-2 agonists are contraindicated.**
- deal with any non-neurological vital needs of the patient:
  - nutritional and fluid support.
  - broad-spectrum antimicrobials if fractures are present.
  - cleaning and dressing of wounds.
  - protect patient from self-inflicted trauma and effects of prolonged recumbency.
- maintain normothermia – monitor for hyperthermia, as may occur after head trauma:
  - treat vigorously by:
    - clipping off hair
    - ice water.
    - fans.
    - antipyretics such as flunixin meglumine.
- maintain normotension and keep blood pressure as normal as possible:
  - isotonic fluids.
  - hypertonic saline
  - colloids:
    - plasma and synthetic colloids.
  - whole blood transfusion is warranted in haemorrhagic shock.
- avoid hypoxaemia:
  - maintain a patent airway.
  - provide supplementary oxygen via nasal or tracheal insufflation.
- control pain with a multimodal approach:
  - flunixin meglumine.
  - constant-rate infusions of butorphanol/lidocaine and opioids (fentanyl or morphine).
- regulate blood glucose and maintain nutrition:
  - enteral and parenteral nutrition when anorexic for more than 12–24 hours.
- prevention and treatment of brain swelling:
  - elevate the head/neck by a few degrees if recumbent.
  - support the head and neck and feed from chest height if standing.
  - hyperosmolar fluids reduce cerebral oedema:
    - hypertonic saline as a continuous intravenous infusion or as boluses.
    - 20% mannitol intravenously.
- antioxidant therapy:
  - 0.1 mg/kg q24 h of dexamethasone.
- Magnesium sulphate, 50 mg/kg i/v (equivalent to 25g in a 500 kg horse) in first 5–10 litres of intravenous fluids:
  - inhibits several aspects of the inflammatory cascade.

### Prognosis

- highly variable:
  - recumbency is a poor indicator.
- stabilisation of neurological signs within 24–48 hours.
- gradual improvement over next week and slower after that.
- plateau in improvement usually means there will be minimal further recovery.
- midbrain syndrome warrants a poor prognosis.
- uncomplicated cerebral syndrome has a good prognosis.

## Poll impact (Figs. 1.30–1.33)

### Definition/overview

- characteristic pattern of brain injury when horse rears over backwards:
  - caudal skull hits an overhead structure or the ground.
  - huge amount of force exerted on the base of the skull.
- usually fractures the paramastoid processes or occipital condyles:
  - possible fractures of the petrous temporal bones or parietal bones.
- hyperextension of the head relative to the neck at impact:
  - consequent contraction of the rectus capitus ventralis muscles:
    - fracture basilar bones from the base of the skull (horses <5 years old).
    - fracture fragments may lacerate adjacent blood vessels:
      - profuse haemorrhage into the retropharyngeal area or guttural pouches.
- fracture fragments may damage adjacent neurological tissue including:
  - brainstem or outflow from it.
- vestibular ataxia or facial nerve deficits may be apparent due to:
  - haemorrhage into the middle or inner ear.
  - fracture of the petrous temporal bone.

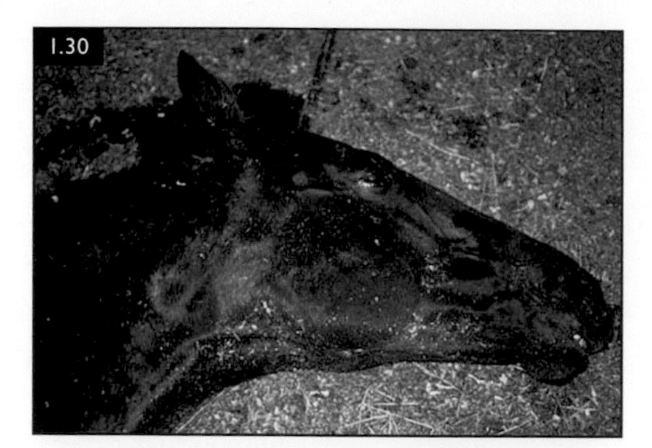

FIG. 1.30 Horse with a basisphenoid fracture.

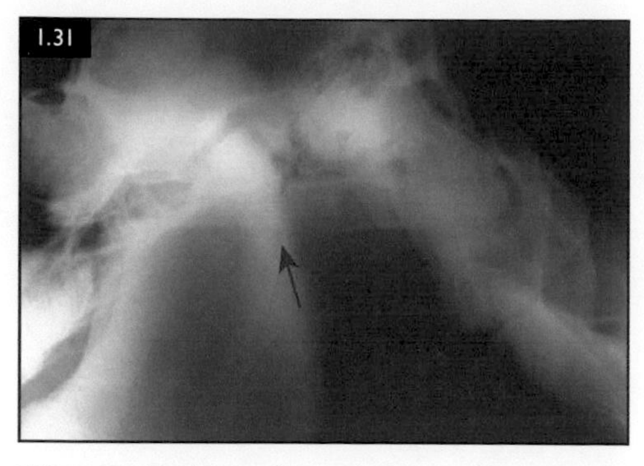

FIG. 1.31 Radiograph depicting a basioccipital fracture (arrow). Note the ventral displacement of the basisphenoid bone.

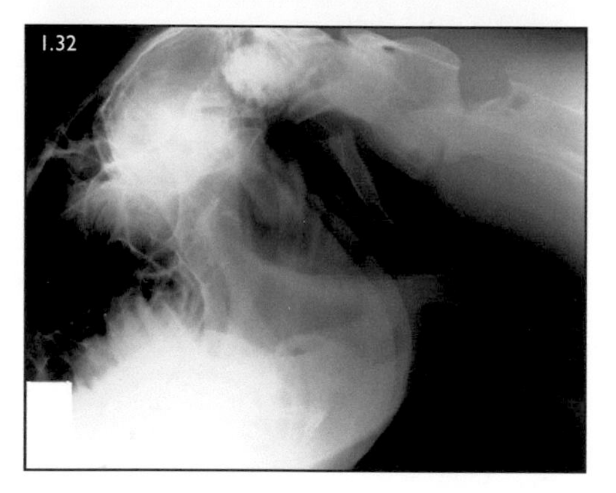

FIG. 1.32 Radiograph of a displaced fracture of the basisphenoid bone. This horse had reared up and flipped over in the show ring, resulting in temporary unconsciousness followed by seizures, ataxia and bilateral epistaxis.

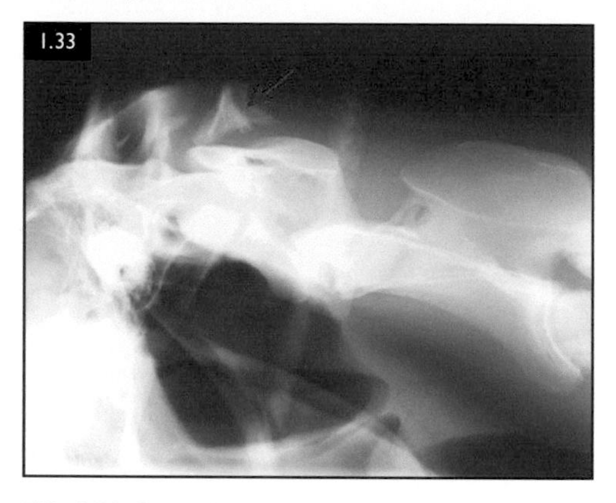

FIG. 1.33 Comminuted, displaced fracture of the occipital protuberance (arrow) because of falling over backwards.

- cerebellar signs are rare:
  - cerebral and optic nerve damage, as described in the last section, can occur.
- treatment similar to that for acute head injury, as described above.
- prognosis depends on the severity, complexity and location of neuronal injury:
  - poor to guarded.

## Intracarotid injection

### Definition/overview

- inadvertent injection is common:
  - close apposition of the jugular vein and carotid artery in caudal third of neck.
- usually leads to generalised seizure, followed by recumbency and degrees of coma.
- contralateral facial twitching and wide-eyed apprehensive appearance may precede seizure.
- animals may rear and strike violently.
- viscous or irritant drugs cause more rapid signs of onset.
- water-soluble drugs – recovery usually occurs:
  - horse standing in 5–60 minutes and completely normal in 1–7 days.
- insoluble and oil-based drugs – recovery is usually unsatisfactory:
  - epilepsy, stupor and coma are common and may necessitate euthanasia.
- treatment is not usually necessary with water-soluble drugs:
  - effects are self-limiting and of short duration.

- violent horses should be sedated with diazepam (50–100 mg i/v):
  - placed in a padded stall.
  - given dexamethasone (0.02–0.05 mg/kg i/v).
- do not give mannitol in the first 24 hours.

## Facial nerve trauma (Fig. 1.34)

### Definition/overview

- one of the most common peripheral nerve injuries.
- site of damage and severity determines:
  - whether some, or all, of the facial muscles are affected.
  - prognosis for the return of facial nerve function.
- unilateral facial paralysis is more common:
  - deviation of the muzzle away from the lesion.
  - drooping of the ipsilateral ear, eyelid and lip.
  - reduced flaring of the ipsilateral nostril during inspiration.
  - inability to close the eyelid and reduced tear production:
    - causes exposure keratitis.
- bilateral facial nerve paralysis:
  - dysphagia and accumulation of feed between teeth and cheeks.
- full facial paralysis:
  - injury proximal to vertical ramus of mandible:
    - fractures of the vertical ramus of the mandible, stylohyoid bone or the petrous temporal bone.
- unilateral facial paralysis without direct injury to nerve:
  - medullary lesions involving the facial nucleus.
  - polyneuritis equi.
  - idiopathic facial paralysis.
  - haemorrhage into middle or inner ear.
  - guttural pouch mycosis and parotid lymph node abscessation.
- distal facial nerve damage usually caused by direct injury from external trauma, head entrapment or prolonged lateral recumbency:
  - often where nerve crosses the mandible or zygomatic arch.
- treatment with:
  - phenylbutazone (3–5 days), flunixin meglumine (3–5 days) or dexamethasone (2–3 days) at standard doses.
  - topical anti-inflammatory cream, such as 1% diclofenac sodium.
  - section of the nerve can be repaired surgically if the ends can be identified.
- prognosis is determined by the site/severity of damage and is guarded to poor.

## Spinal cord trauma

### Definition/overview

- commonly suspected neurological disorder presented to equine practitioners.
- both musculoskeletal and/or neurological disorders may be encountered.

### Aetiology/pathophysiology

- injury is a two-stage process:
  - primary mechanical insult to neurons, axons and microvasculature.

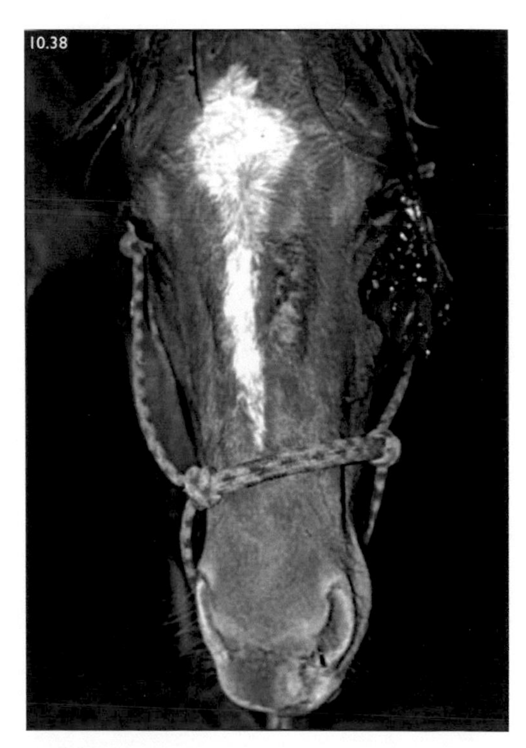


**FIG. 1.34** Skull fracture causing trauma to the left facial nerve. Note the muzzle deviation to the right.

  - secondary injury related to ischaemia, oxidative stress, excitotoxicity and inflammation.
- causes of vertebral trauma +/− spinal cord trauma include:
  - traumatic incident (e.g. a fall).
  - occasionally secondary to other pathology such as neoplasia.
- cervical vertebrae are common sites:
  - occipitoatlantoaxial region in foals.
  - cranial cervical site in rotational or 'head-butt' traumas in adults (Fig. 1.35).
  - caudal cervical and cranial thoracic sites are most common areas for vertebral fractures in the adult horse (Figs. 1.36–1.38).
- neurological signs are:
  - associated with fractures of the vertebral body, arch or articular processes.
  - not associated with fractures of thoracic dorsal spinous processes.

## Clinical presentation

- neurological abnormalities not always present.
- variability in the syndromes that result, depending on area damaged and severity.

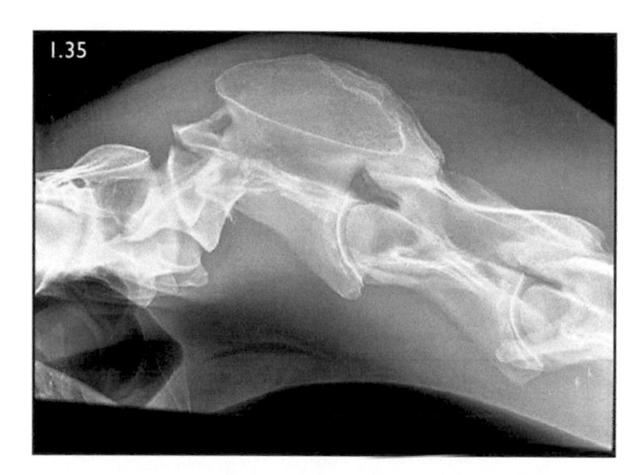

**FIG. 1.35** A lateral–lateral radiograph of the cranial cervical vertebrae (C1–C2) in a 10-year-old eventer that sustained a rotational fall over a fence and head-butted the ground. The horse showed immediate signs of neck stiffness, low head carriage, somnolence, grade 2 quadrilateral ataxia and obvious soft-tissue swelling in the dorsal cranial neck. The radiograph shows evidence of a fracture of the dens (odontoid process) (type IIa), with bone fragmentation and malalignment of C1–C2.

**Cervical spine**

- recumbent with lesion at C1–C3:
  - difficulty raising their head off the ground.
- recumbent with lesion at C4–T2:
  - able to lift the head and cranial neck.
- unilateral lesion:
  - able to lift the head when affected side is on the ground.
- possible tetraplegia, or tetraparesis and ataxia.

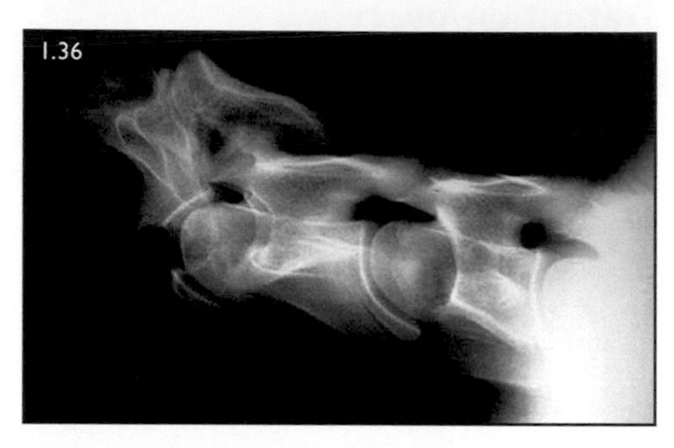

**FIG. 1.36** Displaced cervical (C5) fracture.

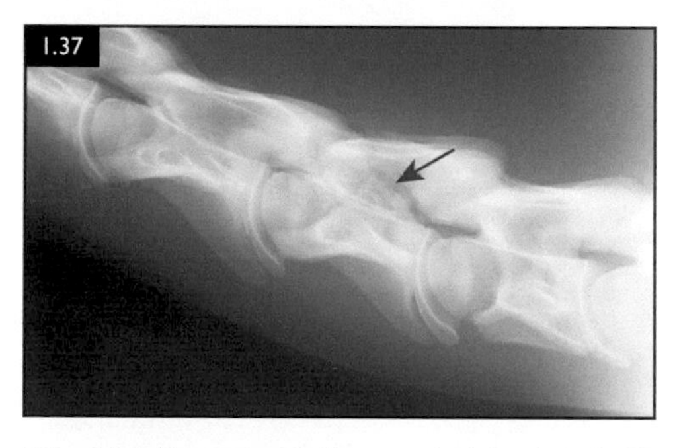

**FIG. 1.37** Fracture of the cervical body of C5 (arrow) that occurred as a result of a fall.

**FIG. 1.38** Foal with symmetrical grade 3 ataxia in all four limbs. Note the swelling in the mid-caudal neck. The foal had a fracture of C5, as displayed in 10.40.

- C1–C6 lesions:
  - muscle tone and spinal reflex responses typical of UMN lesion:
    - normal to increased muscle tone.
    - hyperreflexive spinal reflexes in all limbs.
    - inability to empty the bladder due to increased urethral sphincter tone.

**Cervicothoracic spine (C6–T2)**

- LMN forelimb signs (hypotonia and hyporeflexia).
- UMN hindlimb signs (hyperreflexia and hypertonia).
- conscious perception of pain may be reduced in all four limbs.
- UMN bladder may also be present.
- grey matter lesions between T1 and T3:
  - may result in signs of Horner's syndrome.

**Thoracolumbar** (T3–L2) (Fig. 1.39)

- T3–T6 lesions:
  - paraplegia or paraparesis and ataxia.
- usually bilateral signs.
- normal to exaggerated spinal reflexes and hypertonia of the hindlimbs.
- conscious proprioceptive deficits in the hindlimbs.
- forelimbs are unaffected.
- UMN bladder dysfunction.
- level of hypalgesia on neck or back indicates cranial extent of lesion.
- strip patches of sweating when thoracolumbar spinal nerve roots are damaged.

**Lumbar spine**

- lesions at L1–L3:
  - normal or hypertonic and hyperreflexic hindlimbs (Figs. 1.40, 1.41).
- lesions at L4–S2:
  - hypotonia and hyporeflexia of the hindlimbs.
- bladder distended but sphincter tone normal.
- tail and anal tone normal.

**Lumbosacral spine** (L3–S2).

- paraplegia and paraparesis of the hindlimbs:
  - often flaccid paralysis.
  - hyporeflexia or areflexia.
- no forelimb involvement.
- lesion S1/S2L:
  - bladder shows signs of LMN dysfunction.
  - paraphimosis and reduction in tail tone are also possible.

**Sacrococcygeal** (S3–Coc5)

- Cauda equina syndrome with flaccidity of the tail and paraphimosis:
  - loss of sensation to the entire perineum, and urinary incontinence.
  - unable to urinate or defaecate.

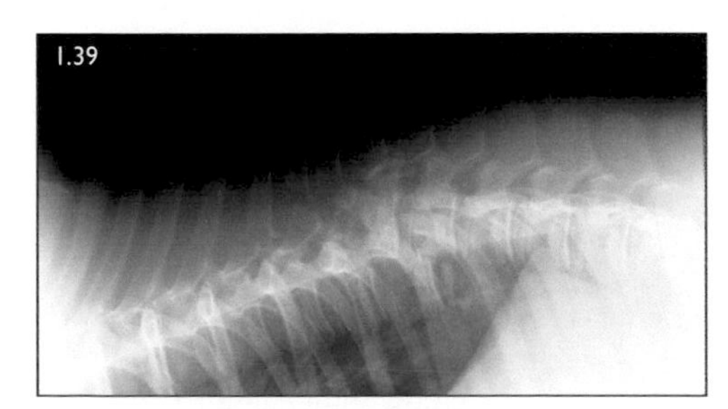

**FIG. 1.39** A markedly displaced fracture of the thoracic spine secondary to a high-speed collision.

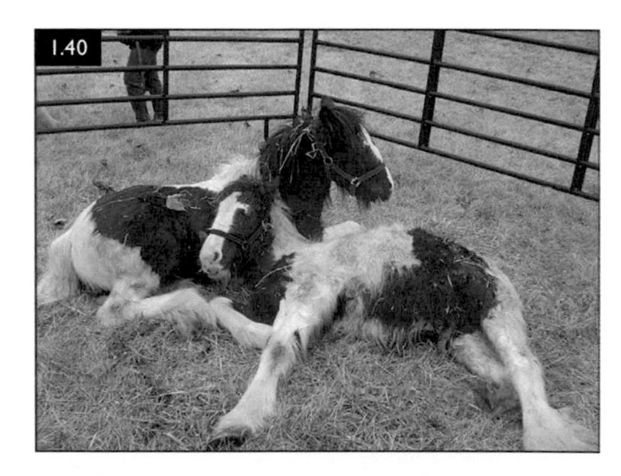

**FIG. 1.40** Two foals that were in a trailer accident and unable to stand. Note the paraphimosis in the foal in the foreground. This foal had a fracture of the 3rd lumbar vertebra (see Fig. 1.41). The foal in the background is sitting sternally and was able to 'dog sit'. This foal had a fracture of the 8th thoracic vertebra.

**FIG. 1.41** Transection of the spinal cord at the 3rd lumbar vertebra of the foal in Fig. 1.40.

- no signs of ataxia or proprioceptive deficits.

## Diagnosis

- neurological examination will allow neuroanatomical localisation of the lesion(s).
- plain radiography:
  - confirms vertebral trauma or displacements of vertebral components.
- CT useful for cervical and cranial thoracic fractures:
  - establish true extent of pathology.
  - limited by horse size and availability of facility.
- CSF:
  - acute changes (<24 h) include:
    - diffuse blood contamination, high red blood cell (RBC) count, normal to high white blood cell (WBC) count and high protein concentration.
  - chronic changes include:
    - normal to slightly increased WBC count, normal to increased RBC count, increased protein concentration and xanthochromia.
  - none of the changes are specific for vertebral fracture.

## Management

- similar to that for cerebral trauma.
- good nursing care is essential, especially for recumbent patients:
  - bladder and rectal evacuation.
- stable fracture and patient can stand with assistance:
  - water tank and/or slings can be used for support (Fig. 1.42).
  - may help minimise secondary complications and hasten recovery.

**FIG. 1.42** Use of a sling in the management of a yearling with a fracture at C7.

## Prognosis

- guarded to poor for return to use.
- healing often results in vertebral malalignment.
- delayed callus formation and degenerative changes can result in delayed permanent spinal compression.
- repeated neurological examinations are the best way to assess the prognosis:
  - longer the recumbency, the poorer the prognosis.

# Peripheral nerve disorders

## Definition/overview

- trauma is the primary cause.
- injections, tumours, abscesses or parasitic invasion of the nerves are also possible.

## Aetiology/clinical presentation

**Suprascapular**

- subsequent to collision between horses or with inanimate objects.
- paralysis of the infraspinatus and supraspinatus muscles:
  - outward bowing of the scapulohumeral joint when bearing weight.
  - neurogenic atrophy over a few months:
    - prominence of the scapular spine.
  - circumduction of limb during protraction.
- known as 'Sweeney' (Fig. 1.43).

**Radial** (Fig. 1.44)

- vulnerable to trauma over the lateral aspect of the elbow joint.
- prolonged anaesthesia or recumbency without sufficient padding.
- humoral fractures or fractures of the first rib.
- effects on the limb depend on location of damage to the nerve:
  - at or near elbow joint (high paralysis):
    - dropped elbow.
    - failure of limb protraction.
    - flexion of all distal limb joints.

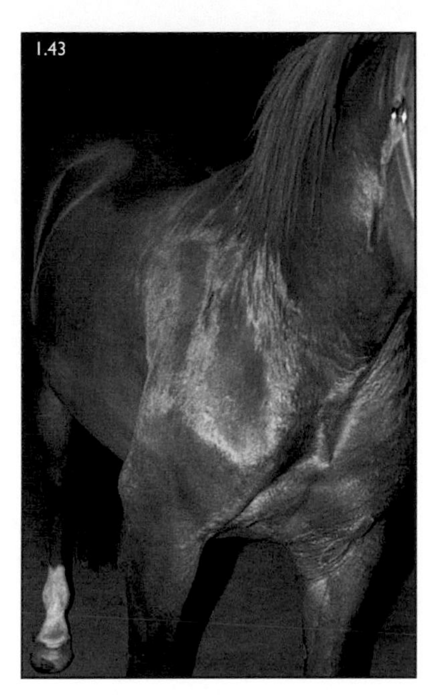


FIG. 1.43 Suprascapular nerve injury secondary to a collision with another horse.

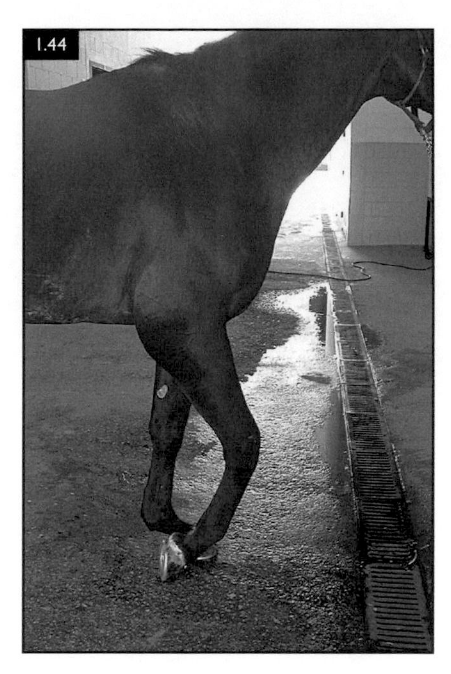


FIG. 1.44 Radial nerve paralysis.

- inability to bear weight on the limb.
- foot knuckled over at rest.
- distal nerve lesions:
  - flexion of carpus, fetlock and pastern joints.
  - can bear weight if lower limb held in extension.
- triceps reflex decreased or absent.
- sensory deficits are variable between patients.

**Brachial plexus**

- caused by shoulder trauma, deep penetrating axillary wounds and traction on fetal forelimbs during dystocia relief.
- deficits in motor innervation to the biceps, coracobrachialis (musculocutaneous), pectoral, subscapularis and triceps muscles:
  - complete flaccidity of the forelimb and inability to bear weight.
- absent triceps and biceps reflexes.
- elbow abduction • dropped shoulder.
- inability to flex the elbow.
- hyperextension at rest.
- signs of radial paralysis and entire forelimb desensitisation.

**Musculocutaneous, median, ulnar**

- usually resulting from brachial plexus injuries.
- results of ulnar neurectomy produce the most pronounced abnormalities:
  - decreased carpal and fetlock flexion, and residual effects such as stumbling.
- gait alterations may take up to 3 months to disappear:
  - may partly be due to crossing over of fibres between the musculocutaneous and median nerves.

**Femoral** (Fig. 1.45)

- caused by:
  - abscesses, tumours and aneurysms in the region of the external iliac arteries.
  - penetrating wounds of the caudal flank.
  - fractures of the pelvis and femur.
  - prolonged anaesthesia or recumbency.
  - spinal cord lesions at L4/5 including in EPM.
- paralysis leads to quadriceps muscle atrophy and inability to extend the stifle.
- extensor paralysis:
  - affected limb resting in a flexed position.

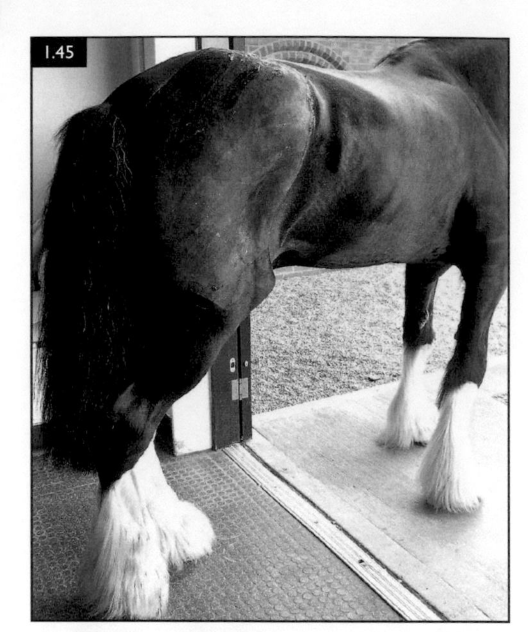

**FIG. 1.45** This Shire horse was presented with a 3-week history, following a traumatic episode in a field, of right-hind lameness and rapid profound quadriceps muscle atrophy. Subsequent clinical examination confirmed this was due to femoral nerve damage.

  - ipsilateral hip in a lower position than the contralateral hip.
  - no weight-bearing on the affected limb during locomotion.
- absent or reduced patellar reflex but a normal flexor reflex.
- bilateral cases lead to difficulty in rising from the ground and a crouching posture.
- prognosis is guarded.

**Tibial**

- uncommon.
- paralysis causes flexed tarsus and flexed, or partly knuckled, fetlock:
  - results in the hip being held lower on the affected side.
- over-flexion of the limb when walking:
  - foot raised higher than normal and dropped straight to ground (stringhalt-like).
- analgesia of caudomedial aspect of the leg.

**Peroneal**

- vulnerable to injury where it crosses the lateral condyle of the femur.
- injury from kicks and lateral recumbency but rarely is the nerve severed.
- paralysis leads to:
  - extension of the tarsus and flexion of the fetlock and interphalangeal joints.
  - horse holds the limb extended caudally at rest:
    - dorsum of the hoof resting on the ground.
  - hoof is dragged along the ground during locomotion:
    - horse can bear weight on the limb when placed manually cranially.
  - hypoalgesia on the craniolateral aspect of the hock and metatarsal regions.
- support and protect the limb whilst animal is given time to improve:
  - most cases, the nerve is not severed, and the horse eventually improves.

**Sciatic**

- damaged by pelvic fractures, especially ischial fractures.
- spinal cord lesions affecting L5–S3:
  - signs of urinary bladder paralysis etc. urinary bladder paralysis, gluteal atrophy and extensor weakness.
- foals:
  - deep injections caudal to the proximal femur.
  - osteomyelitis of the sacrum and pelvis.
- total sciatic paralysis leads to an abnormal gait and posture:
  - at rest, extension of the stifle and hock:
    - limb hanging behind the horse.
    - fetlock and interphalangeal joints partly flexed.
    - dorsum of the hoof on the ground.
  - foot manually advanced and placed on the ground ventral to the pelvis:
    - horse can support weight with some flexion of the hock and take a stride.
- muscles of the caudal thigh and the entire limb distal to the stifle atrophy.
- degrees of hypoalgesia over most of the limb, except for the medial thigh.
- treatment of primary problems may resolve the paralysis.
- prognosis is poor if the nerve is severed, even with surgical repair.

**Post-foaling/obturator paralysis**

- may be observed following foaling with or without a history of dystocia:
  - nerve courses along the medial side of the ilial shaft.

- may be compressed by foal during parturition or subsequent to post-parturient oedema, haemorrhage and swelling.
- signs range from mild stiffness to paraplegia.
- unilateral damage:
  - may present as abduction and circumduction of affected limb at walk.
- bilateral damage:
  - great difficulty in rising and may require support of a body sling.
- prognosis for post-foaling paralysis:
  - fair, with recovery in approximately 50% of cases.
  - prolonged recumbency leads to complications and often euthanasia.

### Management

- medical management involves:
  - reduction of inflammation:
    - 3 days of systemic dexamethasone, flunixin or phenylbutazone.
    - cold water or ice pack application for first 24 hours if localised swelling.
  - relief of musculoskeletal pain:
    - confine in deeply bedded stalls.
  - prevention of secondary medical disorders:
    - turn recumbent animals 6–8 times per day:
      - prevent decubital ulcers and pressure myopathy.
    - support in slings may help maintain strength in the opposite limbs.
  - provision of adequate nutrition.

## *Toxin-associated disorders*

## Botulism

### Definition/overview

- rapidly progressive neuromuscular disease which is often fatal.
- toxins produced by *Clostridium botulinum:*
  - Gram-positive spore-forming anaerobic bacterium.
  - found in animals, soil and organic material.
- horses are very susceptible, with individual and group outbreaks.

### Aetiology/pathophysiology

- 8 botulinum neurotoxins (BoNTs) are produced by different strains:
  - geographical distribution and clinical presentation variation in BoNT types.
- three pathogenic mechanisms:
  - **Adult botulism**: ingestion of pre-formed toxin:
    - most common form of botulism in Europe.
    - caused by ingestion of contaminated forage:
      - animal carcasses.
      - decaying vegetable matter (big-bale hay/silage, grass clippings).
  - **Foal botulism:** ingestion of spores:
    - rare.
    - toxico-infectious botulism or 'shaker foal syndrome'.
    - few days to months old.
    - proliferation and toxin production in the GI tract.
  - **Wound botulism:** contamination by spores – rare:
    - anaerobic environment:
      - umbilical/injection abscess.
      - castration site.
      - injection abscess.
    - allows germination of spores and production of BoNT in host.
- toxin irreversibly binds pre-synaptically at peripheral cholinergic neuromuscular junctions:
  - prevents release of acetylcholine at the junction.
  - causes flaccid paralysis.
  - improvement in clinical signs only with regeneration of new motor end plates.

### Clinical presentation (Table 1.7, Figs. 1.46, 1.47)

- severity and rate of progression associated with the amount of toxin ingested.
- clinical signs within hours to 1–16 days of intoxication.
- symmetrical muscle fasciculations starting at the triceps:
  - progresses to larger muscle groups resulting in recumbency.

**TABLE 1.7** Clinical signs of botulism in horses

| |
|---|
| **Alimentary** |
| Mild colic due to non-strangulating obstructions |
| Reduced prehension of food and ability to swallow |
| Reduced tongue tone (see Table 1.8) |
| Reduced borborygmi and ileus |
| **Musculoskeletal** |
| Weak tail tone |
| Poor anal sphincter tone |
| Low head carriage (with associated head oedema and respiratory stridor in some cases) |
| Muscle fasciculations |
| Stumbling and tripping (due to weakness, not proprioceptive deficits) |
| Increased recumbency |
| **Ocular** |
| Mydriasis |
| Ptosis |
| Slow pupillary light response |
| **Respiratory** |
| Dysphonia |
| Respiratory stridor (related to head oedema) |
| Reduced respiratory rate (6–12 breaths per minute) |
| Altered respiratory movement (prominent abdominal muscular effort) |
| Respiratory muscle paralysis |
| **Urinary** |
| Bladder distension |
| Urinary incontinence |

- exaggerated respiratory effort and reduced respiratory rate.
- urine dribbling due to bladder distension and sphincter paralysis.
- normal sensory function.
- GI impactions due to ileus and dehydration.
- dysphagia may lead to dehydration, malnutrition and aspiration pneumonia.
- death is inevitable in untreated cases due to respiratory failure.

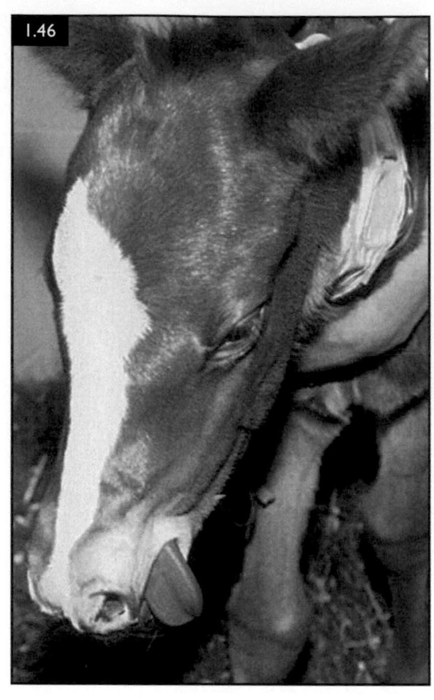

**FIG. 1.46** Weak tongue tone and dysphagia in a foal with type B botulism.

## Differential diagnosis

- other diseases causing profound muscle weakness, dysphagia or muscle fasciculations.
- Equine grass sickness.
- viral encephalitides.
- atypical myopathy.
- EMND.

## Diagnosis

- presumptive diagnosis on history and clinical examination:
  - abrupt onset of diffuse, symmetrical weakness.
  - gradually progresses to recumbency in 1–4 days.
  - normal mentation, concurrent dysphagia and decreased tongue tone.
- dynamic tests may support a clinical diagnosis of botulism (Table 1.8).
- definitive diagnosis requires the detection of the toxin in forage, serum, GI contents or faeces, or a rarely performed mouse bioassay:
  - ELISA and PCR tests for BoNT are available in the USA.

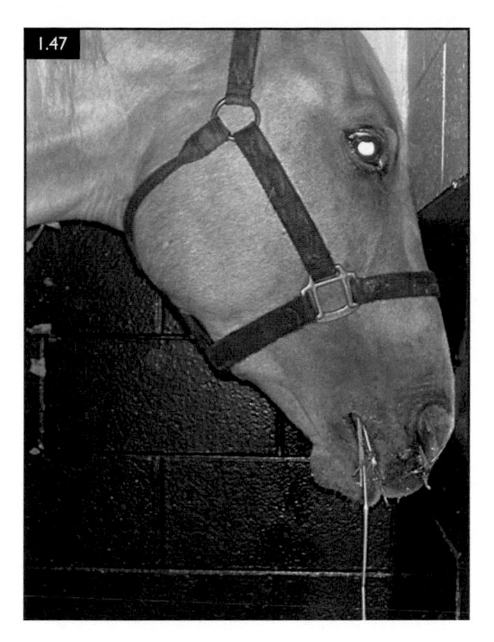


**FIG. 1.47** A mare that was dysphagic and had weak palpebral tone. The mare had botulism secondary to forage poisoning. Eight other horses on the farm were affected.

**TABLE 1.8** Dynamic tests to aid in the detection of a suspected botulism case

| Tongue stress test |
|---|
| • Keep jaws closed with one hand over the bridge of the nose |
| • Retract the tongue from the mouth, between the commissure of the lips |
| • Gently release the tongue |
| • If the horse allows the tongue to hang for more than 2–3 seconds, this is abnormal and indicates a reduced tongue tone due to cranial nerve or neuromuscular dysfunction |
| **Muscle weakness** |
| • Botulism causes weakness, not ataxia; therefore, a neurological examination might help to establish whether there are conscious proprioceptive deficits, or whether clinical signs relate exclusively to muscle weakness<br>• Limb-holding test (holding one forelimb and watching the contralateral limb for muscle fasciculations and buckling) may exacerbate mild muscle weakness |

## Management

- neutralising circulating toxins:
  - treat promptly with specific or multivalent antiserum.
  - >70% survival if antiserum used early in mild cases in adults.
  - clinical signs continue to worsen for a few hours after administration.
  - available in USA but not in all geographical regions.
- excellent nursing care:
  - enteral or intravenous isotonic fluids.
  - enteral feeding:
    - indwelling nasogastric feeding tube.
    - parenteral intravenous nutrition.
  - recumbent animals need regular turning and deep bedding.
  - consider use of slings and hoists.
  - anti-gastric ulcer medication.
- judicious use of antimicrobials in wound botulism:
  - no benefit in adult botulism unless secondary complications (aspiration pneumonia).
  - **do not use procaine penicillin and aminoglycosides as may potentiate effects of BoNT.**

## Prognosis

- poor if clinical signs progress rapidly.
- grave in recumbent/unable-to-rise adults.
- dysphagia cases may start to swallow within 3–14 days.
- return to full muscle strength may take months.

## Prevention

- no vaccines licensed in Europe, but some in the USA.
- vaccination of high-risk mares has reduced the incidence of neonatal botulism.
- good husbandry measures reduce risk of exposure:
  - inspect forage for spoilage and animal carcasses.
  - large round hay or haylage bales are a particular risk and need careful use.
  - strict control of rodents and birds.

# Urea and non-protein nitrogen sources

- urea has been used as a non-protein nitrogen source for adult horses.

- urea is absorbed before reaching the hindgut normally:
  - if it reaches the hindgut it is hydrolysed to ammonia.
- lethal dose is 4 g/kg when ingested orally.
- clinical signs are confined to the nervous system:
  - muscle tremors, incoordination and weakness.
  - death is the result of absorption of ammonia and intoxication.
- diagnosis on clinical signs and history:
  - blood ammonia levels can be measured but influenced by many factors.
- treatment with lactulose (200 ml p/o or per rectum [p/r] q4–6 h) may be attempted.

## Monensin

- Ionophore antibiotic used as a feed additive for growth promotion (cattle) and coccidiostatic effects (poultry).
- toxicity results from abnormal levels of potassium or calcium within the cell, leading to cell death:
  - heart is the primary organ of toxicity.
  - horses very sensitive to toxicosis, with a median lethal dose ($LD_{50}$) of 2–3 mg/kg.
- inadvertent consumption leads to several different toxic syndromes (dose related):
  - **peracute** toxicity:
    - progressive severe haemoconcentration.
    - hypovolaemic shock and death within a few hours of ingestion.
  - **acute** toxicity:
    - ataxia.
    - progressive muscle weakness.
    - tachycardia
    - hypotension.
    - dyspnoea
    - polyuria.
    - anorexia
    - abdominal pain.
    - intermittent profuse sweating.
    - clinical signs for 1–4 days before death.
  - **surviving horses:**
    - signs of unthriftiness.
    - decreased athletic performance.
    - cardiac failure.
- diagnosis on basis of history of exposure to contaminated feed and clinical signs:
  - no clinicopathological test is useful.
- no specific antidote:
  - early and aggressive fluid therapy to combat haemoconcentration and shock.
  - acid–base and electrolyte abnormalities must be corrected.
  - mineral oil/activated charcoal to decrease absorption and evacuate bowel.
- **Glycosides should never be used in acutely affected horses.**
- **Calcium should not be administered to acutely affected horses.**
- prognosis is guarded to poor:
  - recovering animals may have permanent cardiac damage.

## Organophosphate poisoning

- usually due to:
  - overzealous application of insecticides, acaricides or anthelmintics.
  - accidental contamination of feed or water.
- clinical signs result from inhibition of acetylcholinesterase activity:
  - uncontrolled muscarinic, nicotinic and CNS effects of acetylcholine.
  - hyperexcitability, frequent urination, sweating, muscle tremors and colic.
  - followed by weakness, recumbency and death due to respiratory insufficiency.
- treatment consists of:
  - atropine (0.1 mg/kg, half slowly i/v and half i/m).
  - pralidoxime chloride (2-PAM) (20 mg/kg i/m).
  - skin washing and saline purgation to remove residual poison.
- delayed neurotoxicity unusual in horses, but transient laryngeal paralysis is reported.

## Mycotoxicosis (Fig. 1.48)

- secondary metabolites of moulds produced during mould growth in grains or forages.
- poor correlation between spore counts/fungal growth amount and mycotoxin contamination.
- absence of mould growth does not mean mycotoxins are not present in the food.
- many mycotoxins of concern:
  - aflatoxin, fumonisin and ergot alkaloids.

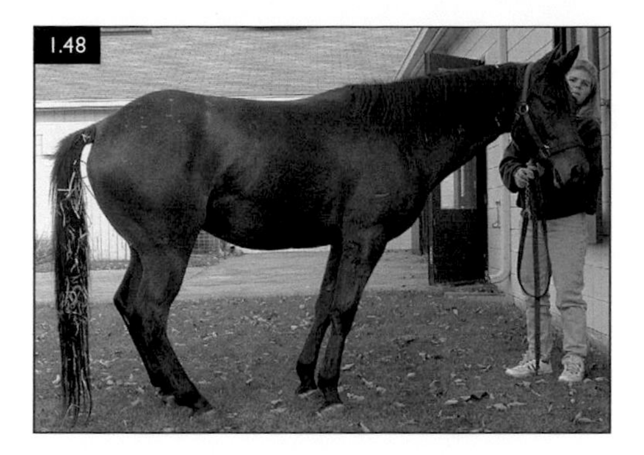


**FIG. 1.48** Mycotoxicosis. One of six affected horses on a farm. At walk there was a marked hypermetric gait of the hindlimbs only and the cases would always stand with all four limbs underneath themselves. Mould was noted in two bales of forage that were fed to these horses.

## Tremorgenic mycotoxicosis (ryegrass staggers, paspalum staggers)

### Definition/overview

- several mycotoxin syndromes characterised by staggering:
  - ryegrass staggers is the most described.
  - reported in the USA, Europe and Australasia.

### Aetiology/pathophysiology

- tremorgenic mycotoxins associated with reduced effects of inhibitory amino acids:
  - increased pre-synaptic neurotransmitter release and prolonged depolarisation.
  - synaptic transmission at the motor end plate is facilitated.
- **Perennial ryegrass** (*Lolium perenne*) is the major forage supporting the endophytic growth of *Neotyphodium lolii*:
  - produces tremorgenic mycotoxins known as lolitrems A, B, C and D.
  - lolitrem B is considered the major toxin:
    - syndrome commonly seen in the late summer or autumn.
    - toxin is concentrated in the seed.
- **Paspalum staggers** due to ingestion of dallis grass or hay (*Paspalum dilatatum*):
  - grown in southern regions of the USA.
  - invaded by *Claviceps paspali* and subsequently produces an ergot sclerotium.
  - contains toxic alkaloids known as paspalitrems.
  - inhibition of gamma-aminobutyric acid (GABA) is the probable mechanism of toxicosis.

### Clinical presentation

- action-related diffuse, intermittent and mild muscle tremors.
- symmetrical vestibular ataxia, stumbling and falls.
- excitement and blindfolding markedly exacerbate the signs.
- condition not usually fatal.

### Differential diagnosis

- fluphenazine reaction.
- reserpine reaction • EPM.
- WNV encephalomyelitis.
- Lead toxicosis.

### Diagnosis

- rule out other potential neurological disorders.
- detection of the toxins in the pasture.

### Management

- no antidote for either syndrome.
- removal of suspect forage results in recovery:
  - 3–7 days for perennial ryegrass.
  - 1–3 weeks for paspalum staggers.
- kept in quiet secure place during recovery, plus good nursing care.

### Prognosis

- good provided the source of the toxin is removed and supportive care is provided.
- chronic ryegrass staggers may lead to residual CNS lesions.

## Leucoencephalomalacia

### Definition/overview

- highly fatal neurological disease caused by ingestion of mycotoxins.
- worldwide occurrence.

### Aetiology/pathophysiology

- ingestion of corn contaminated with *Fusarium verticillioides* (formerly *Fusarium moniliforme*) or *Fusarium tricinctum*:

- produce the toxins (B1, B2 and B3).
- Fumonisin (B1) interferes with sphingolipid metabolism:
  - disrupting endothelial cell walls and basement membranes.
  - liquefactive necrosis and malacia of white matter of one or both cerebral hemispheres results.
  - other organs may be affected, especially the liver.
- growth of fungus increases in cool, humid conditions:
  - toxicosis most common in the late autumn and early spring.

### Clinical presentation

- on average, 3 weeks after initial ingestion of contaminated corn.
- initial signs related to cerebral disease and last 1–3 days:
  - depression, circling, aimless wandering and blindness.
  - no fever.
  - recumbency, coma and death.
- morbidity reported between 14% and 100%.
- recovery can be prolonged, and survivors may have permanent neurological problems.

### Differential diagnosis

- EPM
- verminous migrans.
- hepatic encephalopathy
- Rabies.
- trauma
- cerebral abscess.

### Diagnosis

- clinical presentation and history of possible exposure to contaminated feed:
  - multiple animals may be affected.
- definitive diagnosis is difficult ante-mortem.
- liquefactive necrosis at necropsy is diagnostic.

### Management

- no specific treatment.
- supportive care and removal of contaminated feed.
- elimination of toxins by use of activated charcoal (2.2 kg/500 kg body weight in 4 litres of water q12 h for 24–36 hours).

### Prognosis

- poor – mortality rates of 40–84%

## Metaldehyde toxicosis

### Definition/overview

- uncommon and may result from inadvertent ingestion or malicious poisoning.

### Aetiology/pathophysiology

- polycyclic polymer of acetaldehyde used in slug and snail baits.
- exact mechanism is unclear, but proposed:
  - increased levels of excitatory neurotransmitter.
  - decreased levels of inhibitory neurotransmitter.
- exact toxic dose is unknown.

### Clinical presentation

- within 15 minutes of ingestion.
- ataxia, hyperaesthesia, muscular twitching and agitation.
- convulsive spasms, stringhalt-like limb movements, tachycardia and tachypnoea, and profuse sweating.
- death within a few hours, often accompanied by violent convulsions.

### Diagnosis

- clinical signs and known exposure to metaldehyde.
- clinicopathology is non-diagnostic and post-mortem lesions non-specific.
- testing can be carried out on stomach contents or serum for metaldehyde.

### Management

- no antidote is available.
- control convulsions.
- mineral oil or activated charcoal to minimise absorption.
- supportive care including treatment of dehydration and acidosis.

### Prognosis

- variable and horses that recover do not suffer from any sequelae.

## Lead toxicosis

### Definition/overview

- rare cause of neurological disease associated with inadvertent ingestion of lead.

### Aetiology/pathophysiology

- global contaminant from:
  - nearby mines and smelting operations:
    - forage containing more than 80 ppm can cause chronic lead poisoning.
  - discarded lead–acid batteries.
  - ashes remaining after combustion of older buildings.
  - lead-based paint on buildings (usually built before 1960).
- solubilised in acidic environment of stomach and readily absorbed from proximal small intestine.
- bivalent lead acts like calcium in body and is dynamically bound in bone matrix.
- toxic dose depends on age of the animal:
  - younger animals absorb 10–20% of ingested lead, versus 1–2% in adults.
- chronic lead exposure leads to primarily neurological and haematopoietic damage.
- high levels of lead deposition in the CNS results in CNS toxicity including acute cerebellar haemorrhage and oedema from capillary dysfunction.

### Clinical presentation

- central and peripheral nerve dysfunction:
  - laryngeal hemiplegia
  - ataxia.
  - muscle fasciculations.
  - hyperaesthesia.
  - depression.
- GI tract upset including dysphagia (secondary aspiration pneumonia) and weight loss.
- interference with haematopoiesis:
  - mild to moderate anaemia.

### Differential diagnosis

- EMND.
- anticholinesterase insecticide toxicity (organophosphates and carbamates).

### Diagnosis

- haematology reveals anaemia and increased nucleated RBCs and basophilic stippling.
- blood lead concentration >0.35 ppm with appropriate clinical signs.
- liver and kidney lead concentrations of >10 ppm at post-mortem.

### Management

- enhance urinary excretion with chelation therapy:
  - ethylenediaminetetraacetic acid (EDTA) accelerates excretion from blood and soft tissues.
  - 6.6% solution of calcium disodium EDTA in 5% dextrose or normal saline:
    - administered at 75 mg/kg/day i/v, divided into three doses.
    - 4 days treatment, 2 days off and 4 days treatment.
- concurrent intravenous fluid therapy facilitates excretion of water-soluble chelate and aids hydration.
- effectiveness of therapy evaluated by monitoring changes in lead concentrations within urine and blood.
- blood concentration checked 2 weeks after final chelation; if >0.35 ppm:
  - second course of chelation therapy.
- Thiamine therapy (1 mg/kg i/v) is also used in ruminants with lead poisoning and has proved an effective adjunctive treatment with EDTA.
- supportive care.
- prompt environmental investigation to identify the source of contamination.

### Prognosis

- good with response to chelation therapy.

## Snakebite

- wide variety of snakes with the most common from the families *Crotalidae* (pit vipers) and *Elapidae* (cobras).
- usually cause local effects, with swelling and oedema at the bite site:
  - head and muzzle are commonly affected as well as limbs.
- venom from *Elapidae* is mainly neurotoxic, with minimal local effects:

- envenomation is extremely uncommon in the horse:
  - initial excitement and hyperaesthesia.
  - followed by generalised weakness.
- not usually fatal to adults but can be in young foals.
- treatment includes supportive care, tetanus prophylaxis, broad-spectrum antimicrobials and local treatment of wounds (Fig. 1.49).

## Tick paralysis

### Definition/overview

- rare condition, with a regional distribution, causing ascending LMN paralysis.

### Aetiology/pathophysiology

- ticks *Ixodes holocyclus* and *Dermacentor andersoni:*
  - *Ixodes* paralysis reported in Australia.
  - *Dermacentor* paralysis reported in North America.

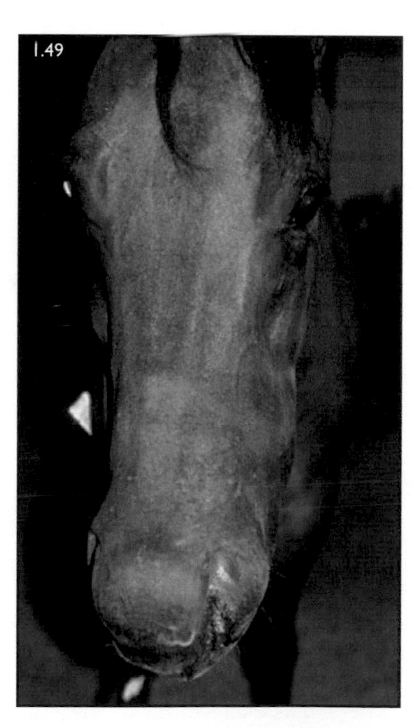

**FIG. 1.49** Rattlesnake bite on the nares of a mare in Arizona, USA.

- tick salivary toxin causes inhibition of acetylcholine release at the neuromuscular junction:
  - results in generalised flaccid paralysis.

### Clinical presentation

- ascending paresis followed by paralysis.
- rapid progression and horses can die within 24 hours.

### Differential diagnosis

- Botulism.

### Diagnosis

- characteristic clinical signs in a specific geographic region.
- identification of ticks on the body.

### Management

- remove all ticks after careful examination of whole body:
  - clipping may be required.
- supportive care similar to botulism.
- prevention of disease involves avoiding tick-infested areas and regular inspection.

### Prognosis

- good with *Dermacentor* infestations, provided ticks are removed.
- poor with *Ixodes* infestation.

## Locoweed intoxication

### Definition/overview

- geographically important cause of neurological disease.
- chronic ingestion of *Astragalus* and *Oxytropis* genera.

### Aetiology/pathophysiology

- *Astragalus* and *Oxytropis* spp. found in western North America and Australia:
  - perennial plants that contain a variety of toxins including:
    - swainsonine and swainsonine *N*-oxide.
    - cellular vacuolation and dysfunction from inhibition of mannosidase.
  - initially eaten when food is scarce, but some horses acquire a taste for it:
    - horses are more susceptible than other species.

## Clinical presentation

- signs 2 weeks to 2 months after ingestion commences.
- continuous ingestion is required to cause disease.
- abrupt onset of signs of a diffuse CNS disorder:
  - periods of depression alternating with frenzy and overreaction to stimuli.
  - gait abnormalities, head-nodding, visual impairment and dysphagia.
  - rapid weight loss, emaciation and death.
- other signs such as abortion and limb deformities in foals.

## Differential diagnosis

- EPM
- WNV encephalomyelitis.
- trauma
- hepatic encephalopathy.
- aberrant parasite migration (*Halicephalobus deletrix*).
- Rabies.

## Diagnosis

- clinical presentation and history of possible chronic ingestion of locoweeds.
- identification of cytoplasmic vacuolation in lymphocytes.

## Management

- no effective treatment for chronic intoxication.

## Prognosis

- mildly affected horses recover in 1–2 weeks if the source is removed.
- no recovery from chronic poisoning with continued clinical signs.

# Nigropallidal encephalomalacia

## Definition/overview

- plant toxicosis only reported in horses.

## Aetiology/pathophysiology

- Yellow star thistle (Fig. 1.50) (*Centaurea solstitialis*):
  - aggressive weed found in USA in California, Oregon and Idaho.
- Russian knapweed (*Acroptilon repens*, previously classified as *Centaurea repens*):
  - broader distribution in intermountain states of the USA.
- not palatable – most poisonings occur in the spring when the plant is young and green:
  - remains toxic when dry and can occur after ingestion of contaminated hay.
- up to twice the horse's body weight has to be consumed before signs of toxicity:
  - weeks to months of consumption before neurological signs.
- inhibition of dopamine release by the toxic agent repin.

## Clinical presentation

- young horses (<3 years old) more frequently.
- acute onset:
  - lack of coordination of facial and oral movements.
  - effective eating or drinking not possible.
- head oedema
- yawning
- ataxia.
- muscle tremors.
- hypertonicity of the lips and tongue.
- involuntary lip-twitching.
- death usually due to starvation, dehydration or aspiration pneumonia.

**FIG. 1.50** Yellow star thistle. (Photo courtesy WA State Noxious Weed Control Board.)

### Differential diagnosis

- EPM
- dental abnormalities.
- fractured jaw
- glossitis.
- pharyngeal abscess/neoplasia.
- oesophageal obstruction
- botulism.
- guttural pouch pathology.
- lead toxicosis.
- arboviral encephalitis.

### Diagnosis

- clinical signs and history of chronic exposure:
  - check pasture and hay.
- MRI successfully used for ante-mortem diagnosis.
- histopathology of brain.

### Management

- no specific treatment for the disease other than supportive care.
- remove from source.

### Prognosis

- poor and complete recovery is not seen.

## Other plant poisonings

### Bracken fern and horsetail poisoning

- *Pteridium* (bracken fern) or *Equisetum* (horsetail) ingestion.
- eaten at times of shortage or in hay.
- contain a thiaminase that causes thiamine deficiency.
- clinical signs weeks after ingestion begins and may continue after removal:
  - most frequently, anorexia, bradycardia and ataxia (severe and all 4 limbs).
  - infrequently – forebrain disease such as blindness and head-pressing.
- thiamine administration (0.5–1 g p/o q12 h) – rapid resolution of clinical signs.

### Solanaceous plants

- *Datura* spp. (thorn apple, jimsonweed), *Atropa belladonna* (deadly nightshade), *Duboisia* spp. (corkwoods), *Solanum nigrum* (black nightshade).
- contain either tropane (atropine-like) or solanum groups of alkaloids:
  - cause neurotoxicity when ingested.
- unpalatable to horses and poisoning occurs when plants included in hay.
- clinical signs:
  - anorexia
  - depression.
  - excessive urination and thirst.
  - diarrhoea
  - mydriasis.
  - muscle spasms
  - convulsions.
  - *Solanum nigrum* toxicity (most toxic):
    - colic, ataxia, weakness, tremors and convulsions.
- treatment with physostigmine or neostigmine (short-acting cholinesterase:
  - 1–5 mg s/c q1–4 h used to effect is the treatment of choice.

### Toxic alcohols

- umbelliferous plants (hemlock-type) contain higher alcohols including:
  - enanthotoxin and cicutoxin.
  - very poisonous and toxins are concentrated in the root and stem.
  - common throughout USA and Europe in wet areas.
- poisoning causes salivation, mydriasis, colic, delirium and convulsions.

## *Miscellaneous neurological conditions*

## Cerebral abscess

### Definition/overview

- rare and sporadic disorder.

### Aetiology/pathophysiology

- secondary to bacterial infection:
  - elsewhere in the body.
  - extension of local disease processes such as sinusitis, rhinitis, otitis media, tooth root abscesses and periocular lesions.
- bacterial agents relate to common causes of septicaemia or those that colonise the skin or respiratory tract:
  - *Streptococcus equi* subsp. *equi*, *S. zooepidemicus*, *Actinobacillus equuli*, *Klebsiella* spp. and *Pasteurella* spp.
- concurrent meningitis may be present.

### Clinical presentation

- behavioural changes such as depression, wandering or unprovoked excitement.

- acute or insidious onset and usually no pyrexia.
- variable severity over a period of days.
- circling and head/neck turned towards side of lesion.
- contralateral impaired vision, deficient menace response and decreased facial sensation are early signs.
- progression to recumbency, unconsciousness, seizures and signs of brainstem compression.

### Differential diagnosis

- EPM
- trauma.
- arboviral encephalitis.
- Leucoencephalomalacia.
- hepatoencephalopathy.
- uraemic encephalopathy
- Rabies.

### Diagnosis

- difficult ante-mortem.
- history of recent bacterial infection:
  - check for septic focus elsewhere.
  - bacterial culture of appropriate specimens, including blood.
- changes in CSF depend on the degree of meningeal or ependymal involvement:
  - xanthochromia and moderate elevation in protein levels.
- CT and MRI.

### Management

- acute, severe and rapidly progressive signs with brain oedema are likely:
  - corticosteroids (dexamethasone, 0.1–0.5 mg/kg i/v q24 h):
    - controversial but used by some clinicians.

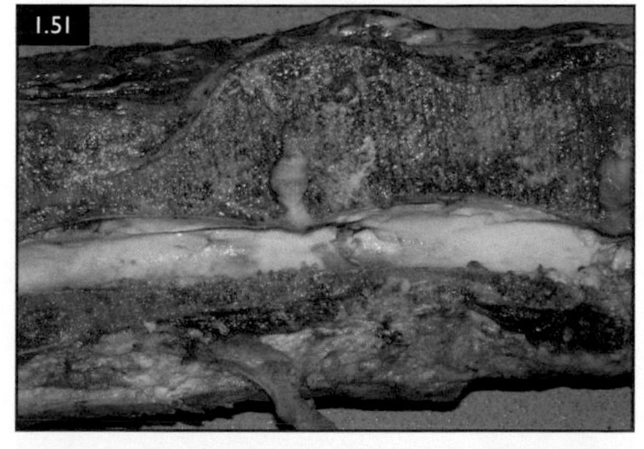

**FIG. 1.51** Vertebral osteomyelitis causing spinal cord compression. *Streptococcus equi* was isolated from the bone.

  - osmotic agents:
    - mannitol or hypertonic saline.
    - diuretics (furosemide 0.75–1.0 mg/kg i/v q12–24 h).
    - DMSO (1 g/kg as a 10–20% solution q12–24 h).
    - counteract life-threatening rises in intracranial pressure.
- prolonged antimicrobials:
  - penicillin (20,000 IU/kg i/v q6 h) and gentamicin (6.6 mg/kg i/v q24 h).
  - initially until culture/sensitivity results.

### Prognosis

- poor as medical therapy is unlikely to be successful.

## Spinal abscessation and vertebral osteomyelitis

### Definition/overview

- very rare and usually associated with vertebral osteomyelitis or discospondylitis.

### Aetiology/pathophysiology

- haematogenous spread of infectious agents in the newborn or extension from local wounds may lead to vertebral osteomyelitis.
- progression of the vertebral infection leads to paravertebral abscess, meningitis, vertebral collapse and spinal cord compression.
- many pathogens isolated including:
  - *Streptococcus equi* subsp. *equi* and *Rhodococcus equi* (Fig. 1.51).
  - *Actinobacillus equuli*, *Escherichia coli* and *Klebsiella* spp. associated with septicaemia in the young foal.

### Clinical presentation

- localised spinal pain.
- fever, stiffness and sensory deficits with variable paresis.
- may rapidly progress to recumbency.
- neurological signs depend on the presence and location of spinal cord compression.

### Differential diagnosis

- spinal trauma or neoplasia
- meningitis.
- arboviral encephalitis.

## Diagnosis

- clinical signs and history.
- positive findings on radiography, scintigraphy, CT or ultrasonography.
- haematology consistent with inflammation.
- CSF normal or consistent with spinal compression.
- repeated blood cultures (taken during pyrexia) useful for identifying pathogens.
- nuclear scintigraphy or myelography to identify a lesion or assess compression.

## Management

- long-term (3–6 months) broad-spectrum antibiotic therapy:
  - culture/sensitivity data.
  - relapses are common.

## Prognosis

- poor if neurological abnormalities are present.

# Polyneuritis equi (neuritis of the cauda equina)

## Definition/overview

- neuritis of the cauda equina with CN deficits:
  - facial, trigeminal and vestibulocochlear nerves are most commonly affected.

## Aetiology/pathophysiology

- immune-mediated response against the myelin of cranial and sacrococcygeal extradural nerve roots.
- may be prior bacterial or viral infections.

## Clinical presentation

- adult horses with LMN deficits at the level of the cauda equina:
  - obstipation and urinary incontinence.
  - absent tail tone (Fig. 1.52).
  - paralysis of anal sphincter (Fig. 1.53) and:
    - rectum, bladder, urethral sphincter and vulva or penis.
  - CN signs are not always present.
- insidious onset with progression over several weeks:
  - may become static at a certain level of severity.

## Differential diagnosis

- EHV encephalomyelitis.
- arboviral encephalitis
- Rabies.
- EPM
- trauma
- EMND.

## Diagnosis

- exclusion of other possible causes.
- antimyelin antibodies can be detected in CSF.
- muscle biopsy from the sacrocaudalis dorsalis lateralis muscle is test of choice.

## Management

- no specific treatment.
- supportive therapy includes:
  - manual evacuation of the rectum.
  - provision of a soft diet.
  - urinary bladder decompression.

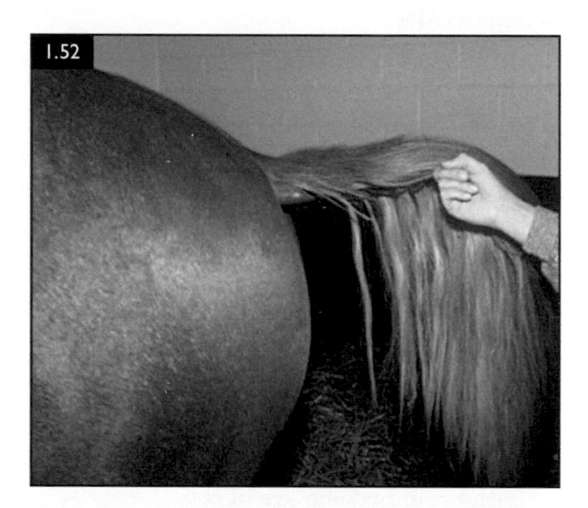


**FIG. 1.52** Poor tail tone secondary to polyneuritis equi.

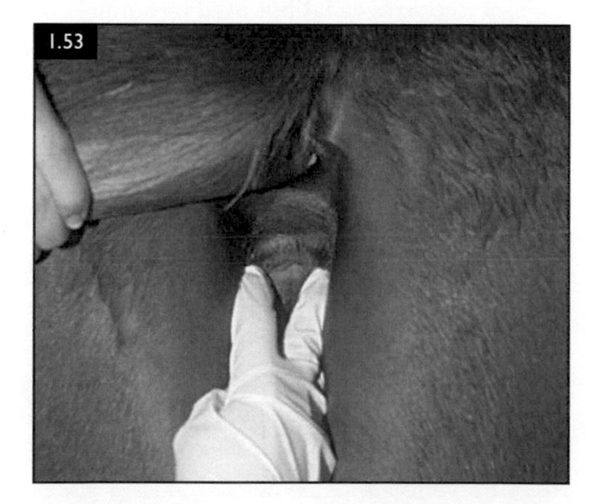


**FIG. 1.53** Poor anal tone secondary to polyneuritis equi.

- corticosteroids (dexamethasone [0.05–0.2 mg/kg q24 h] or prednisolone [1 mg/kg p/o q12 h]).
- monitor carefully for complications including:
  - GI tract impactions.
  - oesophageal obstruction.
  - urinary tract infections.
  - antimicrobial therapy may be necessary to treat secondary infections.

### Prognosis

- poor as rarely responds to treatment.

## Cholesterol granuloma (cholesteatoma)

- incidental finding in the choroid plexuses of up to 20% of older horses.
- larger size in the lateral ventricles where more likely to cause clinical signs.
- brownish nodular thickenings and with abundant cholesterol crystals (Fig. 1.54).
- minority of cases cause compression of brain tissue or an obstructive hydrocephalus.
- insidious onset of clinical signs:
  - altered behaviour
  - depression.
  - somnolence
  - seizures.
  - ataxia
  - weakness.
  - unconsciousness.
- no effective treatment.

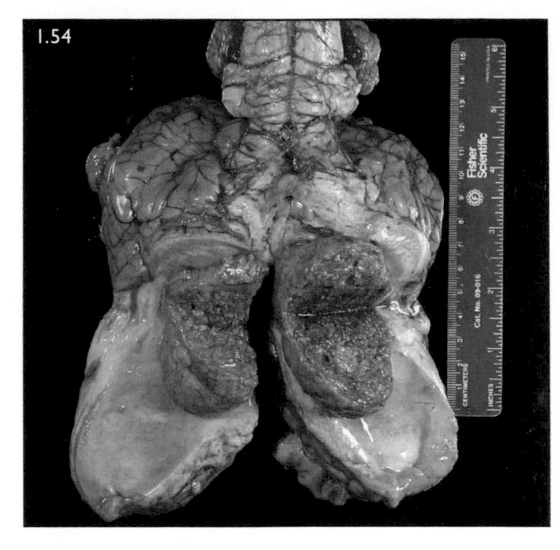

FIG. 1.54 Cholesterol granulomas in the lateral ventricles of a 20-year-old mare. The mare had a history of seizures. (Photo courtesy M Saulez.)

## Post-anaesthetic myelopathy

- reported particularly in young heavy horses, but other breeds have been affected:
  - post-inhalation anaesthesia in dorsal recumbency:
    - systemic arterial hypotension and local venous congestion.
  - compression of caudal vena cava by abdominal viscera.
- difficulty standing to tetraplegia with flaccid paralysis and anaesthesia of hindlimbs:
  - dog-sitting position progressing to lateral recumbency until euthanasia.
- necropsy reveals haemorrhage and congestion of the meninges and spinal cord:
  - degrees of malacia of grey matter over at least several spinal cord segments.
  - sites anywhere from the caudal cervical to caudal sacral spinal cord.
  - ischaemic and hypoxic neuronal damage.
- hopeless prognosis.

## Stringhalt

### Definition/overview

- exaggerated upward flexion of hindlimb (hock).
- **true stringhalt** is sporadic and probably caused by a neuropathy.
- **Australian or pasture-associated stringhalt** is usually bilateral and can occur in groups of horses on pasture.

### Aetiology/pathophysiology

- **true stringhalt:**
  - denervation atrophy of muscles with interference to reflexes that control:
    - posture.
    - coordinate hindlimb muscle contraction.
  - widespread neuropathy elsewhere in body:
    - cause of neuropathy is unknown.
  - some cases may have myopathies or spinal cord disease.
  - can occur following some neurological diseases:
    - EMND (bilateral).
    - equine protozoal myelitis.

    - trauma and bandaging to the dorsal hock and metatarsal region:
        - trauma +/– adhesions to tendons dorsolateral hock/cannon.
        - acquired resetting of neuromuscular spindle trigger.
    - tarsal arthropathy.
- **Australian or pasture-associated stringhalt:**
    - groups of horses on weed-infested pasture in drought conditions.
    - false dandelions/flatweed (*Hypochoeris radicata*).
    - European dandelion or mallow plants.
    - damage to large-diameter axons in long peripheral nerves:
- motor fibres damaged to skeletal muscle:
    - muscle fibre denervation and atrophy:
        - laryngeal and distal hindlimb muscles.
    - muscle spindle afferent fibres damaged:
        - change in resting muscle tone.
        - coordination of muscle contraction.
    - variably produced by feeding plants/extracts, suggesting toxic principle involved:
        - similar syndrome in New Zealand, NW USA, Europe and South America.

## Clinical presentation

- both forms can be sudden in onset.
- one or both hindlimbs can be affected:
    - forelimbs can be affected as well in Australian disease:
        - spasticity, toe scuffing and stumbling also present.
        - left laryngeal hemiparesis may or may not be present (60% of cases).
- hyperflexion of hock:
    - variable severity +/– adduction of leg.
    - at walk usually and not always every stride.
- horse often shows no distress.
    - affected leg up and inside line under body.
    - fetlock may contact ventral abdomen.
    - may remain with leg flexed for few minutes.
    - bilateral cases are rare in true stringhalt and severely affect the horse's gait:
        - peculiar bunny-hopping.
        - unable to stand.
- uncommon at trot
- rare at canter.
- difficulty in backing/turning; exacerbates problem.
- chronic cases have atrophy of the distal limb muscles.
- may worsen with anxiety or cold weather.

## Differential diagnosis

- other mechanical gait abnormalities including:
    - shivering.
    - upward fixation of the patella.

## Diagnosis

- history and clinical signs.
- increased EMG activity in long digital extensor tendon.
- decreased nerve conduction velocity in deep peroneal nerve.

## Management

- box rest inconsistently improves sporadic cases.
- plant toxicity:
    - remove from pasture
    - box rest.
    - recovery can take months or years and may never be complete.
    - oral phenytoin 15–25 mg/kg q12–24 h for 14 days may hasten improvement:
        - mild tranquillisation.
        - extended if clinical signs worsen.
        - EMG nearly normal in long digital extensor tendon afterwards.
        - some cases immediate recovery whilst others take up to 4 months.
- trauma-induced:
    - may resolve with wound repair and gradual exercise.
- botulinum A toxin has been injected into long and lateral digital extensor muscles as a treatment.
- surgical treatments include lateral digital extensor myotenectomy:
    - block lateral digital extensor muscle with LA preop as indicator of success of procedure.
    - standing or general anaesthesia (GA).
    - immediate improvement, or gradually over weeks and months.
    - some cases may worsen after the procedure.
    - often only a salvage procedure in many cases.

### Prognosis

- some Australian stringhalt cases can recover over weeks to a year.
- inconsistent recovery in sporadic cases with rest.
- some sporadic cases respond to surgery, but unpredictably (50%).

## Shivers or shivering

### Definition/overview

- chronic neuromuscular syndrome of variable incidence in horse population.
- generally appears before 5–10 years of age.
- majority of cases (75%) are progressive.
- geldings and tall animals are more susceptible.

### Aetiology/pathophysiology

- unknown aetiology.
- Purkinje cell axonal degeneration in cerebellum:
  - controls slow, learned movements not involved with faster gaits.
  - abnormally elevated muscle recruitment in biceps femoris and vastus lateralis muscles
- not directly related to PSSM.
- probably genetic basis or predisposition, but no specific genetic pattern or test is established.

### Clinical presentation

- susceptible breeds:
  - Draft breeds and their crosses (19% in Belgians).
  - Warmbloods and crosses.
  - sporadically in Thoroughbreds, Hunters and Morgans.
  - rare in ponies.
- spectrum of variable degree and manifestation:
  - maybe intermittent, occasional, inconsistent or latent.
  - difficult to detect in the early stages.
  - not present when standing still.
  - may not be present in first few steps/ turning sharply.
- gradually progressive gait abnormality, especially when backing up.
- not present at trot or faster gaits.
- two primary categories of shivering when backing up:
  - **Hyperflexion:**
    - ♦ hindlimb raised up and away from body with spastic state for several seconds to minutes.
    - ♦ limb trembles in suspension with foot brought rapidly to ground when spasm subsides.
    - ♦ one or both hindlimbs affected.
  - **Hyperextension:**
    - ♦ horse places hind feet further back than normal.
    - ♦ stifle and hock joints hyperextended.
    - ♦ forelimbs also extended when backing begins:
      - stretched or 'sawhorse' stance.
    - ♦ occasionally one or both hindlimbs held out behind in rigid spastic extension.
    - ♦ may stand on toes with heels off ground.
    - ♦ usually both hindlimbs affected.
- other signs can include:
  - trembling of tail while held erect or with sudden jerky extension.
  - milder cases: tenseness or trembling of hindlimbs.
  - severe cases: hyperflexion and shivering in forward walking.
  - occasionally involves muscles of forelimbs, neck, trunk and head:
    - ♦ ears, eyelids, lips and cheeks contract spasmodically.
  - hindlimb weakness in 58% of cases with falls and wide-based stance.
- resistant to hindlimb feet being lifted manually:
  - may hyperflex opposite hindlimb before flexing or abducting hindlimb.
  - cleaning feet and shoeing difficult.
- may be hesitant to lie down, especially when confined, and may lose condition.
- stress/excitement/change in footing may initiate episodes.
- gradual atrophy of hindlimb or generalised muscles with progression.

### Differential diagnosis

- consider other similar gaits:
  - Stringhalt.
  - upward fixation of the patella.

- Fibrotic myopathy.
- EMND
- EPM.
- 'Stiff-horse syndrome': recently reported in horses in Belgium:
  - intermittent stiffness/spasm in axial and pelvic muscles.
  - precipitated by movement/excitement/fear.
  - muscle hypertrophy.
  - immune-mediated GABA deficiency.

## Diagnosis

- history and clinical signs:
  - diagnostic where clinical signs are clear.
- milder cases are difficult.
- rule out other causes of lameness.

## Management

- no effective treatment.
- may improve with turnout and exercise:
  - keep in regular work and constant turnout.
- regression with pain, stress or stabling.
- physiotherapy and acupuncture of hindlimbs.
- diet high in fat and low in starch, similar to that used for treating PSSM (see Book 1):
  - may decrease frequency of muscle spasms.
  - vitamin E supplements have been suggested.

## Prognosis

- many cases can perform at high athletic level.
- early-age onset decreases prognosis and may have quicker progression.
- dressage and driving horses may be more affected by problem.
- short-term prognosis (few years) for performance is fair.
- longer term, over many years up to 50% of cases progress to severe level:
  - difficulty in backing and hyperflexion at walk.
  - drop in performance.

# Hepatic/hyperammonaemic encephalopathy (see Book 4, page 2)

## Definition/overview

- neuropsychological syndrome characterised by biochemical disturbance of CNS function in patients with liver disease.
- also reported in horses suffering from hyperammonaemia associated with:
  - portosystemic shunts.
  - portal vein thrombosis.
  - renal or GI disease.
  - Morgan foal hepatic mitochondrial defect.
- inhibition of normal brain function by ammonia:
  - exact mechanism still not clear, but possibilities include:
    - presence of pseudo-neurotransmitters.
    - activation of the GABA inhibitory system.
- sudden-onset signs of cerebral dysfunction or a gradual subtle change in behaviour.
- cerebral changes may result in maniacal, psychotic or extreme erratic behaviour:
- compulsive walking or circling, head-pressing and coma.
- diagnosis on clinical signs and history, and confirmed by laboratory changes indicative of liver failure, including elevated levels of blood ammonia.
- treatment of liver failure is covered on page 14, Book 4.
- variable prognosis, but generally poor if liver disease.

# Neonatal maladjustment syndrome in foals (see Book 2, pages 184–187)

# Hyponatraemia

- commonly associated with enterocolitis, excessive sweat loss, renal failure, uroperitoneum and adrenal insufficiency:
  - sodium depletion or water retention.
- clinical signs include:
  - depression, seizures (generalised or focal), ataxia and coma.

- when sodium concentration falls below 115 mmol/l (115 mEq/l).
- severity depends on the degree of hyponatraemia and how quickly it develops.

## Hypernatraemia (salt poisoning)

- uncommon in horses:
  - initial stages of diarrhoea, reflux or renal disease:
    - water loss exceeds electrolyte loss.
  - food and water deprivation with continued losses via skin and respiratory tract.
  - inappropriate oral electrolyte supplementation, especially in foals.
- clinical signs include:
  - severe GI disease.
  - neurological signs:
    - head and neck extension.
    - blindness
    - aggressiveness.
    - hyperexcitability.
    - ataxia.
    - proprioceptive deficits and head-pressing.
- pathophysiology involves deposition of sodium ions in the CNS:
  - when followed by access to ion-free water, leads to cerebral oedema.
  - death due to respiratory failure.
- treatment involves gradual reduction of serum sodium over a few days, supportive care and treatment of cerebral oedema.

## Hypocalcaemia (see Book 4, page 23)

- uncommon cause of weakness.

## Neoplasia of nervous tissue

- extremely rare in horses (excluding pituitary neoplasms).
- secondary tumours of a lymphoid origin are the most common.
- primary neoplasms are very rare and usually involve peripheral nerves:
  - neurofibromas and neurofibrosarcomas.
- clinical signs are related to the area of the brain or spinal cord involved:
  - usually due to compression from an expanding mass.

## Juvenile idiopathic epilepsy

### Definition/overview

- also known as benign epilepsy of foals.
- seizures in foals of a few days to several months of age.
- most affected foals are Arabians or Arabian crosses of Egyptian lineage.

### Aetiology/pathophysiology

- unknown:
  - autosomal dominant mode of inheritance suspected.
  - lowered seizure threshold increasing sensitivity to temporary stimuli.

### Clinical presentation

- early onset of signs (first few days or months of life).
- self-limiting and resolves by 1–2 years of age.
- seizure activity can be quite variable.
- focal seizures with abnormal mouth movements.
- facial tremors with head-twitching.
- generalised seizures with entire body muscle spasm, recumbency and thrashing.

### Differential diagnosis

- meningitis
- hypoglycaemia.
- arboviral encephalitis.

### Management

- anticonvulsants initiated at first signs and continued for several weeks:
  - phenobarbital (12 mg/kg p/o or i/v loading dose and 6 mg/kg p/o or i/v q12 h).
  - serum phenobarbital concentrations should be monitored and maintained between 10 and 30 µg/ml.
- stable in an area where less likelihood of injury during seizure.

### Prognosis

- usually resolves with age.
- persistent seizures may result in permanent neuronal damage and seizure foci.

## Acquired epilepsy

### Definition/overview

- adult-onset recurrent seizure activity.

### Aetiology/pathophysiology

- rapid and excessive abnormal neuronal activity from the cerebral cortex that results in involuntary alterations in motor activity, consciousness and autonomic functions.
- recurrent events.
- no evidence of familial adult-onset epilepsy in horses.
- caused by an intracranial epileptogenic focus/foci acquired during post-natal life:
  - may be years between the focus forming and the onset of seizures.
  - paroxysmal neuronal changes spread out from the focus.

### Clinical presentation

- transient changes in horse's behaviour usually precede attacks.
- partial seizures have relatively localised signs such as unilateral facial-twitching, limb-twitching, compulsive running in a circle or self-mutilation:
  - complex partial seizures are relatively common in foals.
  - referred to as 'chewing-gum fits' because the foals show persistent jaw chomping and lip-smacking.
- generalised seizures involve the whole cerebral cortex, and result in:
  - widespread tonic–clonic muscle activity and a loss of consciousness.
- three distinct periods in seizures:
  - **pre-ictal** phase or aura: anxiety or restlessness.
  - **ictal phase:**
    - recumbency.
    - tonic–clonic muscle contractions.
    - eyeball movements.
    - dilated pupils.
    - ptyalism
    - trismus.
    - opisthotonos.
    - excessive sweating.
    - lordosis or kyphosis.
    - uncontrolled urination and defecation.
  - **post-ictal phase:** disorientation and blind for a period of minutes or hours.

### Management

- controlling seizures:
  - minimise further neuronal damage, self-inflicted injury or human injury.
  - **do not attempt if dangerous for human health.**
  - benzodiazepines:
    - diazepam, 0.1–0.2 mg/kg i/v in foals, 0.02–0.08 mg/kg i/v in adults.
    - reduce the electrical activity responsible for the seizure.
    - increase seizure threshold.
    - short half-life and multiple doses may be required.
    - may accumulate in foals, causing respiratory depression.
  - intravenous phenobarbitone.
- preventing seizures:
  - phenobarbitone, bromide, phenytoin and primidone.
  - reduce doses gradually after 2 months without attacks.
  - **do not ride horse until seizure-free without medication for a minimum of 6 months.**
- control any underlying disease.

### Prognosis

- highly variable because of the idiopathic nature of the disease:
  - fair in general, but some horses do not respond well to medication.
- horses with uncontrolled intermittent seizure activity are unsafe to ride.

## Narcolepsy–cataplexy

### Definition/overview

- sudden onset of excessive daytime sleepiness:
  - usually accompanied by cataplexy – sudden and profound loss of muscle tone.
- two different syndromes:
  - fairly common transient condition of foals of light breeds induced by restraint.
  - rare but persistent condition with familial predisposition:
    - Miniature horses
    - ponies.
    - Suffolk Punches.

## Aetiology/pathophysiology

- unclear but a familial predisposition is suspected in certain breeds.
- episode may be stimulated by a specific stimulus.
- abnormalities of the neuropeptides linked to the regulation of sleep.

## Clinical presentation

- intermittent episodes are characterised by:
  - sudden lowering of the head and buckling of the fetlocks.
  - occasional collapse and rapid eye movement (REM) sleep.
- between episodes, animals are clinically normal.

## Differential diagnosis

- other causes of collapse: syncope and seizures.
- any disorder preventing an animal lying down (sleep deprivation):
  - musculoskeletal problems and pain.

## Diagnosis

- history, clinical signs and exclusion of other causes:
  - **sleep-deprivation narcolepsy** has a classic history:
    - horses cannot (due to orthopaedic pain) or will not (due to anxiety/herd hierarchy) lie down to sleep.
  - **vasovagal reflex** has a characteristic history:
    - horse collapses and falls backwards (usually conscious throughout) when bridled or when a dental gag is opened while attached to the head.
    - drop in blood pressure as the mouth opens, leading to hypotensive collapse (syncope).
  - **cardiac syncope:**
    - horses do not usually lose consciousness.
    - more likely to suffer an episode during exercise.
    - cardiac abnormalities detected on electrocardiography and/or echocardiography.
- intravenous administration of physostigmine salicylate (0.1 mg/kg i/v) may elicit signs of narcolepsy within minutes in some individuals:
  - this response is not consistently found in all animals with narcolepsy.
  - **sudden death reported with intravenous injection of physostigmine.**

## Management

- avoid inciting causes.
- tricyclic antidepressant imipramine (0.5–2.0 mg/kg i/m, i/v or p/o q6–12 h) may improve clinical signs in some animals.

## Prognosis

- not a life-threatening condition, but animals are unsafe to ride.
- some foals may outgrow the condition.

# Equine degenerative myeloencephalopathy (EDM)

## Definition/overview

- diffuse, non-compressive, symmetrical degenerative disease of the spinal cord and brainstem.
- incidence could be as high as 23–45% of horses that present with spinal cord disease.
- variable incidence worldwide; more prevalent in North America.
- diagnosed in horses of either sex up to 3 years of age.

## Aetiology/pathophysiology

- familial hereditary basis in some breeds, including:
  - Morgans, Appaloosas, Standardbreds and Paso Finos.
- vitamin E deficiency early in life has been suggested:
  - when clinical signs become apparent, vitamin E levels may be normal.

## Clinical presentation

- individuals or groups (related) may be affected.
- usually, insidious onset:
  - majority of affected horses show signs in the first year of life.
- symmetrical ataxia, paresis and dysmetria:

  - hindlimb signs worse than forelimb (Fig. 1.55).
- clinical signs may stabilise for several months or may progress to cause recumbency.

## Differential diagnosis

- Cervical vertebral stenotic myelopathy.
- EPM
- WNV encephalomyelitis.
- trauma.
- spinal cord impingement secondary to neoplasia/abscess.

## Diagnosis

- clinical presentation and exclusion of other possible causes.
- low serum vitamin E levels may be present but are inconsistent.
- histopathological evaluation of the spinal cord at post-mortem.

## Management

- oral supplementation of vitamin E (5000–7000 IU/day p/o).
- ample green forage.

## Prognosis

- remission or recovery has not been reported.

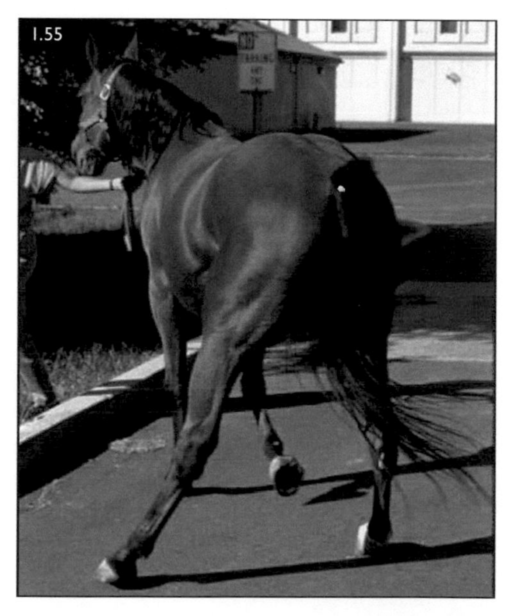

FIG. 1.55 Symmetrical ataxia of the hindlimbs in a horse with EDM.

- treatment is aimed at stabilisation.
- affected animals and parents should not be used for breeding.

# Equine motor neuron disease (EMND)

## Definition/overview

- sporadic, acquired neurodegenerative disorder of the somatic LMNs.
- mean age at onset of signs is 9 years.

## Aetiology/pathophysiology

- unclear:
  - chronic persistent lack of access to antioxidants.
  - absence of access to pasture.
  - chronic vitamin E deficiency.
  - individual predisposition.
- majority of cases reported from north-eastern USA:
  - other countries, occasional reports.
- clinical signs result from oxidative damage to the somatic ventral motor neuron cells:
  - neurogenic atrophy of the type 1 muscle fibres.
  - animals are unable to fix their stay apparatus.
  - signs are only evident when 30% of motor neurons cells die or are dysfunctional.

## Clinical presentation

- depends on duration of the disease.
- neuromuscular weakness due to generalised denervation muscle atrophy.
- **subacute form:**
  - trembling
  - muscle fasciculations.
  - shifting of weight in the hindlimbs.
  - abnormal sweating and frequent recumbency (Fig. 1.56).
  - gradual loss of muscle mass for 1 month prior to clinical signs.
  - low head carriage but appetite and gait are unaffected.
- **chronic form:**
  - usually, but not always, cases that have stabilised from the acute form.
  - poor performance, fatigue and abnormal gaits.
  - muscle atrophy.
  - tail head is raised in both forms.

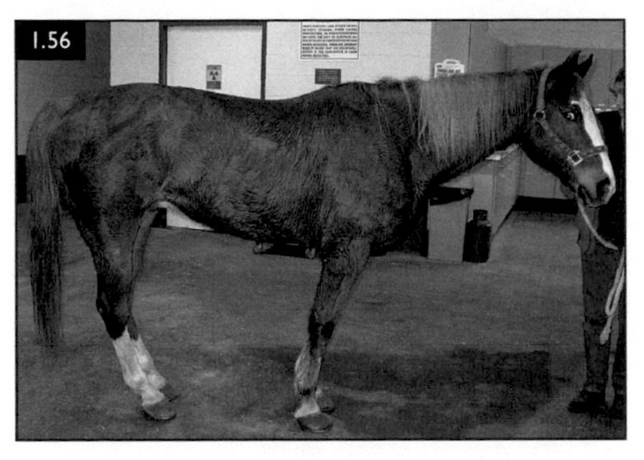

FIG. 1.56 EMND in a Tennessee Walking Horse. Note the 'camped under' posture and full-body sweating. This horse was very weak in all four limbs and muscle fasciculations were noted before the animal would lie down.

- ○ 30% of cases have black dental tartar on the incisors (Fig. 1.57).

## Differential diagnosis

- lead toxicosis
- mycotoxin ingestion.
- trauma
- laminitis.
- atypical myopathy
- colic.

## Diagnosis

- clinical signs and laboratory findings:
  - ○ subacute form:
    - ♦ muscle enzyme activity may be moderately elevated.
    - ♦ plasma vitamin E levels are usually low (<1.0 μg/ml):
      - – CNS, peripheral nerves, muscle, liver and adipose tissue levels also low.
- confirmation:
  - ○ muscle biopsy of the sacrocaudalis dorsalis medialis muscle (Fig. 1.58).
  - ○ biopsy of a ventral branch of the spinal accessory nerve.

## Management

- no treatment.
- vitamin E at 5000–7000 IU p/o per horse per day.
- combined with an increase in fresh grazing.
- supplement all animals in a stable where a case has occurred.

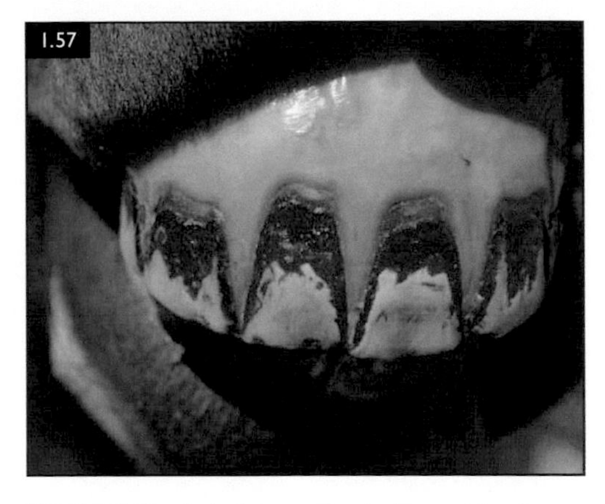

FIG. 1.57 Black tartar on the incisors of a horse with EMND.

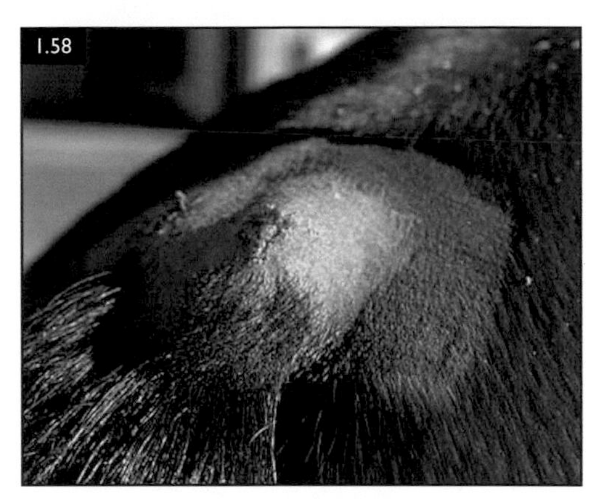

FIG. 1.58 Biopsy site of the sacrocaudalis dorsalis medialis muscle.

## Prognosis

- poor for return to performance and long-term survival.
- affected animals should not be ridden.

# Cervical vertebral stenotic myelopathy (CVSM) (see also Book 1)

## Definition/overview

- often referred to as 'wobbler's syndrome'.
- syndrome characterised by progressive ataxia resulting from spinal cord compression.
- very common cause of ataxia.
- two types of disease:
  - ○ **dynamic stenosis (type 1):**

- instability when the neck is flexed or extended.
- transient decrease in the diameter of the cervical vertebral canal.
- usually at the intercervical spaces C3–C4 and C4–C5 (Fig. 1.59).
- usually in young horses (6–18 months of age), particularly Thoroughbreds.
- may have other developmental abnormalities in the vertebrae:
  - physeal enlargement.
  - malformation of the vertebral canal.
  - extension of the dorsal aspect of the vertebral arch.
  - angulation between adjacent vertebrae.
  - osteochondrosis of the articular processes.
- **static stenosis (type 2):**
  - narrowing present regardless of the position of the neck.
  - more frequent at C5–C6 and C6–C7.
  - any age of horse, but usually mature adults:
    - older horses with progressive degenerative joint diseases of the articular facet joints.
- some overlap can occur between the two types of CVSM:
  - type 1 in a young horse with arthritic changes.
  - mature horses with developmental abnormalities which present clinically later in life.

## Aetiology/pathophysiology

- unclear:
  - combination of a developmental abnormality influenced by genetics and environmental factors such as diet, growth rate, workload and acute trauma.
- spinal cord may be compressed by:
  - malformation, malarticulation and instability.
  - bony changes of the vertebral body and articulations.
  - changes to surrounding soft tissues.
- compression of the UMN and general proprioceptive tracts in cervical spinal cord:

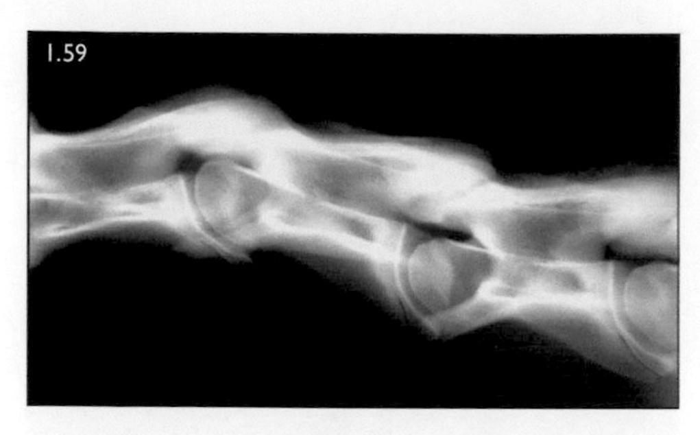


**FIG. 1.59** Cervical vertebral instability of C3–C4. Note the abnormal angulation.

  - occasionally the LMN and nerve roots at the cervical intumescence.

## Clinical presentation

- proprioceptive deficits, weakness and abnormal limb placement in all four limbs.
- ataxia in all four limbs:
  - hindlimbs more obviously abnormal (one grade more severe).
  - some mildly affected horses may only be detectable clinically in the hindlimbs.
- usually, symmetrical signs or mildly asymmetrical:
  - occasionally type 2 cases may be markedly asymmetrical.
- wide-based stance, abnormal limb placement and delayed repositioning of limbs.
- moving their necks is usually normal in type 1 but can be painful and stiff in type 2:
  - related to presence of osteoarthritis in the articular facet joints.
  - neck pain leads to reluctance to flex or laterally bend the neck.
  - holding the neck stuck in a lowered position.
  - inability to bend the neck when being asked to circle.
  - neck muscle atrophy.
- more subtle signs may be present with no ataxia or neck pain:
  - rearing, stopping at fences, reluctance to flex the neck, bring the head and neck up into contact or perform lateral work.
  - stumbling or falling in unexpected circumstances.

- reluctance to be ridden downhill or over drop fences.
- clinical signs are often worsened by specific clinical tests.

## Differential diagnosis

- EPM • trauma • EDM.
- EHV-1 myeloencephalopathy.
- Rabies • viral encephalitides.

## Diagnosis

- neurological examination will localise the lesion to the cervical spinal cord.
- standing radiographs of the cervical spine are helpful in many cases (Fig. 1.60).
- semi-quantitative scoring system can be used to assess the radiographs:
  - angulation of the cervical articulations.
  - minimum sagittal diameter or ratios of vertebral canal:
    - ratios ≤ 0.50 at C4–C6 or ≤ 0.52 at C7 suggestive of stenosis.
  - encroachment of caudal vertebral physis into vertebral canal ('ski-jump' lesion).
  - abnormal ossification of the physis.
  - caudal extension of the dorsal arch.
  - presence of degenerative joint disease.
- clinical relevance of changes can be unclear or not present on standing radiographs.
- myelography is sometimes required to confirm a diagnosis (Figs. 1.61, 1.62).
- percutaneous ultrasonography of the cervical facet joints is useful to assess them for pathology and guide intra-articular injections.
- advanced imaging of the cervical neck with CT and MRI is now available in some countries, including under standing sedation, and is the definitive technique.

## Management

- medical therapy is aimed at reducing cell swelling and oedema around the spinal cord:
  - box rest and systemic or local anti-inflammatory drugs.
- yearlings:
  - restricted exercise and diet:
    - feeding grass hay and a vitamin/mineral supplement.

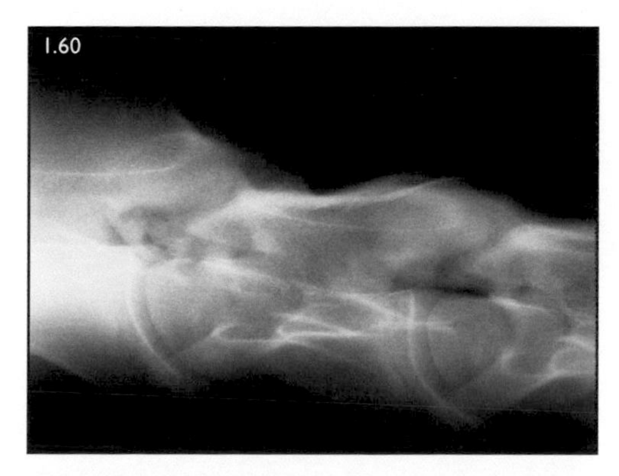


**FIG. 1.60** Osteochondrosis at C2–C3, with fracture of the articular facets.

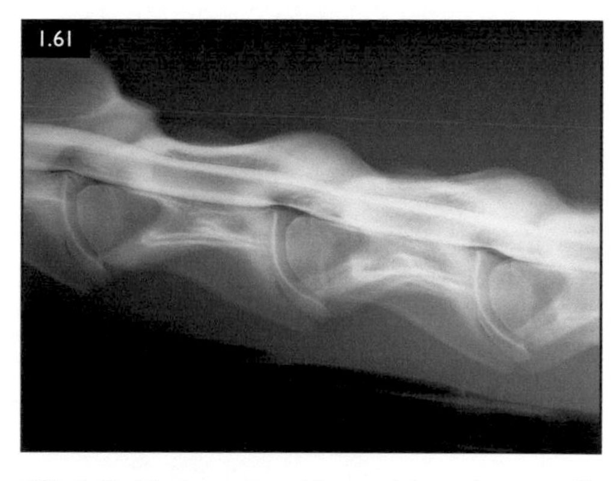


**FIG. 1.61** Myelogram performed in a horse with dynamic cervical vertebral instability. Note the thickness of the ventral and dorsal dye columns at C3 and C4. No compression is noted in this neutral view.

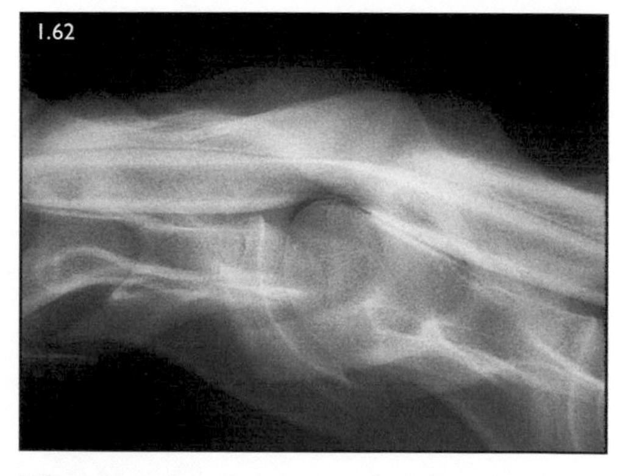


**FIG. 1.62** A flexed view revealed that the dorsal contrast narrowed more than 50% compared with the same site in the neutral view in Fig. 1.61. This myelogram confirms a compressive lesion at C3 and C4.

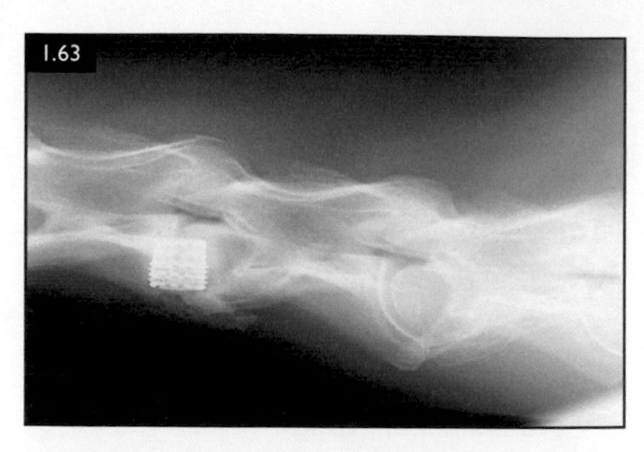


**FIG. 1.63** Fusion of the vertebral bodies of C3–C4 in a horse with wobbler's syndrome.

- intra-articular injection of corticosteroids into the facet joints with type 2 CVSM:
  - may relieve some of the clinical signs for variable periods.
- surgical therapy aims to fuse adjacent vertebrae to eliminate movement in the affected articulation (Fig. 1.63):
  - improvement of one to two grades is expected, but not guaranteed.

## Prognosis

- depends on the age of the horse, severity and duration of the neurological signs, and expectations of the horse's future performance.
- ataxia is often not severe enough to warrant euthanasia, but controversy surrounds whether they can be ridden:
  - horses that are severely ataxic can be dangerous to themselves and humans and should be euthanased.
- best long-term prognosis is obtained from surgical stabilisation.

CHAPTER 2

# Eyes

## Anatomy of the equine eye

- horse has a large eye.
- lateral globe placement and horizontal ovoid pupil allow a total visual field of 350°:
  - narrow blindspots exist:
    - immediately in front of the nose and directly behind the hindquarters.
  - both eyes are used to look at distant objects until within 1.0–1.3 metres (3–4 feet):
    - horse then turns head and looks with only one eye.
- number of unique anatomical features:
  - complete bony orbit, which gives added protection to the eye.
  - well-developed extraocular muscles:
    - make manipulation of the eyelids and globe challenging.
  - globe is slightly deviated medially and ventrally in the neonatal foal:
    - attains the normal adult position by 1 month of age.
  - iridocorneal angle (ICA) is easily visible temporally and nasally.
  - equine pupil is bordered by granula iridica (corpora nigra) (Fig. 2.1):
    - exaggerated prominence of posterior pigmented epithelium layers of the iris.
    - more prominent dorsal pupillary margin than ventrally.
    - may decrease the light through the pupil.
  - equine lens has prominent Y sutures.
  - persistent hyaloid artery remnants are common in the equine neonate:
    - may contain blood in the first few hours after birth.
    - usually disappear by 3–4 months of age.
  - paurangiotic fundus that contains 40–60 small retinal vessels:

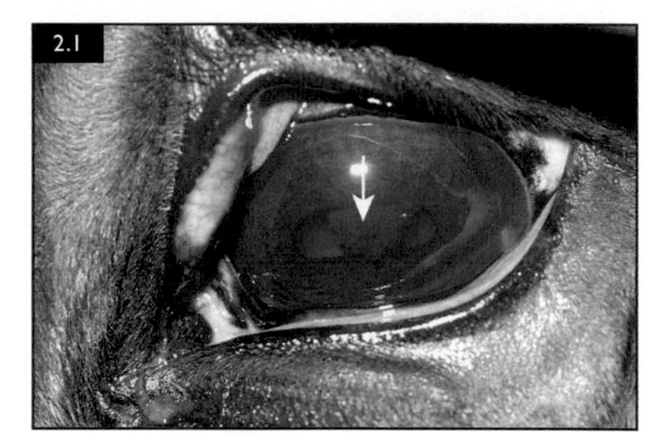

**FIG. 2.1** The granula iridica (arrow) is most prominent dorsally.

    - radiate from the edge of the optic nerve head (ONH).
    - extend only 1.0–2.5 disc diameters from the optic disc.
  - variations in the normal equine fundus are numerous:
    - primarily related to coat and eye colour.
    - optic disc appears as a horizontal, oval to round, salmon-pink structure:
      - located slightly temporally in the non-tapetal region (Fig. 2.2).
      - used to estimate size of fundic lesions in terms of disc diameters.
      - rarely myelinated, but when present may appear as:
        - ◇ white to grey streaks radiating from the ONH.
        - ◇ follows course of retinal vessels nasal and temporal to the disc.
    - normal tapetum may be yellow, green or blue:
      - small reddish-brown dots that represent end-on views of choroidal capillaries, or 'stars of Winslow' (Fig. 2.3).

DOI: 10.1201/9781003451921-2

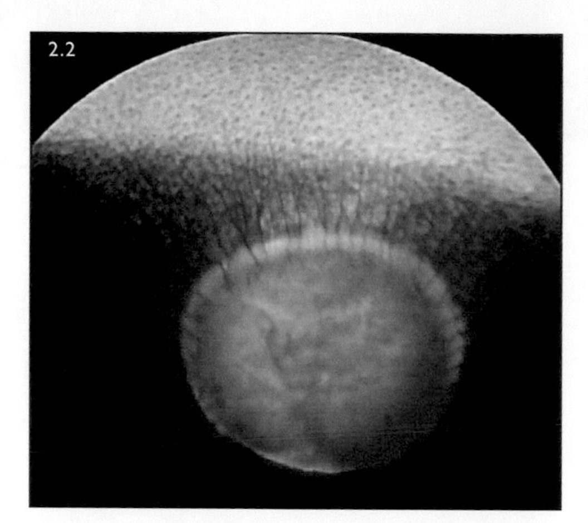


**FIG. 2.2** Normal equine fundus in the optic disc region.

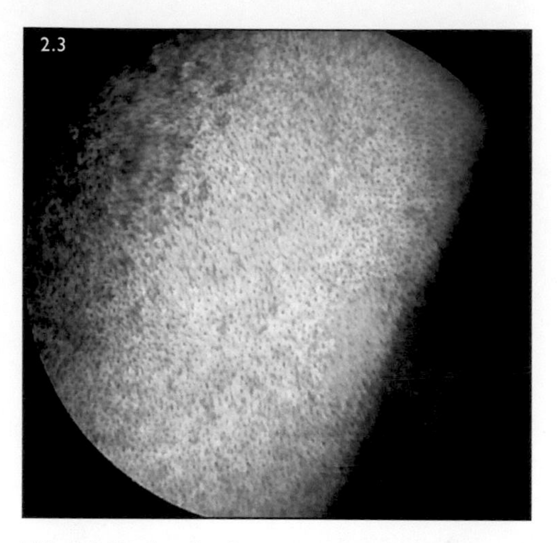


**FIG. 2.3** Equine fundus, tapetal region. The dark foci, called 'stars of Winslow', are end-on choroidal vessels penetrating the tapetum.

- ♦ non-tapetal fundus is typically heavily pigmented:
  - – appearing dark brown or chocolate coloured.
  - – may appear reddish in horses devoid of retinal pigment epithelium (RPE), tapetal cells and choroidal pigmentation (Fig. 2.4):
    - ◇ often seen in horses with blue irises.

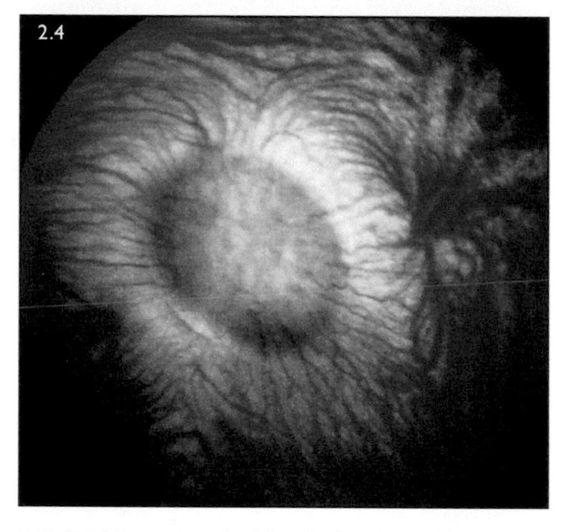


**FIG. 2.4** Normal, subalbinotic fundus. This fundus is characterised by lack of both choroidal pigment and tapetum lucidum. The anatomy of the extensive choroidal vasculature is readily viewed.

## Examination of the eye

- ophthalmic examination completed for all horses undergoing:
  - pre-purchase examination.
  - those exhibiting signs of ocular and/or systemic disease.
- visual impairment may be dangerous to the animal and to humans in contact with it:
  - sudden-onset blindness is likely to be dangerous (Table 2.1).
- ocular disease may affect the use of the animal:
  - recent study reported that 31 of 33 horses that underwent unilateral enucleation returned to their previous performance level immediately postoperatively.

### Ocular examination

- quiet area that can be darkened appropriately.
- chemical restraint, pharmacological mydriasis and auriculopalpebral nerve blocks may be required.
- diffuse and focal light sources, such as a direct ophthalmoscope and transilluminator, respectively, are essential.
- thorough and systematic technique to examine all areas of the adnexa, eye and orbit.
- obtain a full history to include:
  - duration of the problem.
  - any treatment the animal has received and its response.

**TABLE 2.1** Differential diagnoses for sudden blindness

| Abnormal pupillary reflexes (peripheral blindness) |
|---|
| • Optic neuritis<br>• Retinal detachment<br>• Equine recurrent uveitis<br>• Glaucoma<br>• Exudative optic neuritis<br>• Head trauma – optic nerve avulsion<br>• Ocular trauma/intraocular haemorrhage<br>• Retrobulbar granuloma/neoplasia (e.g. cryptococcosis)<br>• Viral encephalomyelitis (e.g. Eastern, Western or Venezuelan equine encephalitis, Borna disease) |
| **Normal pupillary reflexes (CNS/cortical blindness)** |
| • Cataracts<br>• Congenital (hydrocephalus, storage disease)<br>• Metabolic diseases (hypoglycaemia, hepatic encephalopathy)<br>• Toxins (lead poisoning; fiddleneck or horsetail ingestion)<br>• Nutritional (thiamine deficiency)<br>• Head traumatic/vascular (embolus)<br>• Hypoxic – post-ictal; respiratory or cardiac arrest<br>• Infections (toxoplasmosis)<br>• CNS neoplasia or other space-occupying lesions<br>• Idiopathic |

- determine whether primary ocular disease or secondary to systemic disorder.
- perform a general inspection and neuro-ophthalmic examination before sedation, nerve blocks or other diagnostic ophthalmic tests are performed.
- examination:
  - adnexal structures.
  - anterior segment of the eye:
    - conjunctiva
    - cornea.
    - anterior sclera.
    - anterior chamber
    - iris.
    - lens
    - ciliary body.
  - evaluation of the posterior segment of the eye:
    - vitreous and retina.
- other diagnostic tests, such as a Schirmer tear test (STT), fluorescein staining and tonometry, may be required.

## Neuro-ophthalmic examination

- **palpebral reflexes** elicited by touching the eyelids and observing a blink response (Figs. 2.5, 2.6):
  - branches of trigeminal nerve (CN V) for sensory afferent pathway.
  - branches of facial nerve (CN VII) plus orbicularis oculi muscle for motor efferent pathway.
- **menace response** elicited by making a quick threatening motion towards the eye and observing a blink or flinch (Fig. 2.7):
  - both the lateral and medial visual fields.
  - **proper technique is important:**
    - false-positive results if vibrissae are touched or an air current produced.
  - retina and optic nerve (CN II) provide the sensory afferent pathway.
  - branches of facial nerve and orbicularis oculi muscle for motor efferent pathway.
  - menace response is absent in normal newborn foals up to 14 days of age.

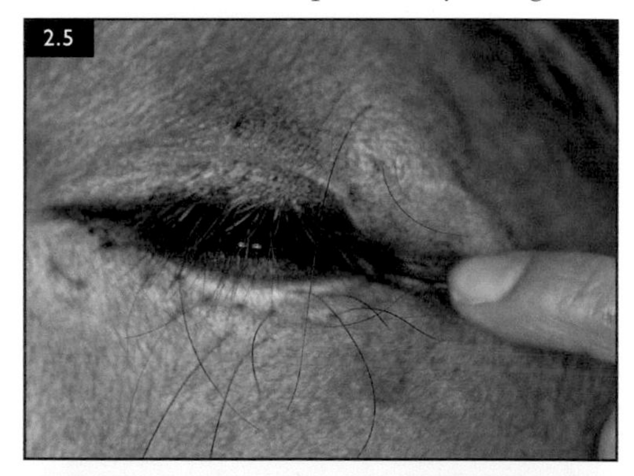


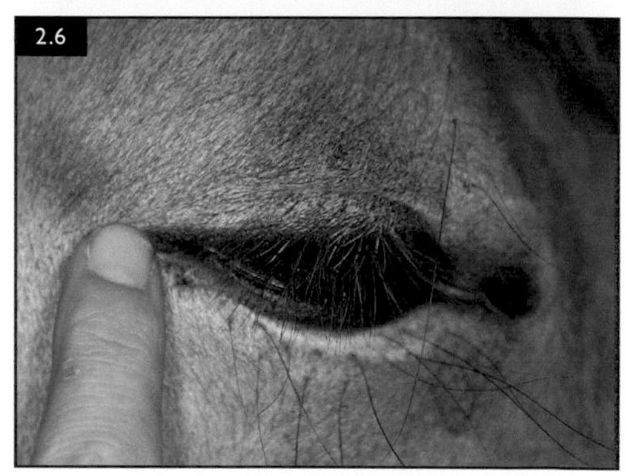


**FIGS. 2.5, 2.6** Palpebral response. Gently tapping the medial (2.5) or lateral (2.6) canthus will stimulate the sensory branches of CN V and generate the afferent stimulus.

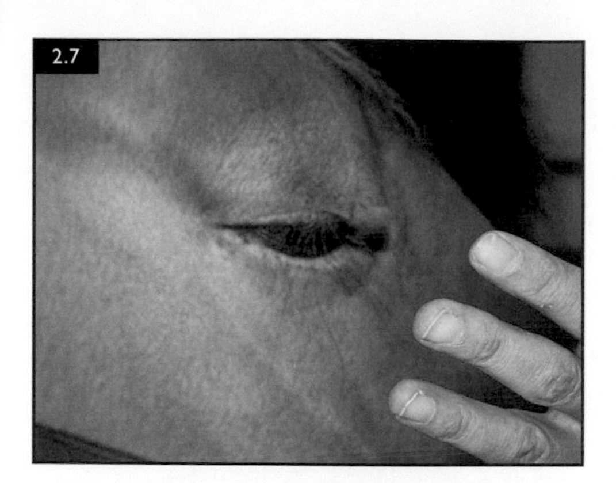

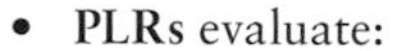

**FIG. 2.7** Menace response. A threatening gesture is made near the eye; the expected response is eyclid closure, often accompanied by movement of the head away from the stimulus.

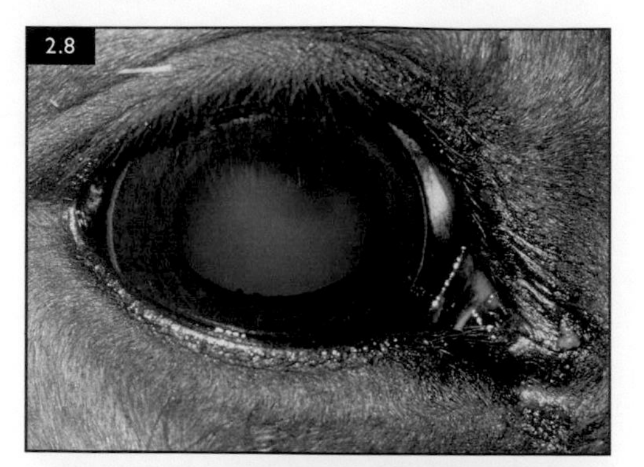

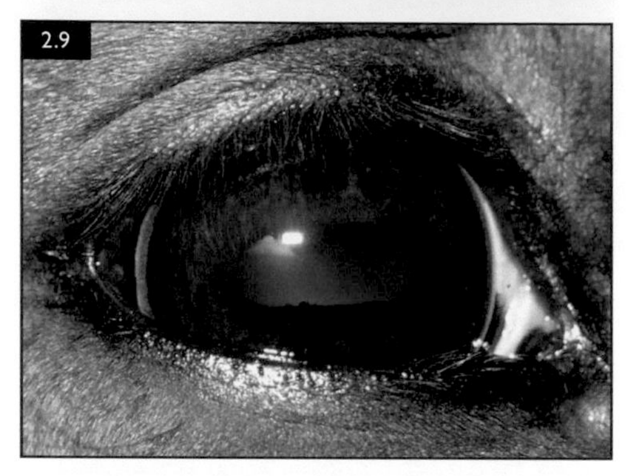

**FIGS. 2.8, 2.9** Pupillary light reflex. (2.8) Normal pupil in ambient light. (2.9) Pupil under effect of direct bright light stimulus.

- PLRs evaluate:
  - retinal function, CN II and the midbrain for sensory afferent pathway.
  - oculomotor nerve (CN III) and iris sphincter muscle for motor efferent pathway.
  - beam of focal light shone into eye:
    - normal pupillary response is constriction of the pupil:
      - **direct PLR** (Figs. 2.8, 2.9).
    - constriction of the contralateral pupil should occur simultaneously:
      - **indirect or consensual PLR.**
      - may require use of an assistant (Fig. 2.10).
  - PLR may be helpful in the evaluation of retinal and optic nerve integrity in an eye with opacities that prevent direct viewing of the posterior segment:
    - corneal oedema, cataract, intraocular haemorrhage.
  - atropine administration, synechiae, iridal colobomas and atrophy affects the PLR.
  - **not a test of visual perception but a subcortical reflex.**
  - normal foals are born with circular pupils and sluggish PLRs:
    - 3–5 days post-partum the pupils become more ovoid and PLRs more rapid.
- **'dazzle' reflex:**
  - subcortical reflex that evaluates the retina, CN II, the rostral colliculus, CN VII and the orbicularis oculi muscle.
  - evaluated using a bright, focal light source shone into the eye causing the horse to squint or blink.

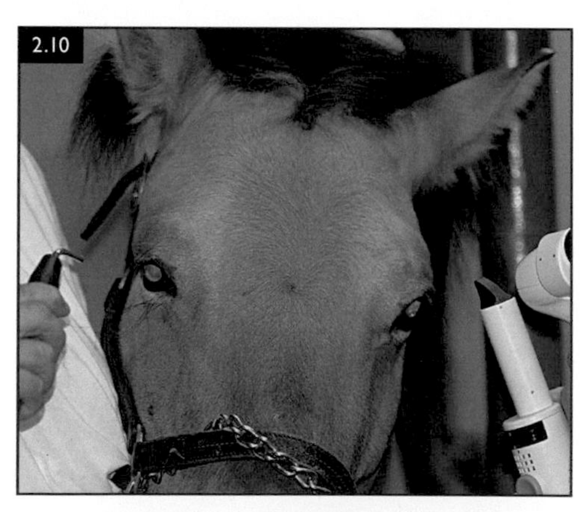

**FIG. 2.10** Evaluation of the indirect PLR. An assistant is required to shine a bright focal light into one eye while the examiner evaluates the response of the contralateral pupil by diffusely illuminating that eye (using distant direct ophthalmoscopy).

- following the neuro-ophthalmic examination, sedation and/or nerve blocks may be performed.

## *Diagnostic procedures*

## Techniques to aid in ophthalmic procedures

### Chemical restraint

- sedation may be required because of the temperament of the patient or ocular pain:
  - alpha-2 adrenoreceptor agonists are used commonly:
    - xylazine (0.3–1.1 mg/kg i/m or i/v).
    - romifidine (12.5–50 µg/kg i/m or i/v).
    - detomidine (10–40 µg/kg i/m or i/v).
  - additional medications may be necessary:
    - butorphanol (0.02–0.05 mg/kg i/m or i/v).
    - acepromazine (0.02–0.05 mg/kg i/m or i/v).
- physical restraint techniques, including a twitch and stocks, may also be required.

### Topical anaesthetic

- 0.5–1% proxymetacaine/proparacaine or 0.5–1% tetracaine.
- applied to desensitise the ocular surface and applied before:
  - collection of conjunctival or corneal scrapings or biopsies.
  - irrigation of the nasolacrimal system (NLS).
  - manipulation of the third eyelid.
  - tonometry.
  - examination of a painful eye.
- equine ocular surface is difficult to anaesthetise completely:
  - multiple drops instilled every minute on 3–4 occasions are required.
  - small syringe (0.1–0.2 ml) can be used to apply directly onto the corneal and conjunctival surface (Fig. 2.11).
  - solution sprayed onto the corneal surface using a tuberculin syringe with needle hub attached:
    - do not get too close to the eye – hub tip is sharp and can damage the cornea.
  - **topical anaesthetic impedes corneal healing and should not be used to treat ophthalmic disorders.**

### Auriculopalpebral nerve block

- paralysis of the orbicularis oculi muscle usually required to allow eyelid manipulation for ocular examination and sample collection, especially when the eye is painful.
- used when placing an SPL system or when the NLS is cannulated or catheterised.
- auriculopalpebral branch of the facial nerve supplies the ipsilateral orbicularis oculi muscle (Fig. 2.12):
  - sedation may be required to complete this nerve block in some horses.

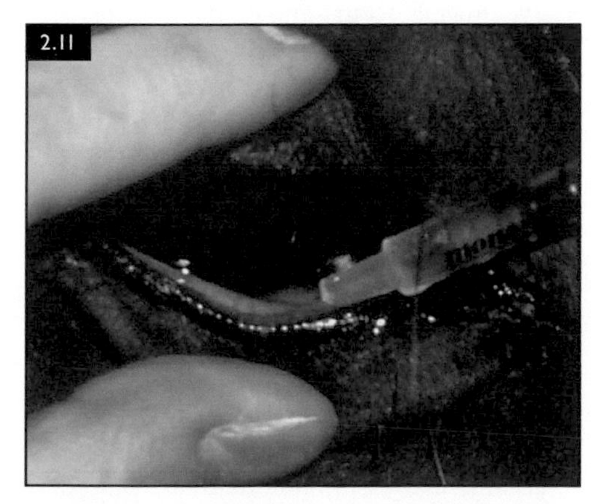


**FIG. 2.11** Administration of topical liquid medication.

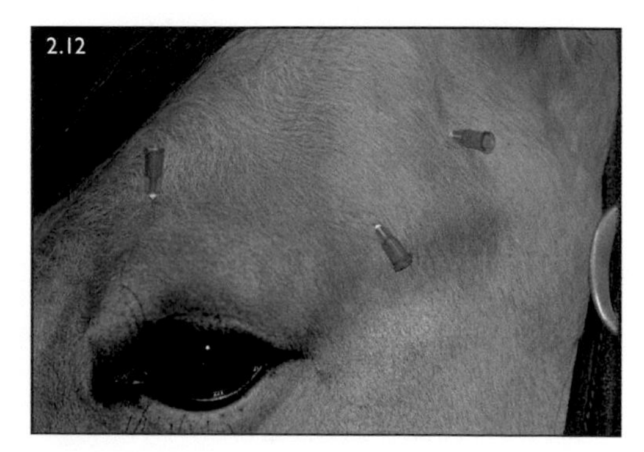


**FIG. 2.12** Location of common peripheral nerve blocks. Sites shown, from left to right: supraorbital nerve block, palpebral nerve block and auriculopalpebral nerve block.

- inject LA over the nerve as it exits the skull at the base of the ear just caudal to the posterior ramus of the mandible and the zygomatic arch:
  - depression can be appreciated in this area, but nerve cannot be palpated.
  - 21–23-gauge 1.5–2.5-cm (5/8–1-inch) needle is inserted into the depression in a dorsal direction, and 5–6 ml of LA is injected.
- facial nerve may also be blocked where it can be palpated as it traverses the dorsal zygomatic arch:
  - 3 ml syringe with a 25-gauge 1.5-cm (5/8-inch) needle and 1–3 ml of LA injected subcutaneously.
  - area of injection may be massaged to facilitate diffusion of the drug.

- successful block will lead to:
  - narrowed palpebral fissure, mild ptosis and eyelid paralysis within 5 minutes.
  - eyelids may remain paralysed for up to 1–2 hours.
  - **block does not provide any sensory nerve analgesia to the eyelid.**

## Supraorbital nerve block

- branch of the ophthalmic division of the trigeminal nerve (CN V).
- blocked as it exits the supraorbital foramen.
  - palpated superior to the orbit in the supraorbital process of the frontal bone.
  - sensory denervation to the majority (middle two thirds) of the upper eyelid.
- 22–25-gauge 1.5–2.5-cm needle is introduced over the supraorbital foramen:
  - 2–3 ml of 2% lidocaine hydrochloride is infiltrated.
  - another 2–3 ml is deposited subcutaneously as the needle is withdrawn (see Fig. 2.12):
    - some variable motor paralysis of the upper lid.

## Other nerve blocks

- lacrimal, zygomatic and infratrochlear nerves are blocked occasionally:
  - all branches of the ophthalmic division of the trigeminal nerve.
  - provides sensory denervation to the lower eyelid.

## Retrobulbar nerve block

- frequently used as an adjunct to general anaesthesia or in standing enucleation:
  - allows a lower depth of anaesthesia to be used during eye surgery.
  - postoperative analgesic purposes.
  - helps control nystagmus and enophthalmos during light anaesthesia.
  - reduces the need for neuromuscular blockage.
  - reduces risk of bradyarrhythmia and hypotension associated with oculocardiac reflex.
- three techniques for injection can be used:
  - direct injection into the orbital cone using a 6.25-cm (2.5-inch) 22-gauge needle:
    - inserted perpendicular to the skin in the orbital fossa just posterior to the dorsal orbital rim.
  - 10-cm (4-inch) 18-gauge needle may be inserted:
    - 1 cm caudal to the lateral canthus.
    - advanced in a ventromedial direction parallel to the medial canthus.
    - 10–12 ml of 2% lidocaine hydrochloride (or 2% mepivacaine) is injected for both techniques.
  - four-point block may also be used:
    - use a 7.5-cm (3-inch) 20-gauge needle.
    - inserted into each quadrant.
    - inject 5–10 ml of 2% lidocaine hydrochloride per site.
- syringe is aspirated to ensure that the needle is not in a vessel prior to injection.
- cycloplegia occurs and ocular reflexes are lost within 20–30 minutes (depends on LA used).
- risk of orbital haemorrhage, optic nerve damage or neuritis, and globe penetration.

# Basic ophthalmic tests

## Schirmer tear test (STT)

- measures basal and reflex tear production.
- conduct before instilling fluorescein stain, topical anaesthetic or ocular medication.
- commercially available strips are placed in the middle to lateral third of the inferior conjunctival fornix for 1 minute.

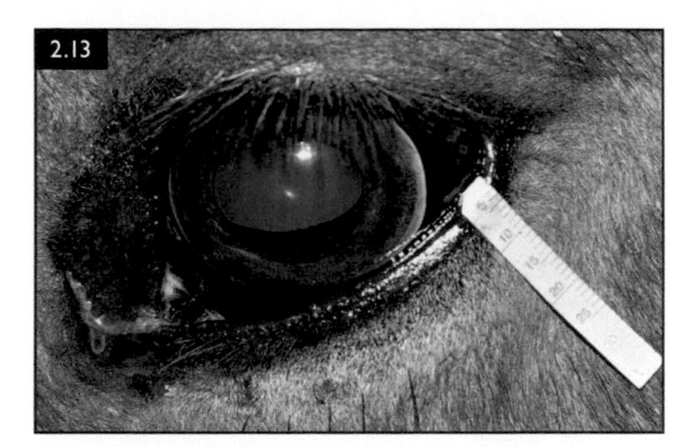


**FIG. 2.13** Schirmer tear test. The test strip is placed in the lateral lower conjunctival sac. The test strip shown contains a convenient millimetre ruler and is impregnated with a blue dye, which travels along the strip with the tears for accurate measurement.

- strip is removed, and the amount of tear-wetting is measured in mm/minute.
- dye-impregnated strips also available; can facilitate measurement (Fig. 2.13).
- results may vary:
  - normal horse usually 15–35+ mm/min.
  - 18–27 mm/min in donkeys.
  - sedative agents and topical tropicamide will reduce this reading.
  - STT value of <10 mm/min is considered abnormal:
    - consider clinical signs (e.g. mucoid discharge, conjunctival hyperaemia).

## Fluorescein staining

- used to evaluate the eye for corneal epithelial defects or ulcerations:
  - will not stain normal corneal epithelium or Descemet's membrane.
  - stains the hydrophilic corneal stroma following a disruption in the epithelium.
- either:
  - wet a fluorescein strip with sterile saline or eyewash solution.
  - touch it to the dorsal bulbar conjunctiva.
  - allow the horse to blink.
  - can lead to false-positive results if the strip touches the cornea directly.
- or:
  - fluorescein solution can be applied to the superior palpebral conjunctiva:
    - single-dose disposable ampoules.

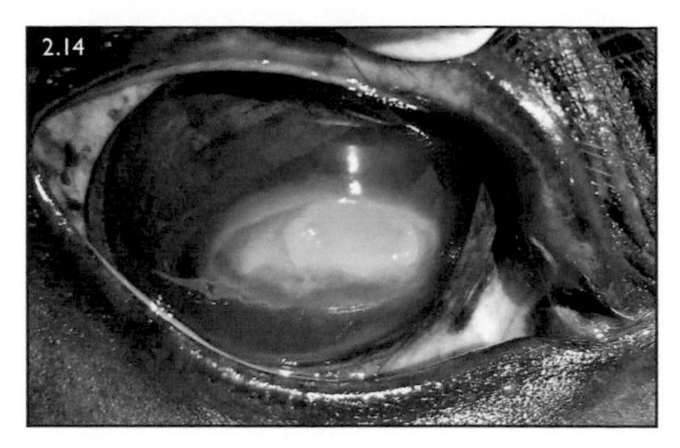


**FIG. 2.14** Corneal ulcer stained with sodium fluorescein stain.

    - or made by irrigating a fluorescein strip with saline eyewash:
      - do not dilute the fluorescein stain too much.
      - may result in a false-negative result.
- areas of ulceration appear as an apple-green, fluorescent lesion (Fig. 2.14).
- identification of lesions is enhanced by:
  - irrigation of the eye with sterile saline to remove excess fluorescein.
  - cobalt blue light source to illuminate the eye.
- other uses of fluorescein:
  - indication of nasolacrimal patency (Jones test):
    - fluorescein appears in the nostril after traversing the NLS following topical instillation (Fig. 2.15).
  - tear film break-up time (average 21.8 ± 10 seconds):

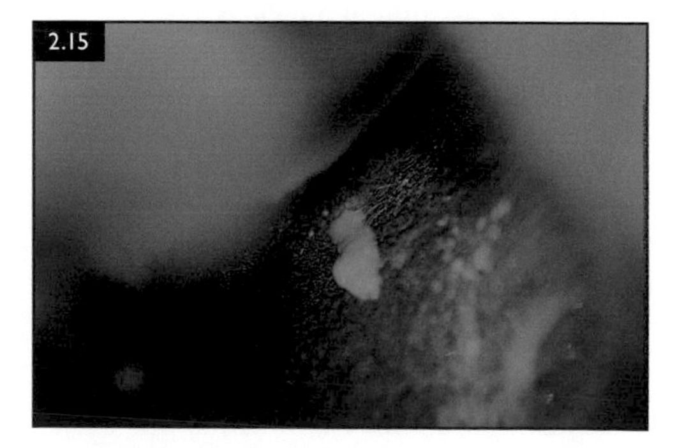


**FIG. 2.15** Close-up photograph of the nasal lacrimal punctum at the floor of the opening of the nares. The opening is highlighted by fluorescein.

    - helpful in diagnosing qualitative tear film disorders.
  - identify corneal perforation and aqueous leak (Seidel test):
    - 'wash-out effect' ('dark' [unstained] fluid stream/rivulet within the fluorescein-stained tear film).

## Tonometry

- measuring intraocular pressure (IOP).
- indirect measurement using:
  - applanation tonometer such as the Tono-Pen™
  - rebound tonometer Tonovet™ (Fig. 2.16).
  - both are relatively expensive but are portable, easy to use and accurate.
  - tip of the applanation tonometer is gently and repeatedly touched perpendicular to the corneal surface (after topical anaesthetic application) until IOP reading is obtained:
    - disposable rubber membrane covers the tip of the tonometer.
    - replaced between animals to prevent the spread of infectious disease.
    - **important to avoid inadvertent pressure on the globe through the eyelids or an artefactually increased IOP will be obtained:**
      - hold eyelids open, by resting your fingers on the bony orbital rim.
- normal IOP range in the horse is 15–32 mmHg.

## Lighted ocular examination

- initial evaluation in a lighted environment.
- general distance examination looking for evidence of:
  - facial asymmetry.
  - globe positioning, size and movement (Table 2.2).
  - abnormal ocular signs.
  - vision loss:
    - photopic (in ambient light) obstacle course, or maze test.
- neuro-ophthalmic examination and basic diagnostic tests are then completed.
- palpation of the orbital rim and globe retropulsion.

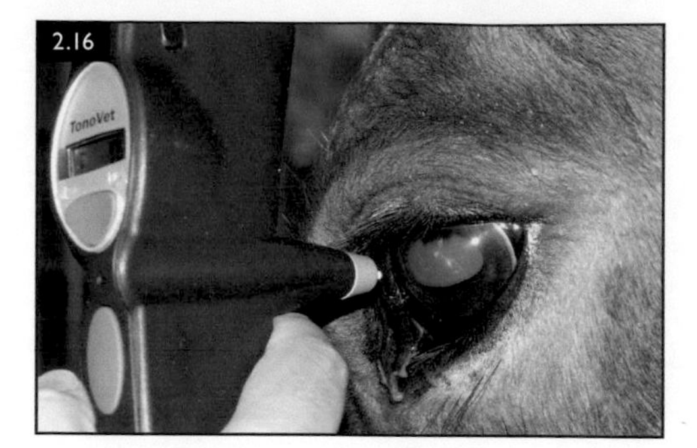


**FIG. 2.16** Rebound tonometer (Tonovet™) in use on a horse.

- examination of the adnexa and anterior segment:
  - diffuse or 'wide-beam' illumination:
    - transilluminator, penlight, direct ophthalmoscope or slit lamp.
    - magnification to aid in the identification of lesions:
      - head loupe, surgical glasses, direct ophthalmoscope or slit lamp.

## Dark ocular examination

- dimmed light.
- examination of the adnexa and anterior segment with diffuse illumination:
  - detect lesions involving the eyelid, conjunctiva, cornea, anterior chamber, iris, lens and anterior vitreous.
- focal light source, narrowed slit beam and magnification are used to identify and evaluate lesions:

**TABLE 2.2** Main causes of enophthalmos and exophthalmos

| Enophthalmos |
| --- |
| • Globe rupture/perforation<br>• Dehydration<br>• Horner's syndrome<br>• Orbital fat loss – starvation<br>• Abducens nerve paralysis/paresis<br>• Orbital fractures<br>• Phthisis bulbi |
| **Exophthalmos** |
| • Orbital abscess – bacterial<br>• Orbital granuloma – fungal (i.e. cryptococcosis)<br>• Orbital tumour (e.g. lymphoma)<br>• Orbital trauma<br>• Retrobulbar extra adrenal paraganglioma |

- slit aperture on direct ophthalmoscopes and, occasionally, penlights.
- slit lamp biomicroscope provides stereopsis and excellent magnification:
  - expensive and requires training to use properly.
- slit beam produces three images inside the eye as it strikes:
  - anterior corneal surface.
  - anterior lens capsule.
  - posterior lens capsule.
  - known as Purkinje–Sanson images.
  - used to determine the depth of ocular lesions.

- examination of the posterior segment:
  - dark environment will allow adequate pupil dilation in most cases.
  - pupil dilation with 1% tropicamide may be necessary in some cases:
    - mydriatic of choice.
    - provides dilation within 20–25 minutes and persists for up to 8 hours.
    - **only after neuro-ophthalmic examination, STT, tonometry and diagnostic sample collection for culture and sensitivity are completed.**
  - direct or indirect ophthalmoscope may be used to evaluate the posterior segment.

## Direct ophthalmoscopy (distant and close)

- **distant direct ophthalmoscopy:**
  - hold the instrument at arm's length from the patient (approximately 50–75 cm) and view the tapetal reflex.
  - dioptre dial allows the observer to set the focus level within the eye:
    - setting of 0 allows quick screening for any opacity present in the eye.
- **close direct ophthalmoscopy:**
  - move to within 2–3 cm of the eye to view the fundus (Fig. 2.17).
  - hand holding the instrument should rest on the horse's head:
    - sudden movement does not then injure the horse's eyes, the examiner or damage the instrument.
  - intensity of light beam is set at a level comfortable for the horse and that illuminates subtle lesions.

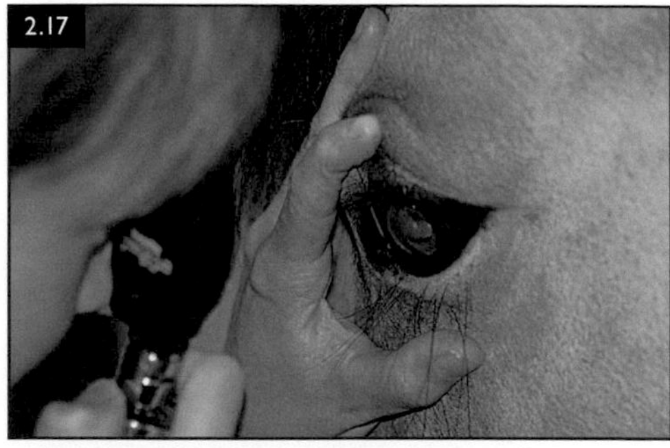

**FIG. 2.17** Direct ophthalmoscopy. With the dial set at 0, bring the fundic reflex into view at arm's length, then move in to approximately 2–3 cm from the eye, at which point the dioptre setting can be adjusted for clearer focus.

  - image generated is real, erect and magnified up to eight times.
  - fundus should be in focus using a dioptre setting of 0 to –3:
    - ONH should be examined closely.
    - rest of the fundus is examined in quadrants.
  - progressively higher positive dioptre strengths are then used to focus on more anterior structures of the eye:
    - +12 D to +8 D for examination of the lens.
    - +15 D to +12 D to evaluate the iris.
    - +20 D to +15 D to examine the external eye and adnexa.
  - advantages of this technique include:
    - greater magnification
    - availability of slit aperture.
    - ability to alter the dioptric strength.
  - disadvantages include:
    - small field of view
    - lack of stereopsis.
    - short working distance from the horse's face.

## Indirect ophthalmoscopy (Figs. 2.18–2.20)

- can be performed using a 20 D convex lens and a transilluminator or penlight.
- fundic image is inverted and reversed.
- 14 D lens provides a more magnified view of the ONH or any retinal lesions.
- magnification is less than with direct ophthalmoscopy (approximately 0.8–2 times).
- more difficult to master the technique.

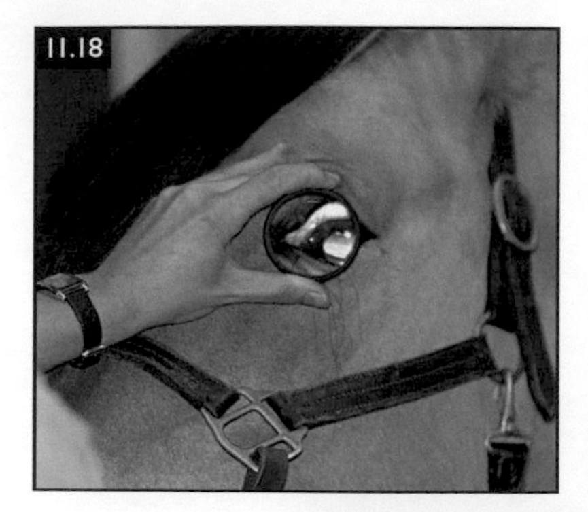


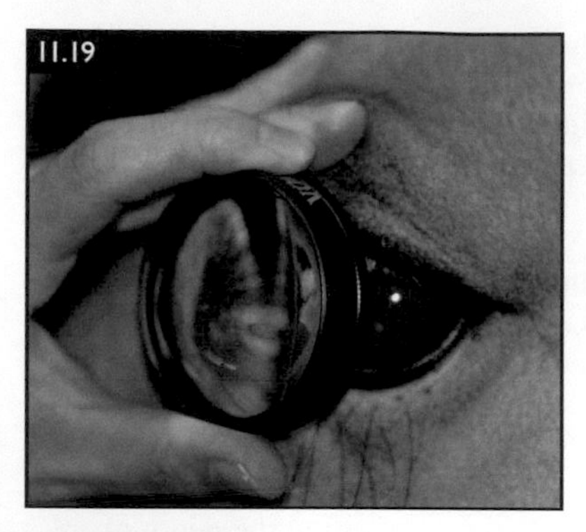


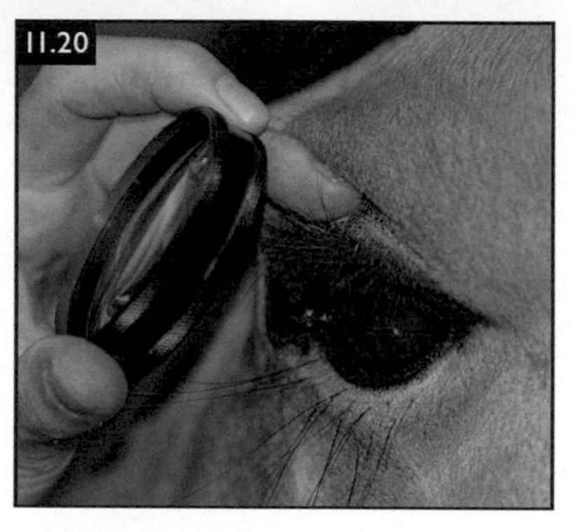


**FIGS. 2.18–2.20** Indirect ophthalmoscopy. (2.18) With the examiner positioned approximately an arm's length from the eye (50–75 cm), the fundic reflex is established with a light source held near the examiner's eye. (2.19) The lens is inserted into the beam approximately 3–5 cm away from the surface of the horse's cornea and the upper eyelid retracted with the examiner's middle finger. The fingers of the hand holding the lens should rest lightly on the animal's head to prevent injury to the eye in case of any sudden movement. (2.20) The lens can be lifted up out of the direct path of the light to re-establish the fundic reflex without taking the finger off the eyelid.

- advantages include:
  - greater field of view.
  - examiner is farther away from the patient.
  - equipment is relatively inexpensive.
- indirect ophthalmoscope headset (binocular indirect ophthalmoscopy):
  - provides stereopsis.
  - allows both hands to be used to manipulate the adnexa.
  - more difficult to master and the instrument is expensive.

## Special diagnostic tests

### Culture and sensitivity testing

- used routinely to diagnose infectious disease:
  - antimicrobial sensitivity results essential in determining appropriate antimicrobial therapy.
- samples for culture must be collected prior to:
  - fluorescein staining.
  - ideally before topical anaesthetic administration:
    - ♦ not possible to collect samples without anaesthesia in many cases.
    - ♦ solutions may inhibit the growth of microorganisms or cause contamination of the sample.
- corneal cultures are indicated in suspected cases of bacterial or fungal keratitis:
  - samples taken from the centre and edge of the corneal lesions (Fig. 2.21).
- conjunctival cultures are rarely indicated as potential pathogens are present in the conjunctiva as normal flora:
  - normal flora of conjunctival sac varies with season and geographical location.
  - includes the bacteria:

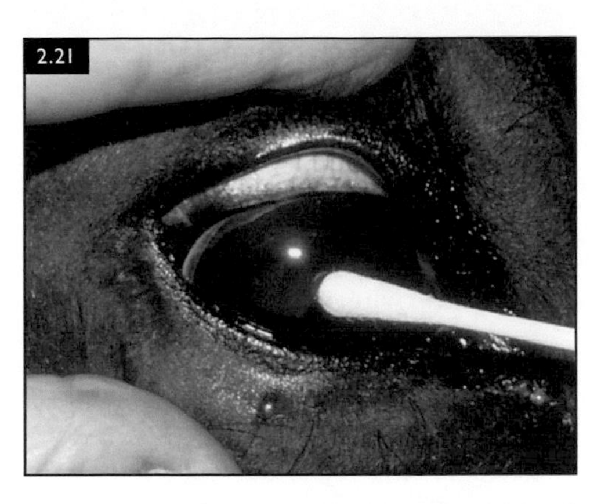

FIG. 2.21 Collection for culture and sensitivity. A sterile culturette can be used to swab the cornea for microbes. Care should be taken to avoid inadvertent contamination from the eyelids.

- ♦ *Acinetobacter* spp., *Staphylococcus epidermidis*, *Corynebacterium* spp. and *Bacillus cereus*.
- ○ fungi:
  - ♦ *Aspergillus*, *Penicillium*, *Alternaria*, *Mucor*, *Absidia* and *Cladosporium* spp.

## Corneal/conjunctival cytology

- cytology samples may be collected following the application of a topical anaesthetic.
- usually from corneal ulcers and abscesses.
- Dacron swab, cytobrush, stainless steel spatula (e.g. Kimura spatula) or the blunt end of a scalpel may be used to collect scrapings for cytological evaluation.
- fungal organisms have an affinity for Descemet's membrane and deeper layers of the corneal stroma:
  - ○ deep scrapings from centre of the lesion plus from the periphery are required.
- place on microscope slides and stain with Gram or Wright–Giemsa for evaluation of the types of cells and organisms present:
  - ○ specialised stains such as periodic acid–Schiff (PAS) stain and Gomori's methenamine silver stain may be used when fungal infection is suspected.

## Nasolacrimal system cannulation and lavage

- indicated if an obstruction or abnormality in the NLS is suspected.
- easily catheterised and flushed in a retrograde manner from the distal nasal opening:
  - ○ 20–60 ml of physiological saline.
  - ○ soft 5- or 6-Fr feeding tube, cat or dog urinary catheters, or polyethylene tubing (Figs. 2.22, 2.23).
- upper or lower eyelid punctum may be cannulated and flushed anterograde with:
  - ○ small cat urinary catheter, or an intravenous catheter with the stylet removed.
  - ○ flushed with 20–30 ml of physiological saline (Fig. 2.24).
  - ○ adding a small amount of fluorescein dye will facilitate identification of the fluid as it exits the nasal or lacrimal punctum.
- debris flushed from the NLS should be collected for cytology and culture and sensitivity testing.

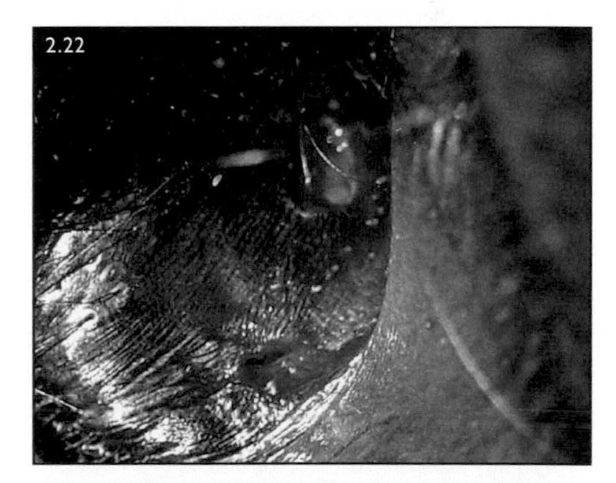

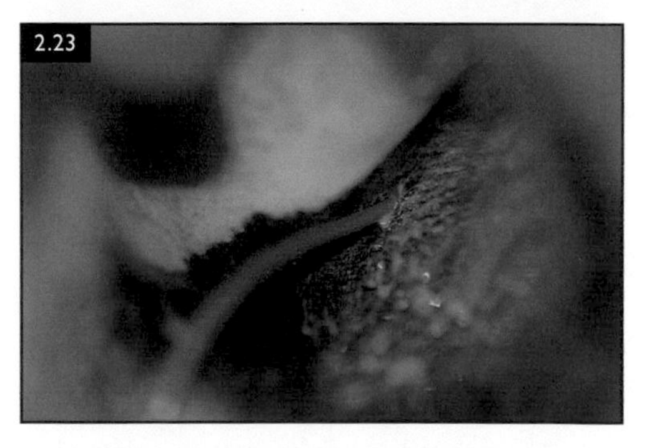

FIGS. 2.22, 2.23 Retrograde flushing of the NLS. (2.22) The distal opening is on the floor of the nasal vestibule. After administration of an anaesthetic gel over the distal duct, the opening can be cannulated with a variety of catheters, such as the 5-Fr feeding tube shown (2.23).

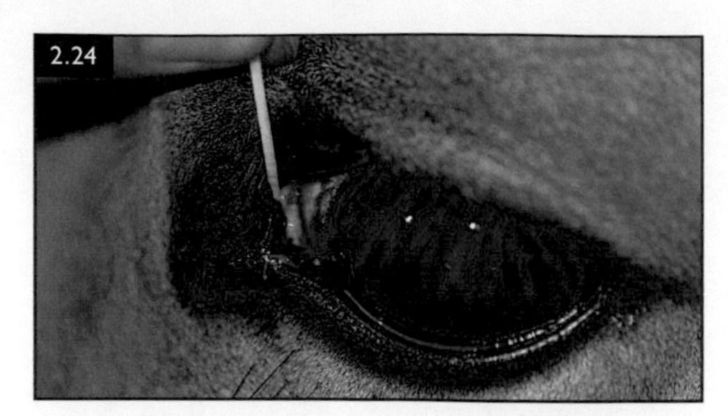


**FIG. 2.24** Anterograde flushing of the NLS. The proximal openings to the nasolacrimal duct (puncta) are located just inside the eyelid margin of the superior and inferior eyelids. Following application of topical anaesthetic to the area, the puncta can be cannulated with an open-ended tom-cat catheter, or an intravenous catheter (without interior needle) as shown here.

### Anterior and posterior chamber paracentesis

- very rarely indicated.
- these procedures should only be performed by a specialist.

### Electroretinography

- electroretinography (ERG) is a test of retinal, not visual, function.
- study of electrical potentials produced by the retina when it is stimulated by light of varying intensity, wavelength and flash duration.
- used to diagnose retinal dysfunction (e.g. Appaloosas suspected of congenital stationary night blindness [CSNB]) and to ensure retinal function prior to cataract extraction.

### Radiography

- useful for an abnormality involving bone or if there is the possibility of a radiopaque foreign body.
- Flieringa ring may be placed at the limbus to establish a landmark for orientation.
- contrast radiography includes:
  - dacryocystorhinography:
    - establish the extent of a congenital or an acquired NLS abnormality.
    - upper lacrimal punctum or distal nasal puncta is cannulated or catheterised.
    - approximately 5 ml of iodine-based contrast medium is injected to outline the NLS.

### Ocular ultrasonography

- valuable, non-invasive diagnostic technique for evaluating the eye.
- sedation and local anaesthesia are usually required.
- 7.5–10-MHz probe is ideal:
  - probe can be applied directly to the cornea, with or without a stand-off.
  - transpalpebral technique may also be used.
- horizontal and vertical sections of both eyes should be evaluated and compared.
- useful in retrobulbar disease, intraocular cysts, lens subluxation or luxation, intraocular haemorrhage, retinal detachment and intraocular masses or foreign bodies.

### Computed tomography/ magnetic resonance imaging

- CT is used primarily for evaluating retrobulbar disease.
- MRI provides excellent visualisation of the eye and orbit.
- available for horses at referral centres.

## Methods of medicating the equine eye

### Topical administration

- provides high concentrations of the medication to the anterior segment of the eye.
- decreases likelihood of associated systemic side effects.
- used for treating diseases of the conjunctiva, cornea, anterior part of the sclera, anterior chamber, iris or ciliary body.
- will not establish therapeutic drug levels in the posterior segment tissues or eyelids.
- available in ointment, suspension and solution forms:
  - topical ophthalmic ointments may be applied to the conjunctival sac of the eye:
    - longer contact time in the eye than solutions or suspensions.
    - lower rate of systemic absorption.
    - may have a greater ability to impair corneal healing.
    - **should not be used:**

- presence, or high risk of globe rupture, globe perforation or intraocular surgery.
- may irritate and damage the internal structures of the eye.

- topical ophthalmic solutions or suspensions may be applied by using a tuberculin syringe and a 25-gauge needle with the tip broken off at the hub:
  - ♦ spray the medication gently into the eye.
  - ♦ helps avoid contamination of the medication bottle.
  - ♦ decreases the likelihood of globe perforation or damage.
  - ♦ solutions and suspensions have the shortest contact time:
    - mix readily with tears and wash away rapidly.
  - ♦ frequent therapy is required to maintain high therapeutic drug levels.
  - ♦ application can be difficult even with adequate physical restraint:
    - strength of orbicularis oculi muscles.
    - horses may become resistant to, and even violent during, administration of ocular medications.
- SPL system recommended in horses requiring:
  - frequent or long-term administration of topical medications.
  - those animals becoming resistant to topical therapy.

## Subpalpebral lavage (SPL) system

- safe placement requires:
  - adequate chemical and physical restraint.
  - auriculopalpebral nerve block.
  - regional anaesthesia and topical anaesthetic instilled into the conjunctival fornix.
  - postoperatively under general anaesthesia.
- site of placement clipped and sterile preparation with dilute povidone–iodine sterile saline solution.
- placed aseptically with sterile gloves.
- various modifications exist for the placement of these systems, but kits are available commercially for ease of use.
- traditionally, SPL systems have been placed in the dorsal or superior conjunctival fornix in the upper eyelid (Figs. 2.25–2.27):
  - require the use of a supraorbital nerve block for sensory denervation of the upper eyelid.
- inferomedial placement of a single-entry tube is also used:
  - preferred by some clinicians as may be associated with lower incidence of complications.
  - footplate located between less mobile inferior eyelid and anterior aspect of the nictitating membrane:
    - ♦ helps protect the cornea from the tube.
    - ♦ makes the tube less likely to migrate out.
    - ♦ proper positioning is relatively easy, and gravity helps maintain the appropriate position.
  - first introduce a gloved finger into the conjunctival sac and slide the provided trochar (attached to tubing) alongside the finger into the inferomedial conjunctival fornix and through the eyelid in a ventronasal direction (Fig. 2.28).
  - haemostats placed over the exit site will provide counterpressure and help the needle exit the skin.
  - commercially available silastic tubing, with an attached footplate, is then threaded through the needle from the conjunctival surface to the skin surface.
  - once tubing exits the sharp end of the needle, both the needle and tubing are pulled through the eyelid.
  - footplate positioned deep in medial aspect of inferior conjunctival fornix lying flat between the lower eyelid and anterior surface of nictitating membrane (Fig. 2.29).
  - piece of tape attached to tubing flush with the exit point from the skin and sutured in place to prevent tube from sliding back (Fig. 2.30).
  - tubing is brought over the poll, braided into the mane and secured to the horse using tape and skin sutures.

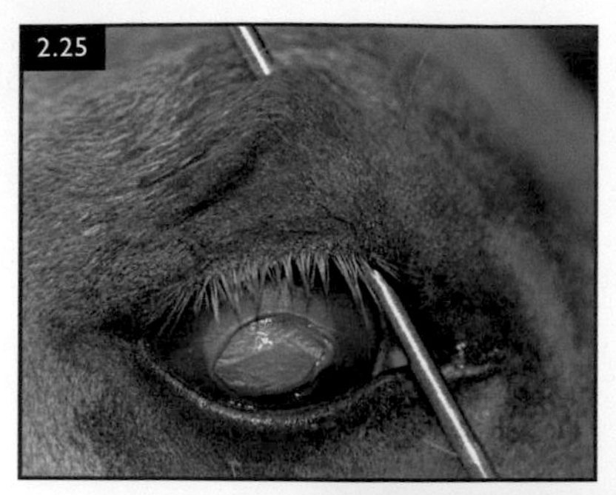


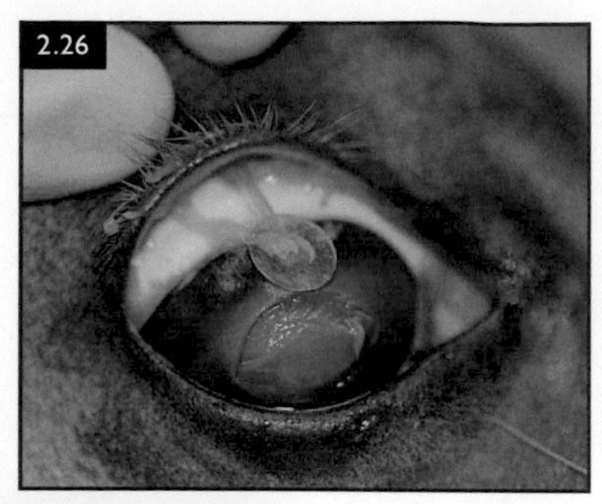


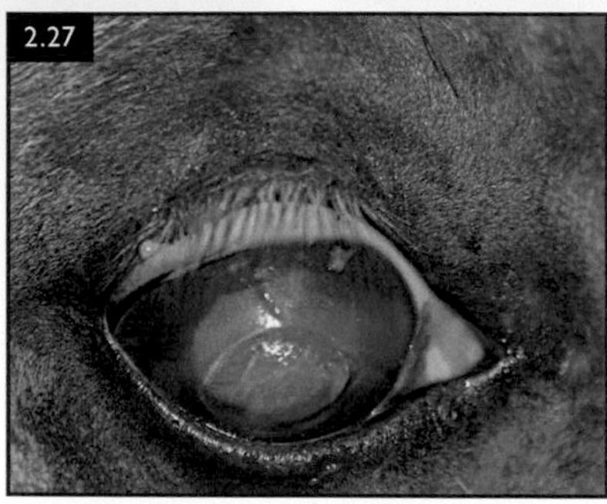


**FIGS. 2.25–2.27** Dorsal placement of an SPL system. (2.25) The needle is tunnelled under the eyelid through the palpebral conjunctival fornix. The lavage tubing is fed through the needle and both needle and tubing are pulled through the hole in the eyelid. (2.26) The footplate is shown as it is pulled up under the dorsal eyelid. (2.27) The lavage entry hole should be at the highest point possible when passing through the conjunctival fornix to avoid inadvertent corneal contact with the lavage footplate.

  - some clinicians tunnel the tubing under the skin in multiple sites over the face in lieu of tape 'butterflies' and skin sutures.
  - delicate silicone lavage tubing can be reinforced by encasing it in intravenous drip-line tubing which is split longitudinally and wrapped around the silicone tubing.
  - injection port, for manual injection, can be attached to the free end of the tubing, secured to a tongue depressor and wrapped in gauze and tape to make it less likely that the end will bend or kink (Fig. 2.31).
  - manual delivery involves injecting approximately 0.15–0.2 ml of medication at the injection port and flushing the drug into the eye using 3 ml of air:
    - air must be injected slowly, to avoid irritation to the corneal or conjunctival surface, and decreasing any discomfort.
  - continuous drip or pump systems can also be used.
- complications associated with SPL systems include:
  - infection of the eyelid.
  - loss of the footplate in the eyelid:
    - no subsequent long-term problems have been reported.
  - conjunctival granuloma.
  - endophthalmitis.
  - iatrogenic trauma to the globe during insertion.
  - plugging or breakage of the tubing.
  - tube displacement or premature removal by the horse.
  - suture loss and injection port damage or loss.
  - poorly placed tubing or tube slippage can quickly produce:
    - corneal irritation or ulceration.
    - allow topical medications to leak into the subcutaneous tissue, leading rapidly to eyelid swelling (chemosis) and severe inflammation.
- SPL systems should be checked daily for complications and to ensure patency.
- systems are well tolerated for extended periods of time.

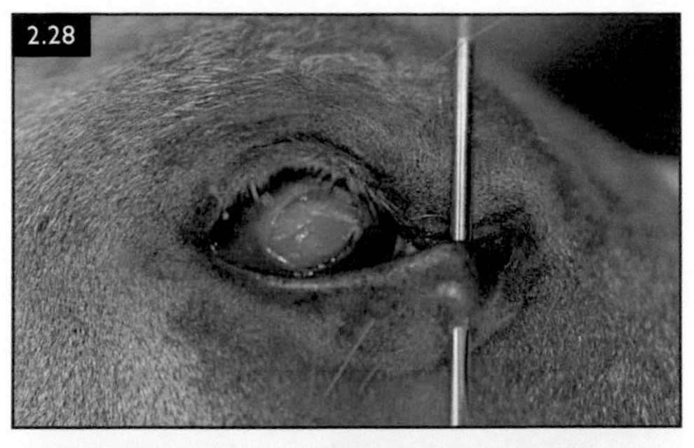

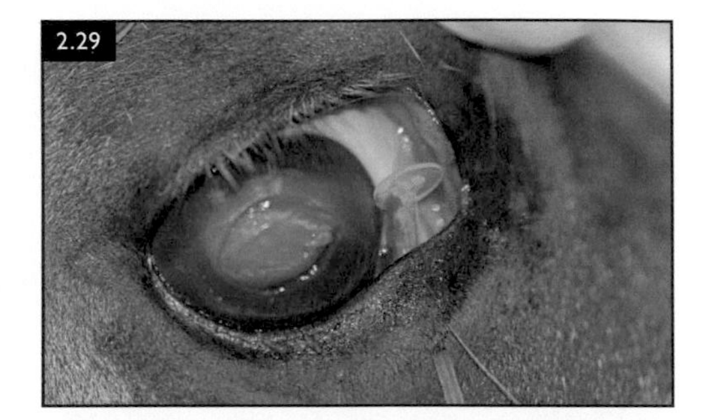

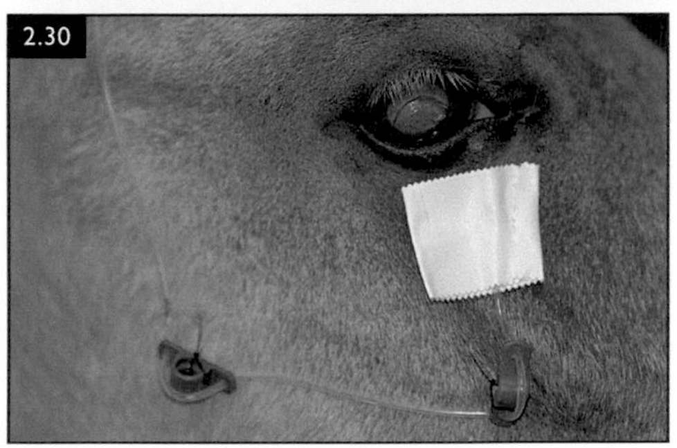

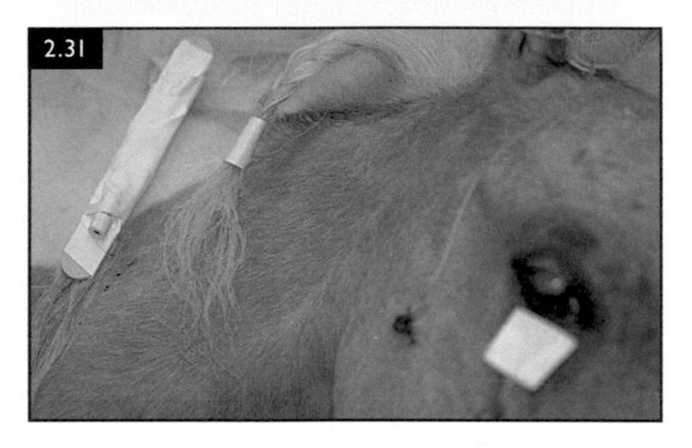

**FIGS. 2.28–2.31** Ventromedial SPL system placement. (2.28) The needle is tunnelled through the medial aspect of the ventral eyelid. Care is taken to avoid the inferior nasolacrimal punctum. (2.29) The lavage footplate is pulled into position. (2.30) The lavage system in place with tubing secured to facial skin at several points to minimise movement. (2.31) The distal aspect of the lavage system, showing the injection port of the tubing secured to a tongue depressor for added support.

- topical ophthalmic suspensions may precipitate and block the tubing.
- ointments should not be used in an SPL system.

## Topical drug reservoirs

- drug-impregnated collagen shields and contact lenses can be used as drug delivery devices (Figs. 2.32, 2.33):
  - pre-soaked in the chosen drug.
  - initial corneal drug levels are high, but they deplete rapidly, thereby limiting the apparent benefit over topical therapy.
  - some drugs are less bioavailable from soaked contact lenses, so drug choice is important.

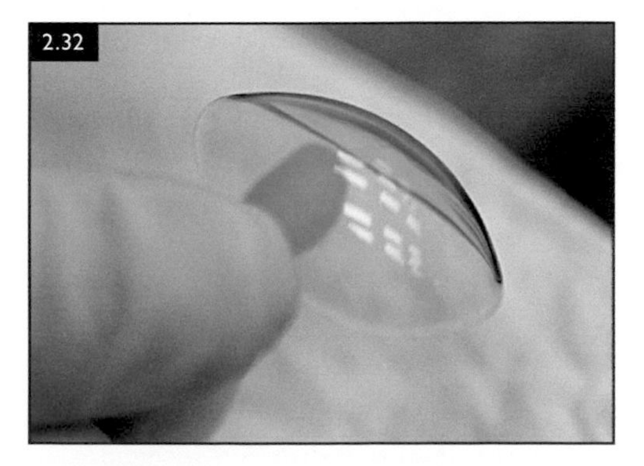

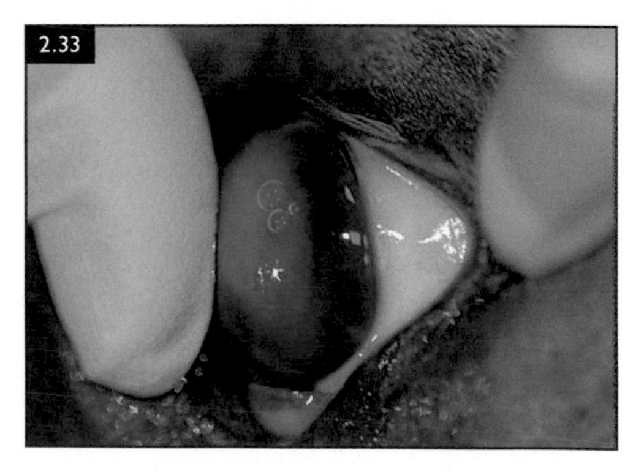

**FIGS. 2.32, 2.33** Contact lens. (2.32) A tinted contact lens may be easier to work with and identify on the eye. (2.33) The ideally sized contact lens will not extend onto the sclera.

## Subconjunctival administration

- Subconjunctival injection may establish much higher medication levels in tissues for a longer period of time than those attained with occasional topical application.
- beneficial in emergency situations to establish high tissue levels of:
  - antimicrobials.
  - repository corticosteroids.
  - therapeutic levels may be maintained for several days.
- low intravitreal levels of the drug may be achieved.
- often used to augment topical therapy.
- injection facilitated by use of topical anaesthesia and an auriculopalpebral nerve block.
- maximum volume of 1 ml may be injected under the bulbar conjunctiva:
  - use a 25–27-gauge needle.
  - bevel of the needle should remain up.
  - hand holding the syringe should rest on the horse's head when injecting.
  - decreases the risk of inadvertent globe perforation or trauma (Fig. 2.34).
  - dorsolateral quadrant is the easiest place to inject:
    - important to inject as close to the lesion site as possible.
    - drug levels are highest in the region immediately adjacent to the injection site.

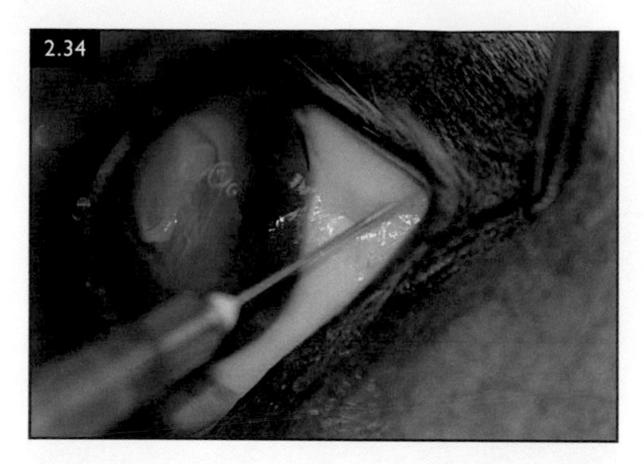


**FIG. 2.34** Subconjunctival administration of a drug. Using a fine-gauge needle (25–30 gauge), the conjunctiva is tented with the needle tip and the subconjunctival space is then expanded with drug.

## Intraocular administration

- Intracameral (into the anterior chamber) or intravitreal (into the vitreous) administration is used occasionally.
- useful in cases of endophthalmitis.
- both are specialist techniques and rarely indicated.
- associated with several severe complications and side effects.

## Parenteral therapy

- most drugs are unable to cross the blood–ocular barrier (BOB):
  - limits the usefulness for intraocular disease.
  - chloramphenicol and sulphonamides penetrate the BOB adequately.
- inflammatory conditions may facilitate penetration of antimicrobials not normally able to penetrate the BOB.
- intravenous administration preferred as higher plasma levels and intraocular concentrations.
- no advantage to continuous infusion over pulsed therapy.
- major drawback is the relatively higher risk of associated systemic side effects.

# NEONATAL AND CONGENITAL OCULAR DISORDERS

- ocular diseases of the foal may be congenital or acquired.
- congenital ocular defects have a reported incidence of 0.5–35% in the horse (depending on the population studied).
- foals may be born with single or multiple ocular defects including:
  - NLS atresia.
  - entropion.
  - eyelid, iris or fundic colobomas.
  - anterior segment dysgenesis.
  - cataracts (most commonly reported).
  - microphthalmia.
  - retinal dysplasia, retinal detachment and CSNB.
- acquired lesions or disorders include:
  - ocular trauma, globe perforation and phthisis bulbi.

- normal foals:
  - decreased corneal sensitivity and lower tear production than adult horses.
  - menace reflex may not appear until 7–14 days post-partum.
    - increases likelihood of acquired ocular disease.
    - especially in systemically ill and/or recumbent foals.
- all compromised neonatal foals should be examined daily for signs of ocular disease:
  - ophthalmic lubricant ointments applied regularly to help protect the cornea from damage.

## Anophthalmos and Microphthalmos

### Definition/overview

- Anophthalmos is a rare condition with complete absence of ocular tissues, resulting in blindness.
- Microphthalmos describes a congenitally small globe:
  - **simple** or pure microphthalmos (or nanophthalmos):
    - small but otherwise normal and visual globe.
  - **complicated** microphthalmos:
    - more common.
    - small globe with multiple anterior and/or posterior segment defects.
    - may affect vision.
- one or both eyes may be affected.

### Aetiology/pathophysiology

- both are congenital anomalies with unknown aetiologies.
- most cases are thought to be spontaneous and idiopathic:
  - suggested causes include intrauterine maternal infections and exposure to toxins.
  - Thoroughbreds have been reported to be at higher risk:
    - inherited component not confirmed.
- disfigured orbit may occur secondary to an absent or smaller than normal-sized globe.

### Clinical presentation

- foals born with an absence of ocular tissue or a small eye:
  - usually obvious on examination (Fig. 2.35):
    - small palpebral fissure.
    - elevated/prominent nictitating membrane.
    - conjunctivitis.
    - facial asymmetry.
    - entropion secondary to failure of the smaller globe to support the eyelids.
    - mild keratitis to severe corneal ulceration.
- vision may be present but usually the eye is blind.
- multiple ocular anomalies may also be present, including:
  - colobomas, cataracts, retinal dysplasia and/or retinal detachment.

### Differential diagnosis

- globe perforation or rupture.
- phthisis bulbi:
  - globe that atrophies secondary to severe ocular disease (Fig. 2.36).

### Diagnosis

- thorough history and ocular examination:
  - differentiate between congenital microphthalmos and acquired diseases.

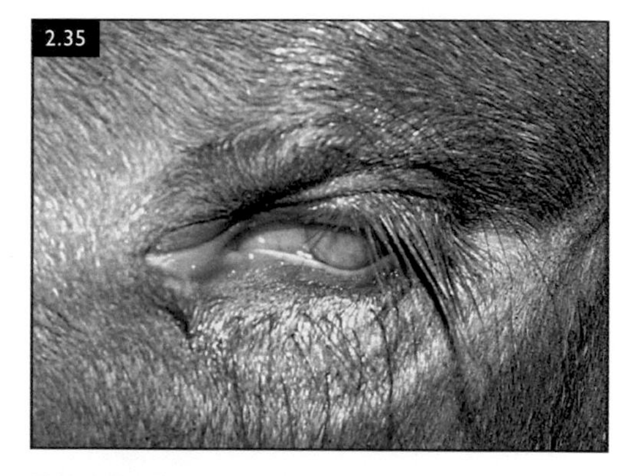


**FIG. 2.35** This neonatal Thoroughbred was born with bilateral microphthalmos and was blind. The left eye had a palpably smaller-than-normal orbit and the picture shows a prolapsed nictitans and a small, pink, residual globe.

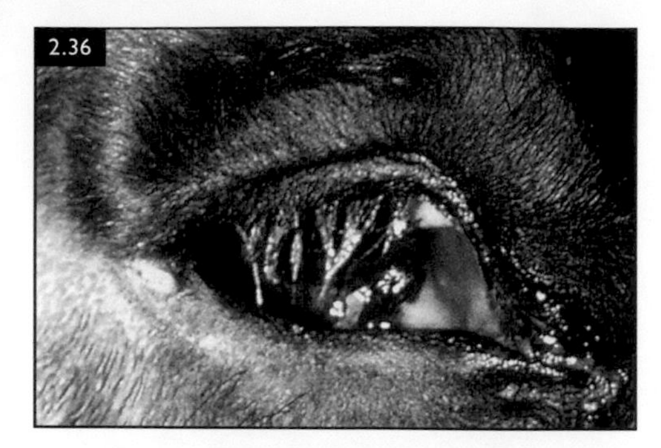


**FIG. 2.36** Phthisis bulbi (shrunken globe), typically associated with severe chronic intraocular inflammation, must be distinguished from microphthalmos in the horse.

### Management

- no treatment.
- secondary entropion present and eye is uncomfortable – surgical correction of former may be indicated.
- unilateral cases with chronic ocular irritation or discomfort, and where the eye is blind:
  - enucleation may be treatment of choice.
- bilateral involvement with severe visual deficits or blindness:
  - euthanasia may be indicated.

### Prognosis

- congenital non-progressive lesions with a variable effect on vision.

## Strabismus

### Definition/overview

- deviation of the globe from its normal position that may be congenital or acquired.
- unilateral or bilateral.
- may be present continually or intermittently.
- normal neonatal foals have a slight medioventral strabismus:
  - changes to normal adult position by 1 month of age.
- bilateral convergent strabismus ('cross-eyes') is termed esotropia.
- divergent strabismus is exotropia.
- vertical upward deviation is termed hypertropia.
- downward deviation is hypotropia.

### Aetiology/pathophysiology

- congenital lesions may be inherited:
  - dorsomedial strabismus and hypertropia seen in CSNB.
  - Saddlebreds have an increased incidence of esotropia.
- due to congenital anomalies or trauma to, or atrophy of, the extraocular muscles.
- periorbital abscesses or tumours can cause deviation of the globe.
- central or peripheral neurological disease may lead to strabismus.

### Clinical presentation

- clinical examination reveals:
  - deviation of one or both globes from the normal position (Fig. 2.37).
  - may be noted as excessive scleral exposure.
  - dorsomedial displacement of the globe may be associated with dim-light vision deficits in animals with CSNB.
  - spontaneously resolving medioventral strabismus in normal neonatal foals.
  - ambylopia (vision loss in the deviated eye leading to a lack of true stereopsis) may be present in congenital or early-onset cases.

### Differential diagnosis

- differentiate primary congenital strabismus from causes of acquired strabismus.

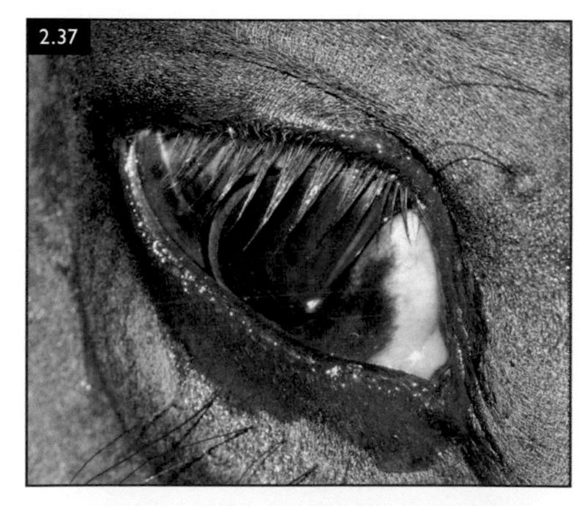


**FIG. 2.37** Enophthalmos and protrusion of the third eyelid in combination with medial strabismus and internal ophthalmoplegia due to a basal skull fracture damaging the parasympathetic fibres of the oculomotor nerve and the abducens nerve.

## Diagnosis

- clinical appearance.
- history and/or additional diagnostic tests such as ultrasonography may help identify the cause of the strabismus.

## Management

- surgical therapy to repair strabismus can be attempted:
  - technically challenging and often unsuccessful.
- acquired strabismus – underlying cause should be treated.
- horses with strabismus associated with CSNB should not be bred.

## Prognosis

- excellent for normal neonates with medioventral strabismus – resolves spontaneously.
- guarded for other causes.

# Entropion

## Definition/overview

- inversion of the eyelid:
  - **primary** congenital ocular disorder:
    - possibly more common in Thoroughbreds and Quarter horses.
    - no inheritance has been proven.
    - weakness of tarsal plate, abnormal positioning of the globe or abnormally small eyelids.
  - **secondary** (acquired) to another ocular disease is more common:
    - enophthalmos resulting from microphthalmos.
    - quite common in premature, dysmature or systemically ill foals:
      - secondary to dehydration and/or cachexia with orbital fat loss.
    - occasionally in adults secondary to:
      - trauma and severe scarring of the eyelids (cicatricial).
      - prolonged blepharospasm caused by chronically painful conditions of the conjunctiva and cornea (spastic).
      - enophthalmos or phthisis bulbi.
- lower and/or upper, but former is much more common.
- unilateral or bilateral.

## Aetiology/pathophysiology

- any type of entropion leads to rolling in of the eyelids:
  - cilia and facial hairs may contact the conjunctiva and/or cornea (trichiasis).

## Clinical presentation

- lower and/or upper eyelid is inverted (Fig. 2.38).
- trichiasis present:
  - corneal irritation and increased lacrimation.
  - blepharospasm may exacerbate entropion.
  - photophobia
  - conjunctivitis.
  - keratitis and/or corneal ulceration.
- other congenital eyelid abnormalities may be present in cases of primary entropion:
  - ankyloblepharon or coloboma.
  - dermoid and ciliary abnormalities.
- other ocular abnormalities may be present in cases of cicatricial or spastic entropion.

## Differential diagnosis

- other causes of ocular pain or irritation:
  - conjunctivitis, keratitis, foreign bodies and corneal ulceration.

## Diagnosis

- based on clinical examination, which confirms inversion of the eyelid margin:
  - determine whether the entropion is primary or secondary.
  - spastic entropion resolves with topical anaesthesia.

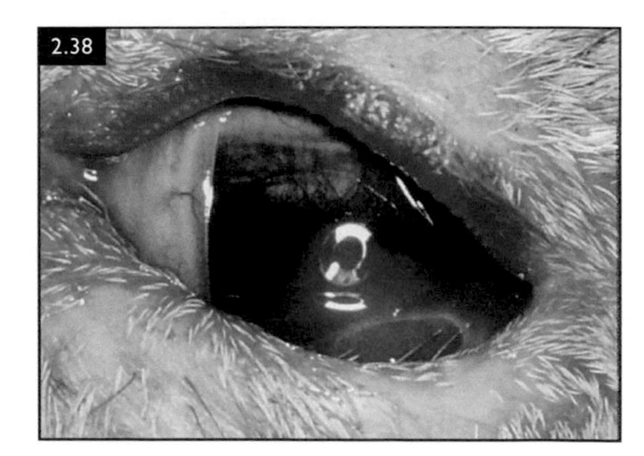


**FIG. 2.38** Primary entropion in a foal as a result of microphthalmos. Note the corneal ulcer present in the ventral aspect of the cornea.

- positive fluorescein staining of corneal ulcer when spastic entropion is secondary.
- cicatricial entropion is diagnosed based on a history of trauma or eyelid surgery.

## Management

- treatment varies with cause, severity and chronicity:
  - primary entropion may be self-correcting in foals.
  - young animals:
    - evert the eyelid margin temporarily by manual manipulation.
    - eliminate the corneal irritation.
    - temporary tacking to create a normal conformation by using either:
      - non-absorbable suture (e.g. 4-0 silk) in vertical mattress pattern.
      - surgical staples.
      - maintained until orbital growth establishes normal eyelid conformation or the cause of the entropion has resolved.
    - adjunctive medical management includes topical eye lubrication.
    - topical antimicrobials and atropine where corneal ulceration is present.
  - any other underlying cause should be treated.
- surgical repair via a modified Hotz–Celsus procedure may be required (Fig. 2.39) if the condition persists.

## Prognosis

- depends on the cause and severity of the condition plus additional ocular pathology.
- guarded to good.

# Coloboma

## Definition/overview

- congenital absence of tissue that is normally present.
- defined by location in relation to line of closure of the optic fissure (6 o'clock position):
  - along the line of closure are typical forms.

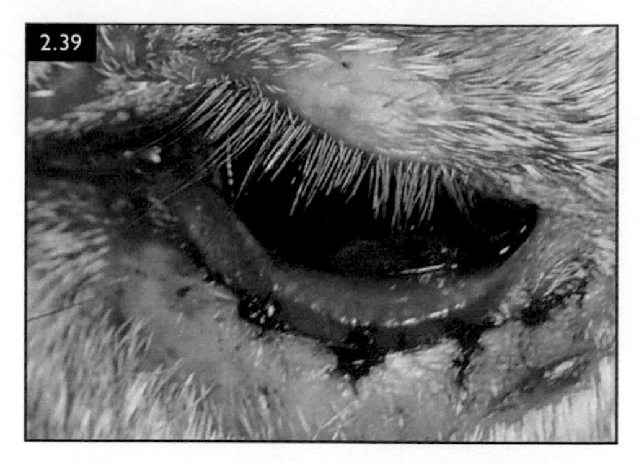


**FIG. 2.39** Inferior entropion has been resolved following a Hotz–Celsus surgical eversion technique.

  - other areas are atypical.
- eyelids, iris, ciliary body, lens, choroid, retina (including RPE) alone and/or optic disc may be affected:
  - may be seen in combination with other ocular disorders.
  - individual incidental finding in some blue-eyed or incompletely albinotic horses.
- rare in the horse.

## Aetiology/pathophysiology

- unknown cause and occur sporadically.
- colobomas occur early in embryogenesis:
  - typical forms involve defective closure of the embryonic fissure.
  - atypical forms lack the development of ocular tissue outside the fissure.

## Clinical presentation

- vary in appearance:
  - small notches in tissue to an almost complete absence:
  - eyelid, iris, ciliary body, lens, choroid, retina and/or ONH.
- eyelid coloboma:
  - trichiasis.
  - secondary keratitis and blepharospasm.
- iris colobomas (Fig. 2.40):
  - dyscoria (abnormal pupil shape).
- retinal or optic nerve colobomas:
  - partial or complete retinal detachment.
  - decreased or absent vision.
- occasionally present:
  - scleral ectasia.
  - other congenital ocular abnormalities:
    - microphthalmos.
    - cataract and retinal dysplasia.

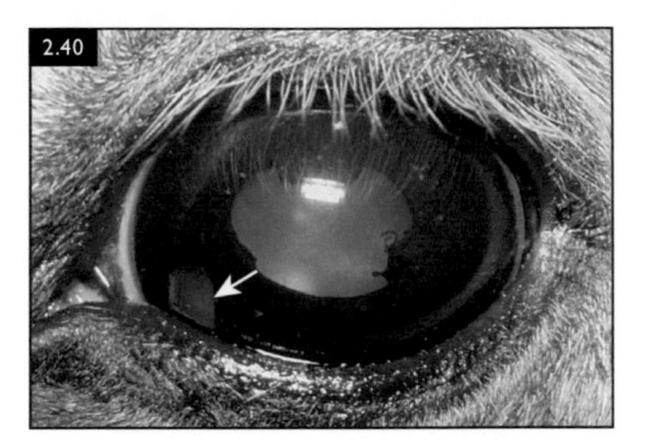


**FIG. 2.40** A coloboma (arrow) is present in the ventromedial iris of this foal. Additional anomalies in this eye include cataract, iris hypolasia and a dysplastic corpora nigra. (Photo courtesy D Ramsey.)

### Differential diagnosis

- eyelid lacerations may resemble eyelid colobomas.
- synechiae may appear similar to iris colobomas.
- lens luxation may mimic lens colobomas.
- other causes of retinal detachment (Table 2.3) must be ruled out.
- glaucoma with optic nerve cupping should be differentiated from optic nerve colobomas.

**TABLE 2.3** Aetiologies of retinal detachment

- Congenital/inherited
  - Associated with retinal dysplasia and/or cataracts
- Systemic infectious diseases (mycoses, lymphosarcoma, toxoplasmosis)
- Neoplasia
- Vitreoretinal traction bands/adhesions (traction detachment)
- Trauma
- Vitreal degeneration
- Cataracts (i.e. uveitic, traumatic or idiopathic)
- Sudden decreases in IOP
- Serous or fluid detachments (vasculitis, uraemia, vascular hypertension)
- Glaucoma
- Extraocular pressure
- Retinal tears/holes (rhegmatogenous detachment)
- Equine recurrent uveitis
- Head trauma or perforating globe wounds
- Idiopathic
- Postoperative complication of phaecoemulsification

### Diagnosis

- clinical identification of a defect in the eyelid, iris, lens, choroid, retina and/or optic nerve that has been present since birth.

### Management

- some eyelid colobomas:
  - defect closed by blepharoplasty using advancing skin and/or conjunctival flaps.
  - eyelashes or facial hair contributing to corneal irritation may be relieved by:
    - cryoepilation.
    - ocular lubrication.
- no therapy for intraocular colobomas.

### Prognosis

- usually, non-progressive:
  - associated risk of retinal detachment with optic nerve or retinal colobomas.
  - **coloboma directly or indirectly causing visual loss renders the horse unsound.**

## Dermoids

### Definition/overview

- relatively uncommon.
- focal congenital masses consisting of displaced normal skin tissue (choriostomas).
- eyelid, nictitating membrane, conjunctiva and/or cornea:
  - most common in the temporal limbus:
    - can involve neighbouring bulbar conjunctiva and cornea.
  - coarse hairs may grow from mass and cause irritation to conjunctiva or cornea.

### Aetiology/pathophysiology

- unknown.
- association of limbal dermoids with iris hypoplasia and cataracts in Quarter horses.
- possibly due to abnormal differentiation of isolated groups of cells early in development or abnormal invagination of ectodermal tissue later in gestation.
- may interfere with blinking, and lead to exposure keratopathy.

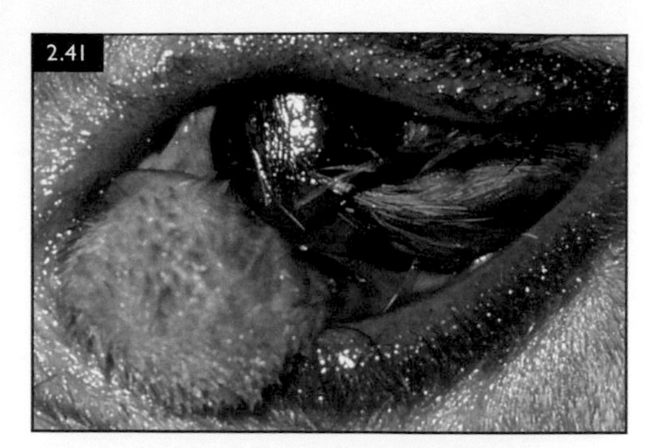


**FIG. 2.41** Dermoid in a foal involving the temporal cornea and conjunctiva and observed separately on the palpebral surface of the nictitating membrane.

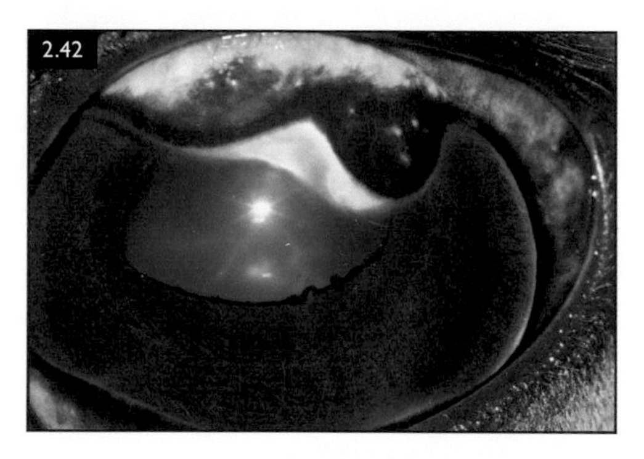


**FIG. 2.42** Extensive dorsal limbal dermoid in the left eye of a foal, associated with anterior segment dysgenesis (note the persistent pupillary membrane running from 5 o'clock on pupil to posterior cornea). This foal had a less obvious dermoid in the right eye.

- degree of visual impairment depends on the extent of corneal involvement and the occurrence or absence of disease secondary to corneal problems.

## Clinical presentation

- pigmented mass with or without hair growth involving the eyelid, nictitating membrane, conjunctiva and/or cornea (Figs. 2.41, 2.42).
- mechanical irritation of the cornea by hair, which leads to epiphora and keratitis.
- vision can be variable.
- Quarter horses may have iridal hypoplasia, cataracts and limbal dermoids.

## Differential diagnosis

- aberrant pigmentation.
- ocular melanomas.
- eyelid, conjunctival and corneal neoplasms.

## Diagnosis

- clinical appearance is usually adequate.
- confirmed by histopathology.

## Management

- left untreated if they are not causing clinical signs.
- with ocular signs, the treatment of choice is surgical resection of the aberrant tissue.
- depending on the location:
  - eyelid resection, conjunctivotomy and/or a superficial to deep keratectomy may be necessary.
  - removal may be followed by reconstructive blepharoplasty.
  - medical therapy post-resection, as for corneal ulcer management.
  - dermoids involving the deeper layers of the corneal stroma:
    - rare, but post-resection may require corneal mechanical support:
      - contact lens, collagen shield or surgical grafting procedure.

## Prognosis

- good for vision.

# Nasolacrimal system atresia

## Definition/overview

- congenital absence of part of the NLS pathway.
- upper or lower eyelid puncta, part of the nasolacrimal duct itself and/or the nasal punctum may be involved.
- imperforate nasal punctum is the most common abnormality.
- unilateral or bilateral.

## Aetiology/pathophysiology

- cause is unknown.

## Clinical presentation

- epiphora may be noted by 4–6 months of age, but not always acted upon.

FIG. 2.43 Profuse mucopurulent discharge emanating from the lower eyelid nasolacrimal punctum due to nasolacrimal duct atresia and dacryocystitis.

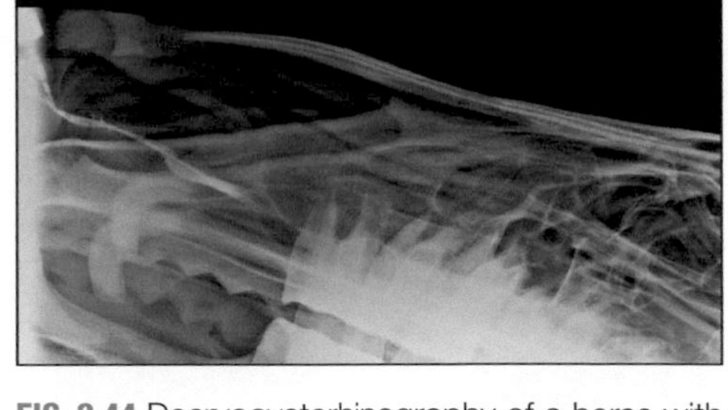

FIG. 2.44 Dacryocystorhinography of a horse with an acquired obstruction of the nasolacrimal duct highlighted by the failure of the contrast to pass up the duct after introduction at the distal puncta.

- secondary infection within the abnormal tract leads to:
  - dacryocystitis and conjunctivitis:
    - copious chronic mucoid/ mucopurulent ocular discharge (Fig. 2.43).
  - discharge often brings attention to a possible problem:
    - many cases may not develop this until 1–2 years of age.
- absence of the eyelid or nasal meatus punctum may be evident.
- chronic epiphora can lead to dermatitis, poor cosmesis and infectious keratoconjunctivitis.

## Differential diagnosis

- acquired NLS obstruction due to:
  - trauma.
  - chronic dacryocystitis.
  - foreign body, parasites, respiratory infection or neoplasia.

## Diagnosis

- confirmed by visual absence of eyelid or nasal punctal openings on ophthalmic examination.
- failure of fluorescein solution to exit the nasal punctum after application to the surface of the globe may suggest a blockage in the NLS.
- inability to cannulate or flush the NLS.
- distension of the mucosa overlying the imperforate nasal punctum in response to irrigation from the ocular punctum.
- samples for culture and sensitivity are recommended:
  - frequently secondary infection (dacryocystitis).
- contrast radiography (dacryocystorhinography) may be useful to determine whether there is associated nasolacrimal duct agenesis (Fig. 2.44).
- CT can also be helpful in diagnosing NLS atresia.

## Management

- treatment involves surgery to relieve the obstruction by creating a new opening:
  - imperforate punctum:
    - the patent, opposite punctum is first cannulated.
    - occasionally, the tip of the catheter may be palpated slightly proximal or distal to the expected location of the punctum.
    - NLS is flushed, which often causes dilation of the eyelid conjunctival or nasal mucous membrane overlying the site of the atretic puncta.
    - once the atretic site is identified:
      - cut down through palpebral conjunctiva or nasal mucosa to establish patency.
      - electrocautery may be necessary for haemostasis.
    - catheter tip is grasped with a haemostat and pulled through the new punctal opening.

**FIG. 2.45** Postoperative photograph of a 2-year-old Clydesdale horse following surgical correction of a congenital distal nasolacrimal duct atresia. A catheter has been placed from proximal to distal and then out through the false nostril before suturing to the face proximally and distally. The catheter was left in place for 4 weeks.

        - ♦ stent of polyethylene tubing or silicone is sutured in place for 3–8 weeks:
            - – prevent closure of new opening while duct and punctum epithelialise and dacryocystitis resolves (Fig. 2.45).
- medical management after surgery to treat secondary bacterial infection and prevent re-obstruction:
    - ○ topical antimicrobials or antimicrobial/corticosteroid preparations and systemic antimicrobials used for several weeks to allow dacryocystitis to resolve.
    - ○ antimicrobial based on culture and sensitivity results.

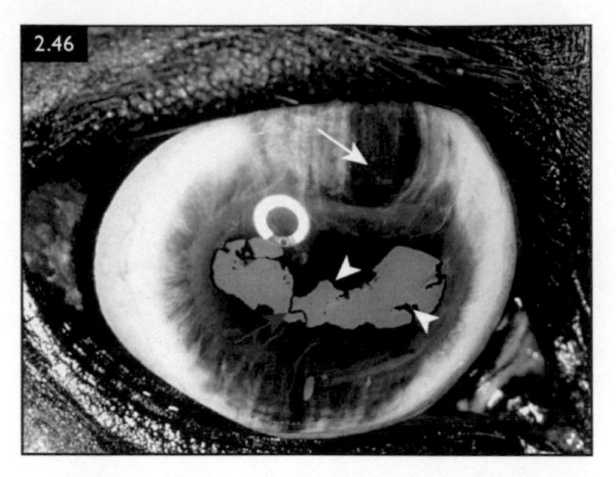


**FIG. 2.46** Heterochromia and iris stromal hypoplasia are apparent in this horse. The 12 o'clock location is typical for iris stromal hypoplasia (white arrow). This eye also has an incidental finding of several areas of posterior synechiae (arrowheads) and persistent pupillary membranes (red arrow).

- nasolacrimal duct agenesis may also accompany eyelid or nasal puncta atresia:
    - ○ requires a more complicated surgical correction at a referral centre.

### Prognosis

- depends on the extent of the atresia.
- surgery can be challenging, and relapses can occur.

## Heterochromia iridis

### Definition/overview

- iris with more than one colour or two differently coloured irides.
- lay terms include:
    - ○ 'wall eye' (partially or completely white and blue iris colour with a brown corpora nigra) (Fig. 2.46).
    - ○ 'China eye' (white iridal colour with brown corpora nigra).
- lack of iris stromal pigmentation in one iris or a portion of an iris.
- normal inherited variation present from birth in:
    - ○ Appaloosa.
    - ○ palomino, chestnut, grey, spotted and white horses.
- may be associated with uveal stromal cysts (iris stromal hypoplasia):
    - ○ iris may bow into the anterior chamber.

- tapetum of affected eyes is often poorly developed or even absent:
  - areas of the fundus may be deficient in pigment (subalbinotic fundus).
- Rocky Mountain horses and other breeds with partial albinism:
  - heterochromia iridis can be associated with multiple ocular defects.
- differentiate from chronic uveitis (increased pigmentation) or uveal melanoma.
- no treatment is necessary.

## Uveal cysts

### Definition/overview

- pigmented cysts arising from the posterior pigmented epithelium of the iris or granula iridica.
- either attached at the pupillary margin or free-floating in the anterior chamber.
- may be unilateral or bilateral.

### Aetiology/pathophysiology

- unknown cause.
- uveal stromal cysts are associated with iris hypoplasia and heterochromia iridis.

### Clinical presentation

- pigmented, spherical, smooth-surfaced structures free-floating in the anterior chamber or attached to the pupillary margin (Fig. 2.47):
  - lightly pigmented and appear translucent.
  - darkly pigmented with a solid appearance.
- may fluctuate in size over time.
- usually no associated clinical signs:
  - rarely, they may obstruct the pupil if associated with the granula iridica.
    - visual impairment, decreased performance level and headshaking.
  - may also rupture:
    - circular area of benign pigment on the corneal endothelium.
  - uveal stromal cysts occur commonly at the 12 o'clock position:
    - iris appears to bow into the anterior chamber and may cause dyscoria.
    - more common in older horses and ponies.

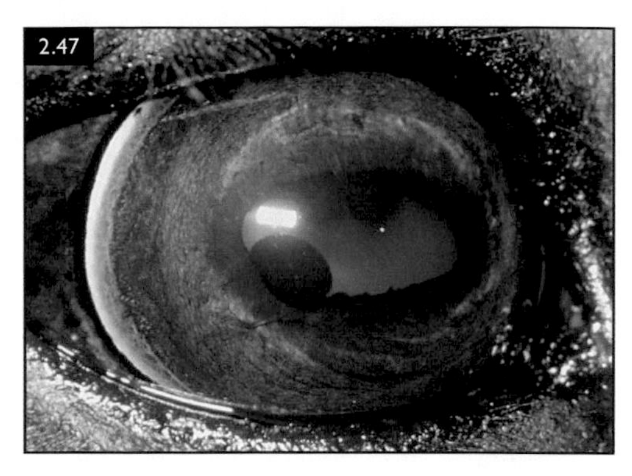

**FIG. 2.47** Cystic dilation of the ventral corpora nigra. Large uveal cysts in this location can impair vision, particularly when the pupil is constricted.

### Differential diagnosis

- intraocular neoplasia (e.g. uveal melanoma).
- hypertrophic granula iridica.

### Diagnosis

- gross appearance.
- using a focused light source may be transilluminated, but this is inconsistent.
- ultrasonography may be necessary to achieve a definitive diagnosis.

### Management

- treatment is rarely necessary.
- needle aspiration.
- ablation with a Nd:YAG or diode laser (Fig. 2.48):
  - large enough to occlude the pupil and obstruct vision.
  - compromise the ICA.

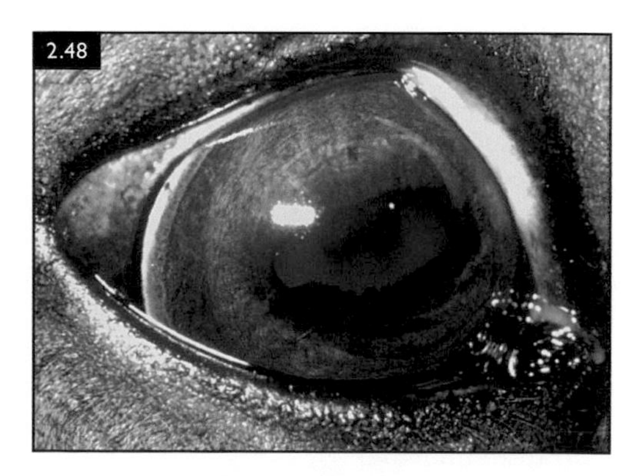

**FIG. 2.48** The eye shown in Fig. 2.47 following ablation of the cyst with a diode laser.

### Prognosis

- excellent prognosis.
- occasionally they are associated with other ophthalmic abnormalities.

## Aniridia and Iris hypoplasia

### Definition/overview

- aniridia is the complete absence of the iris tissue:
  - seen on its own in the American Quarter horse and Belgian Draught horse:
    - can be seen associated with limbal dermoids and/or congenital cataracts.
- iridal hypoplasia is where only a small remnant of iris tissue is present:
  - common in Appaloosas and eyes with heterochromia iridis.
  - usually severe and bilateral.

### Aetiology/pathophysiology

- unknown cause.
- aniridia is inherited in the Belgian Draught horse, and possibly in the American Quarter horse, as an autosomal dominant trait.
- iris hypoplasia involves the failure of ingrowth of mesenchyme and disorderly differentiation of neuroectodermal tissue:
  - lens of an affected eye is often cataractous, or occasionally ectopic or hypoplastic.
  - ICA may be malformed, leading to glaucoma.

### Clinical presentation

- both problems may present with blepharospasm, photophobia or reduced vision.
- PLR may be marginal or absent.
- may exhibit ocular discharge, perilimbal keratitis and/or cataract formation.
- iris hypoplasia may cause the iris to bulge anteriorly.
- lens equator and processes of the ciliary body may be visible in some cases (Fig. 2.49).
- affected areas are often readily transilluminated.
- iris hypoplasia may be seen alone or associated with multiple ocular anomalies.

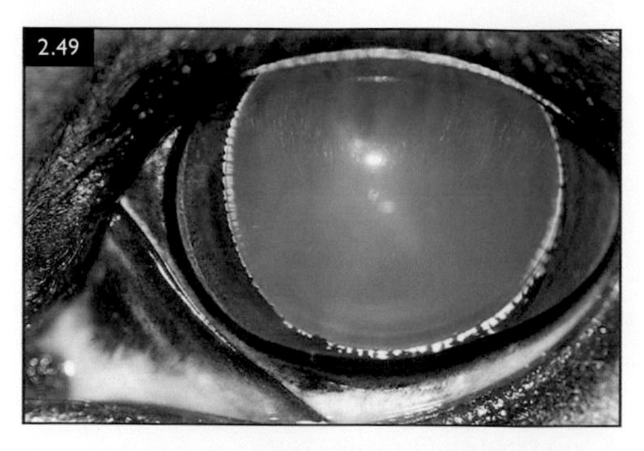


**FIG. 2.49** Aniridia. The almost complete absence of iris exposes the normally hidden ciliary processes attached to the periphery of the lens.

- Belgian Draught horses and Quarter horses:
  - anterior cortical cataracts and/or conjunctival dermoids may also be present.

### Differential diagnosis

- any cause of an abnormally large pupil (mydriasis).
- lens subluxation.

### Diagnosis

- history and clinical appearance.

### Management

- no specific treatment available.
- supportive measures may be useful:
  - face masks or dark-tinted contact lenses to decrease ambient light.
  - increase patient comfort.

### Prognosis

- non-progressive congenital condition,
- affected horses are unsound and should not be bred.

## Congenital ocular anomalies in Rocky Mountain horses

### Definition/overview

- inherited syndrome of ocular lesions in the Rocky Mountain horse.
- increased prevalence of anomalies is found in horses with a chocolate-coloured

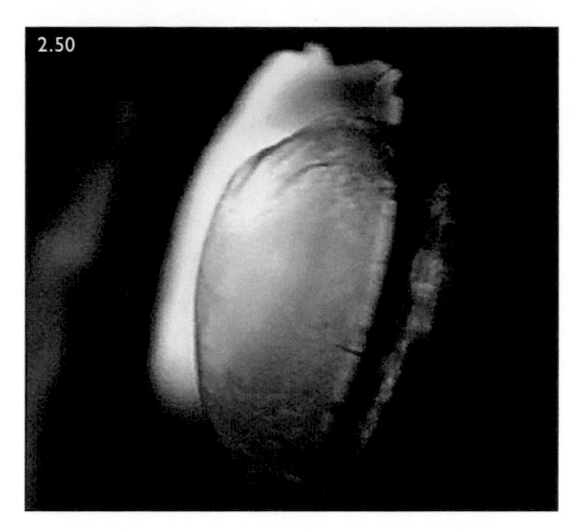

FIG. 2.50 Ciliary body cyst in a Rocky Mountain horse. These are typically located in the medial or lateral aspects of the ciliary body and can be large. (Photo courtesy D Ramsey.)

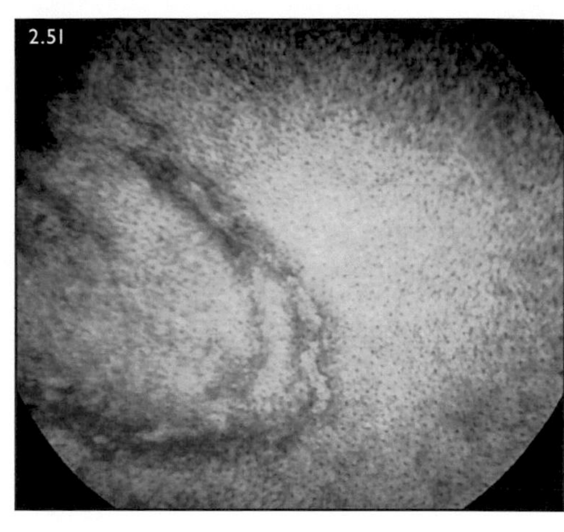

FIG. 2.51 Fundus of a young Rocky Mountain horse demonstrating curvilinear retinal lesions, possibly associated with previous detachment. (Photo courtesy D Ramsey.)

coat and a white or flaxen-coloured mane and tail (partial albinism).

- comparable set of congenital ophthalmic lesions has also been seen in the Kentucky Mountain Saddle horse, Mountain Pleasure horse, Morgan horse and pony, and miniature breeds, in animals with similar colour combinations.
- bilateral, but often asymmetrical.

## Aetiology/pathophysiology

- dominant inheritance with incomplete penetrance.
- same founder stallion was used to produce the Kentucky Mountain Saddle horse, Mountain Pleasure horse and Rocky Mountain horse breeds.

## Clinical presentation

- large, translucent, cystic structures arising from the posterior surface of the iris, ciliary body or the peripheral retina are seen most frequently (Fig. 2.50):
  - usually located in the temporal region.
- single to multiple well-delineated, darkly pigmented curvilinear streaks may be seen in the temporal part of the peripheral tapetal fundus (Fig. 2.51):
  - frequently bilateral, but not symmetrical.
- unilateral or bilateral retinal dysplasia most often seen in the temporal peripheral retina:
  - characterised clinically as linear folds or vermiform streaks.
- additional ocular abnormalities also noted:
  - megalocornea, macropalpebral fissures, abnormal corneal contour with excessive protrusion, excessively deep anterior chamber, miosis, dyscoria, iris stromal hypoplasia and absence of a discernible collarette.
  - readily visible pupillary sphincter muscle, appearing as radially oriented deep stromal strands of iris tissue extending from the pupillary ruff (Fig. 2.52).
  - granula iridica may be
    - hypoplastic
    - appear flattened
    - circumferentially oriented at the pupil margin.
  - PLRs can be decreased or absent.
  - minimal, or no, response to mydriatics in eyes with iris abnormalities.
  - areas of poorly developed or absent pectinate ligaments in the ICA or even areas where multiple strands of pigmented tissue extend from the temporal peripheral iris to the peripheral cornea (goniosynechiae) (Fig. 2.53).

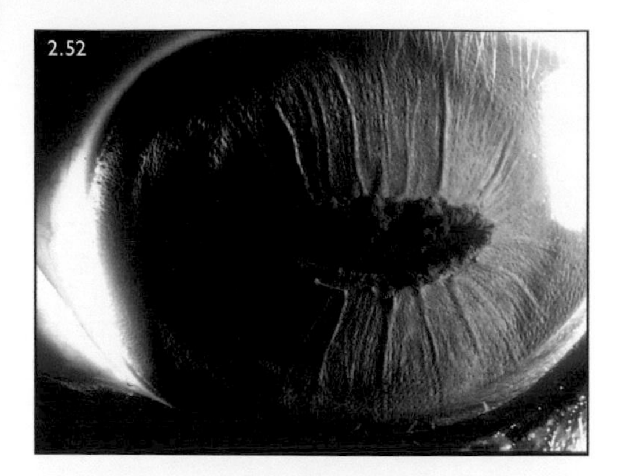

**FIG. 2.52** Rocky Mountain horse with hypoplastic granula iridica, miosis, circumferential corpora nigra, visible sphincter muscle and an absent iris collarette. (Photo courtesy D Ramsey.)

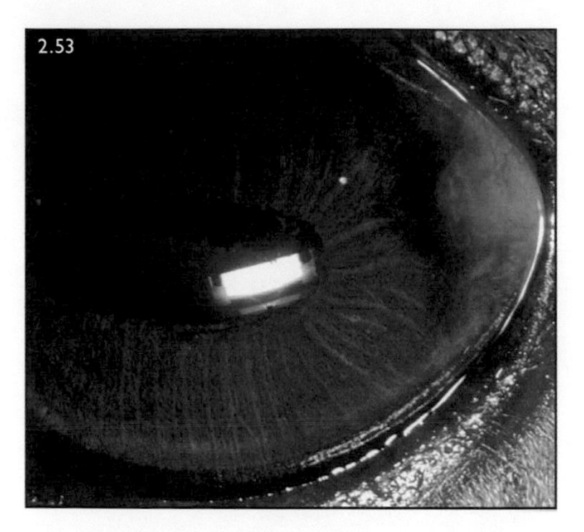

**FIG. 2.53** Goniosynechiae in the periphery of the anterior chamber of a Rocky Mountain horse. (Photo courtesy D Ramsey.)

   - immature lenticular nuclear opacities (cataracts) and ventral subluxation of the temporal part of the lens (Fig. 2.54).
   - abnormal prominence of anterior orbital rim and/or microphthalmos may also be visible.

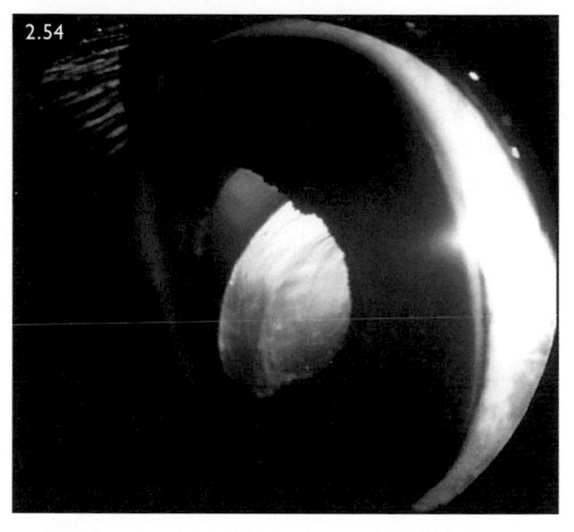

**FIG. 2.54** Lens subluxation with ventral displacement in a Rocky Mountain horse. (Photo courtesy D Ramsey.)

### Differential diagnosis

- non-inherited congenital ocular defects.
- causes of chronic uveitis.
- acquired cataracts.
- other retinal diseases.

### Diagnosis

- based on signalment, history and ophthalmic examination.

### Management

- no treatment is available.

### Prognosis

- non-progressive disease.
- affected animals should not be bred.

## Congenital cataracts

### Definition/overview

- focal or diffuse opacification of the lens that may involve the capsule, cortex and/or lens nucleus.
- unilateral or bilateral, symmetrical or asymmetrical, stationary or progressive and congenital, inherited or acquired.
- variable effect on vision.
- congenital cataracts present at birth, and most are bilateral:
   - may occur alone or in association with multiple ocular anomalies.
   - most common congenital ocular anomaly in the horse.
   - common cause of visual impairment or blindness in foals.

### Aetiology/pathophysiology

- inherited or develop following *in utero* stresses or occur secondary to other developmental abnormalities:

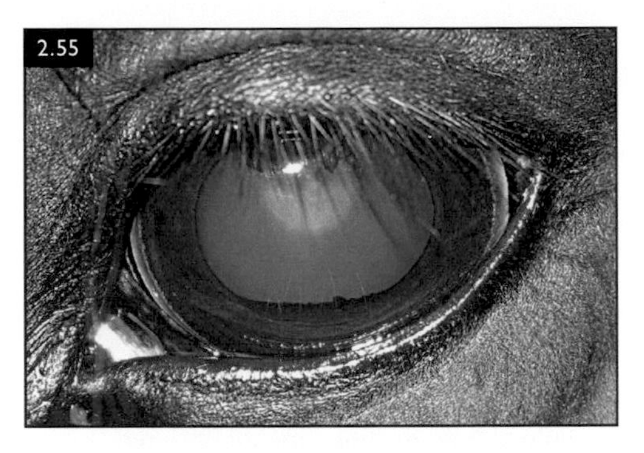

FIG. 2.55 [Nuclear cataracts, present in the centre of the lens, are hereditary in the Morgan breed. Although nuclear cataracts are usually non-progressive, their central position can create vision distortion, particularly when bright light causes pupillary constriction.

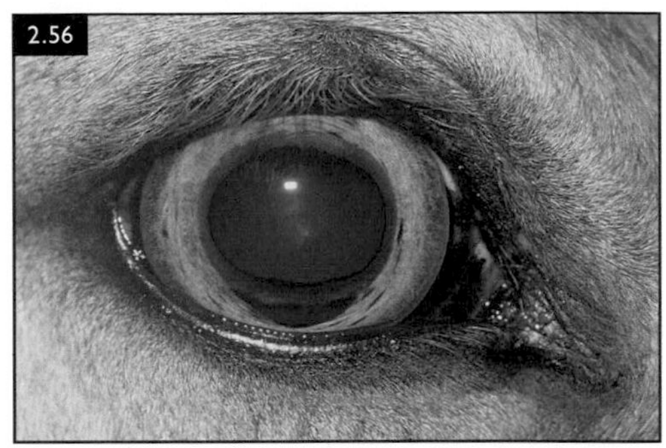

FIG. 2.56 The faint anterior cortical incipient cataract present in this horse was identified on pre-purchase examination and was not associated with apparent visual compromise. The cause was undetermined.

- underlying cause unknown in most cases.
- inherited in Belgian Draught horse and possibly in the Thoroughbred as dominant trait.
- non-progressive, nuclear, bilaterally symmetrical cataracts that do not seriously affect vision are seen in Morgan horses (Fig. 2.55).
- cataracts may be associated with mild microphthalmos.
- persistence of a portion of the tunica vasculosa lentis and/or hyaloid artery can lead to congenital cataract formation.

## Clinical presentation

- lens opacity (cloudy lens) which varies in shape, size and location within the lens (Figs. 2.56, 2.57).
- level of visual compromise is dependent on the degree of opacification.
- other ocular anomalies may be present:
  - synechiae, uveitis, lens luxation and/or retinal disease.
  - anterior segment dysgenesis/multiple ocular anomaly syndrome in Rocky Mountain horses.

## Differential diagnosis

- differentiate from prominent posterior Y sutures, which are very common in foals.
- inherited or acquired cataracts (Table 2.4).

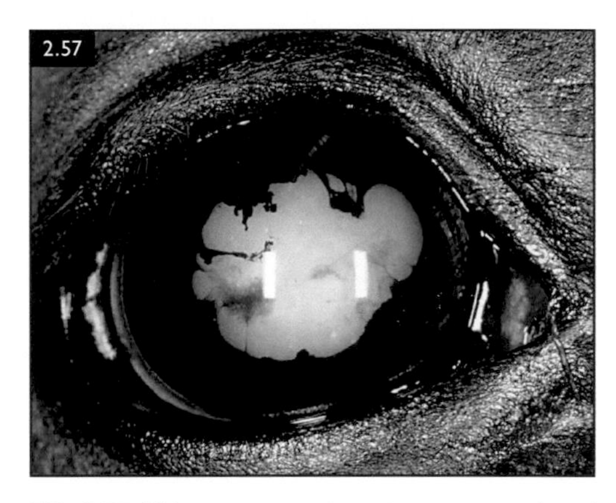

FIG. 2.57 This mature cataract was associated with chronic bouts of recurrent uveitis. Note the posterior synechiae, a typical post-inflammatory finding.

## Diagnosis

- identification of a lenticular opacity on ophthalmic examination.
- menace response may be absent if the cataract is severe.
- ocular ultrasonography and ERG will help identify posterior segment abnormalities, if present, prior to cataract surgery.

## Management

- foals with unilateral or bilateral immature or mature cataracts that interfere with vision should be referred to a veterinary ophthalmologist promptly for evaluation:

**TABLE 2.4** Main causes of cataracts

| |
|---|
| **Congenital** |
| • *In utero* infections/toxins/ocular inflammation/trauma/stresses/faulty nutrition<br>• Persistent hyperplastic tunica vasculosa lentis, persistent hyperplastic primary vitreous, persistent pupillary membranes |
| **Inherited** |
| • Belgian, Thoroughbred, Morgan, Rocky Mountain horse, Quarter horse, Arabian |
| **Acquired** |
| • Chronic uveitis<br>• Glaucoma<br>• Traumatic – blunt or penetrating; whiplash injury<br>• Iatrogenic from surgical trauma/laser therapy<br>• Lens subluxation/luxation<br>• Retinal disease/degeneration/detachment<br>• Neoplasia<br>• Senility (>18 years old)<br>• Exposure to toxins<br>• Systemic metabolic disease<br>• Nutritional<br>• Electrical (i.e. lightning)<br>• Radiation |

- removal by phaecoemulsification is the treatment of choice for cataracts in animals with:
  - visual impairment.
  - good PLRs.
  - no other ocular abnormalities or diseases that may interfere with vision.

## Prognosis

- age of the patient influences the postoperative success rate:
  - cataract surgery in foals less than 6 months of age typically has very good results (approximately 80% success rate).

# Congenital glaucoma (see page 123)

## Definition/overview

- elevation in IOP:
  - abnormal development of the ICA (goniodysgenesis):
    - cause often unknown.
    - idiopathic, inherited or secondary to pre-natal or parturient trauma.
  - obstruction to the outflow of aqueous humour.
- detrimental to normal ocular function and eventually results in optic nerve damage and blindness.
- rare in horses and is often associated with multiple congenital anomalies.
- no breed predisposition.
- ophthalmic findings include dilated pupil, vision impairment/blindness, buphthalmia, corneal oedema, Haab's striae, lens subluxation/luxation, tapetal hyperreflectivity, retinal degeneration/ atrophy and/or optic nerve degeneration/ cupping.
- differential includes other forms of glaucoma and other causes of vision deficits in foals.
- history, clinical appearance and tonometry are used to diagnose glaucoma:
  - examination of the ICA may show malformations.
- medical and surgical treatment (see page 123):
  - affected eyes often become blind and chronically painful.
  - enucleation or evisceration with intrascleral prosthesis is the treatment of choice in these cases.
- early detection and treatment are vital, but the prognosis is poor for vision.

# Congenital stationary night blindness (CSNB)

## Definition/overview

- inherited autosomal or sex-linked recessive trait condition in the Appaloosa and possibly in the Thoroughbred, Paso Fino, Standardbred and Quarter horse.

## Aetiology/pathophysiology

- associated with leopard complex coat-spotting patterns in Appaloosa:
  - single autosomal dominant locus (LP).
  - homozygous (LP/LP) are affected with CSNB.
  - heterozygous (LP/lp) and wild type horses (lp/lp) are not affected by CSNB.

  - Appaloosa horses with a chocolate-coloured coat most affected.
- defect in neural transmission between the photoreceptors and the bipolar cells within the retina has been suggested as the cause of vision problems in horses with CSNB.

### Clinical presentation

- usually exhibit visual impairment or blindness in reduced illumination.
- normal vision in light conditions except in the most severe cases.
- normal ophthalmoscopic examination, including fundic evaluation.
- affected animals may appear disoriented, stare off into space and seek lighted conditions:
  - may have bilateral dorsomedial strabismus, nystagmus and/or subtle microphthalmos.

### Differential diagnosis

- other causes of visual deficits or blindness in the horse.

### Diagnosis

- history, visual impairment in a poorly lit (scotopic) obstacle course and a normal ophthalmoscopic examination.
- confirmed by ERG that shows a large negative A-wave.

### Management

- no treatment available.
- affected animals should not be used for breeding.

### Prognosis

- static, non-progressive disease.

## Optic nerve hypoplasia

### Definition/overview

- congenital underdevelopment of the optic nerve of unknown cause.
- smaller than normal optic disc on fundoscopy:
  - differential diagnosis in the older animal is optic nerve atrophy.
- rare, but when it occurs it is most often seen in the Quarter horse or Appaloosa breeds.
- unilateral or bilateral.
- animals exhibit near normal vision to total blindness, depending on the severity.
- occasionally a single entity, but usually seen with other ocular abnormalities.
- no treatment.
- disease is non-progressive.

## Ocular manifestations of neonatal sepsis in foals

### Definition/overview

- foals with sepsis may exhibit associated ocular signs including anterior uveitis, chorioretinitis, endophthalmitis and panophthalmitis (Fig. 2.58):
  - corneal oedema can be dramatic in foals with anterior uveitis.
  - multifocal haemorrhages, exudates and focal detachments may be seen in the retina.
- ocular lesions occur following breakdown of the brain–ocular barrier and entry of bacteria into the ocular tissues.
- diagnosis following a thorough ophthalmic examination and additional testing for a definitive diagnosis of sepsis.
- treatment includes:
  - topical prednisolone acetate (or dexamethasone) and atropine sulphate if anterior uveitis is present.

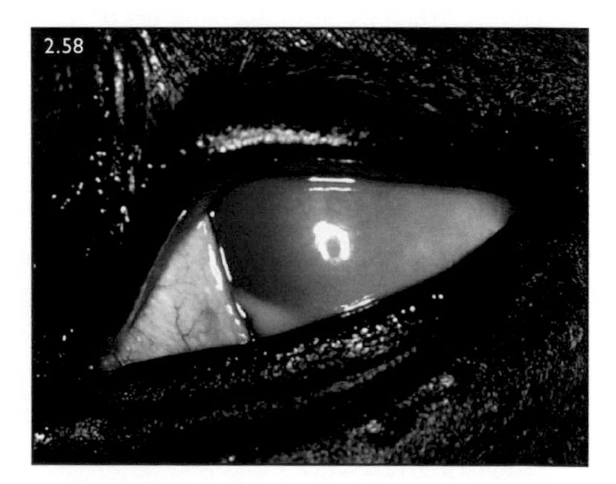


**FIG. 2.58** Endophthalmitis in a septicaemic neonatal foal. Corneal oedema and hypopyon are present in this severely inflamed eye.

  - systemic antimicrobials, based ideally on culture and sensitivity testing results.
  - systemic NSAIDs help control inflammation and counteract the changes associated with endotoxaemic and septic shock.
  - other components of therapy are discussed elsewhere (see Book 2, pages 178–181).
- prognosis for foals with neonatal sepsis is guarded:
  - animals that survive may be left with ocular lesions such as posterior synechiae and chorioretinal scars.

# TUMOURS OF THE EYE, ORBIT AND ADNEXA

## Periocular sarcoids

### Definition/overview

- second most common tumour of the equine eye and adnexa.
- may present as solitary or, more often, multiple masses.
- may be typed clinically as:
  - occult.
  - verrucous (hyperkeratotic fibropapilloma) which can convert to fibroblastic types following intervention.
  - nodular types A and B.
  - fibroblastic types A and B which are especially locally aggressive.
  - mixed.
  - malignant, which are very rare.
- medial canthus and/or upper eyelid are the most affected sites:
  - lower incidence in the lateral canthus and/or lower eyelid.
- locally aggressive, but very rarely malignant.
- usually affect horses 7 years of age or less:
  - no sex or coat-colour predilection.
  - reported higher incidence in Arabians, Appaloosas, Thoroughbreds and Quarter Horses.

### Aetiology/pathophysiology

- cause is uncertain:
  - bovine papilloma virus and C-type retrovirus have been implicated.
- may occur at the site of:
  - previous trauma.
  - areas that come into contact with pre-existing sarcoids.
- strong familial predisposition:
  - genetic or inherited component has not been established.
- direct contact, an arthropod vector (e.g. flies) or fomites may play a role in transmission by translocating sarcoid cells into open wounds.

### Clinical presentation

- variable appearance.
- occult can appear as small, subepidermal, miliary nodules or plaques that are hairless or have altered hair.
- verrucous can appear partially or totally hairless, rough and wart- or cauliflower-like, with miliary nodules and/or thickened areas that may be ulcerated.
- nodular are typically solid, smooth and spherical or ovoid, with a well-defined outline (Figs. 2.59, 2.60):
  - may be entirely subcutaneous.
  - variable amount of the overlying skin may be involved.
- fibroblastic are fleshy masses that resemble granulation tissue ('proud flesh'):
  - may be pedunculated or have a broader base with ill-defined margins.
  - can be ulcerated.
- mixed sarcoids encompass the characteristics of a combination of two or more of the above lesion types.
- malignant sarcoids may be characterised by nodules and cords of abnormal tissue and typically grow at a much faster rate.
- periocular sarcoids may cause secondary corneal ulceration or nasolacrimal duct obstruction.
- tumours themselves are not painful or pruritic.

### Differential diagnosis

- other periorbital neoplasms such as squamous cell carcinoma (SCC), papilloma, fibroma, fibrosarcoma,

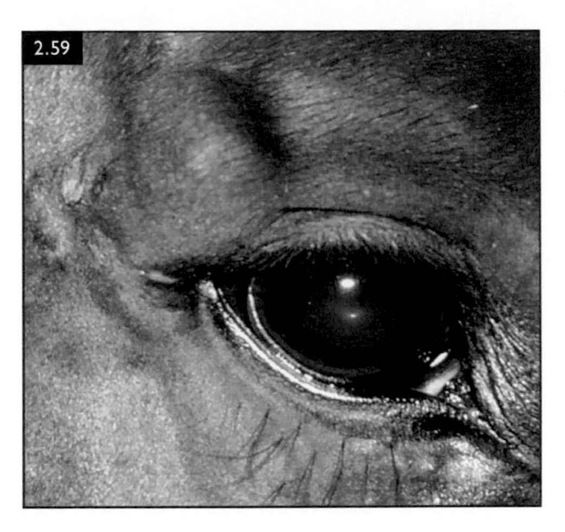

FIG. 2.59 Nodular sarcoids are common in the periocular region of the horse. (Photo courtesy A Gemensky-Metzler.)

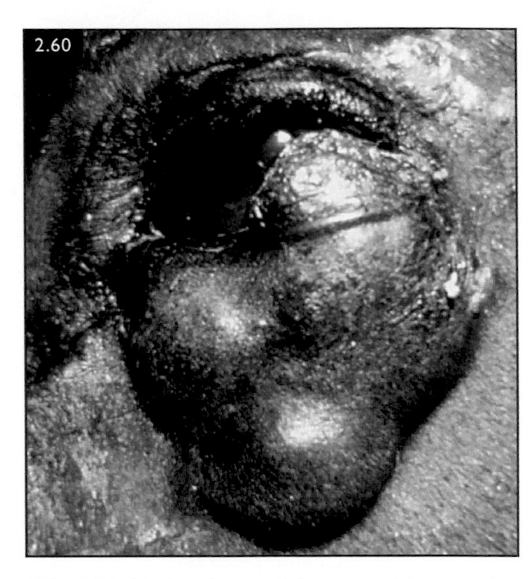

FIG. 2.60 Periocular nodular sarcoids can be extensive. (Photo courtesy American College of Veterinary Ophthalmologists.)

neurofibroma, neurofibrosarcoma, schwannoma, melanoma, myxosarcoma, fibromyxosarcoma and dermoids.
- cutaneous habronemiasis (parasitic granuloma).
- any non-neoplastic granulation tissue (e.g. exuberant granulation tissue/'proud flesh', bacterial granuloma, foreign body reaction).
- dermatophilosis (rain scald).
- subcutaneous or deep fungal infections.
- dermatophytosis (ringworm).
- abscesses, traumatic superficial abrasions or scratches, rub marks, scarring.
- sebaceous cysts, alopecia areata and idiopathic periorbital vitiligo.

## Diagnosis

- clinical appearance is suggestive.
- definitive diagnosis is made based on histopathological examination following tumour removal:
  - **partial excision or biopsy is contraindicated as it may activate the lesion.**

## Management

- depends on:
  - size, location and type of the sarcoid.
  - equipment available and financial constraints of the owner.
- benign neglect is possible as some lesions spontaneously regress (over years).
- scalpel or carbon dioxide laser excision:
  - periorbital sarcoids are highly infiltrative:
    - high rate of recurrence when treated with excision alone.
  - surgical debulking or excision should be accompanied by adjunctive therapy:
    - cryotherapy, radiofrequency or hyperthermia.
    - immunomodulation, chemotherapy or radiation.
- cryotherapy is simple, rapid and inexpensive:
  - useful for verrucous or occult sarcoids that are < 2 $cm^2$.
  - high overall recurrence rate has been reported.
  - depigmentation of the treated area is expected after cryosurgery.
- radiofrequency hyperthermia has only a limited ability to treat deeper tissues.
- Bacillus Calmette–Guérin (BCG) immunomodulation requires injection of 1 $ml/cm^3$ BCG into the tumour tissue:
  - reports of 100% success rates with surgical debulking followed by intralesional BCG administration.
- intralesional cisplatin or 5-fluorouracil (5-FU) chemotherapy may be useful for fibroblastic or nodular lesions.

- interstitial brachytherapy irradiation with a variety of implants including iridium$^{192}$, beta strontium$^{90}$ and strontium$^{90}$:
  - generally, radiation treatment has an excellent outcome.
  - potentially dangerous for handlers.
- enucleation or exenteration may be required for extensive periocular tumours.

### Prognosis

- variable depending on size, location, type of sarcoid and treatment modality chosen.
- some regress spontaneously, but generally the prognosis is guarded:
  - rarely metastasise.
  - often impossible to remove periocular sarcoids completely while allowing for wound closure:
    - leads to high rates of recurrence, especially with excision alone.
- extensive occult and verrucous or mixed sarcoids are the most difficult to treat.
- cases that are unsuccessfully treated several times have a poorer prognosis regardless of tumour type or the treatment modality used.
- lowest recurrence rate has been reported in cases treated with intralesional brachytherapy with iridium$^{192}$.

## Squamous cell carcinomas

### Definition/overview

- most common tumour of the equine eye and adnexa.
- predominately in middle-aged to older horses.
- Appaloosas, Paints, Pintos and draught horses (e.g. Belgians, Clydesdales) at higher risk.
- white, grey or palomino coat colours are predisposed.
- commonly located on the eyelid, nictitating membrane, conjunctiva, cornea and/or limbus, in one or both eyes.
- locally invasive:
  - rarely metastasise to regional lymph nodes, salivary glands and lungs.

### Aetiology/pathophysiology

- cause is unknown, but likely to be multifactorial.
- increased susceptibility:
  - prolonged exposure to ultraviolet (UV) radiation, increased altitude and latitude.
  - non-pigmented or lightly pigmented ocular and periocular structures.
  - exposure to mechanical irritants and papillomavirus.
- unknown pathophysiology but lesions typically progress from non-cancerous plaques → papillomas → carcinomas *in situ* prior to transforming into SCC.
- local invasion and/or metastasise.

### Clinical presentation

- depends on anatomical location and stage of development (Figs. 2.61–2.63):
  - well-circumscribed, small, white, elevated, hyperplastic plaques.
  - raised, rough, irregular pinkish-white warty or cauliflower-like structures with a broad base of attachment.
  - ulcerated and necrotic, with lesions that may bleed easily.
  - can be invasive or infiltrative.
  - tumours involving the nictitating membrane may present as inconspicuous, small lesions on the leading edge (Fig. 2.64):
    - extension of mass to deeper aspects of the third eyelid is common.
    - only appreciated by retropulsing the globe to expose the surface of the nictitans (Fig. 2.65).
  - limbal SCCs often appear as a raised, vascularised, grey-white corneal opacity with associated conjunctival hyperaemia and thickening (Fig. 2.66).
  - SCCs may also invade the orbit, leading to signs associated with retrobulbar masses (e.g. exophthalmos, lagophthalmos, exposure keratitis).

### Differential diagnosis

- depends on the location but should include:
  - granulation tissue, abscesses, habronemiasis, cutaneous onchocerciasis.
  - *Thelazia* infestation, bacterial and fungal granulomas, foreign body reactions.
  - dermoids.

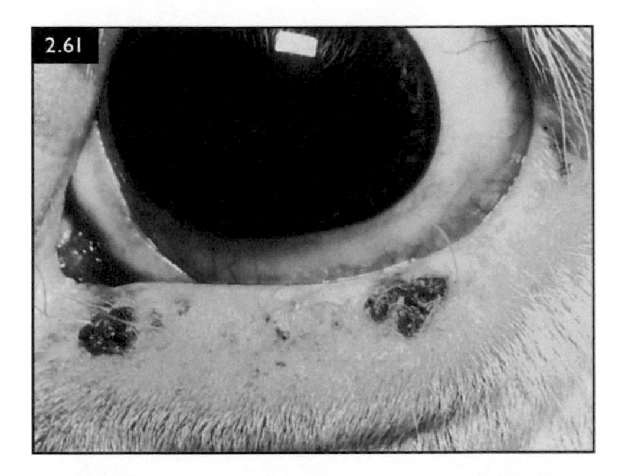

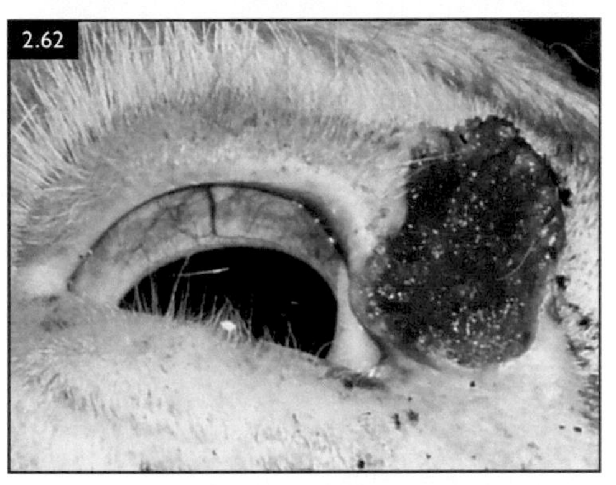

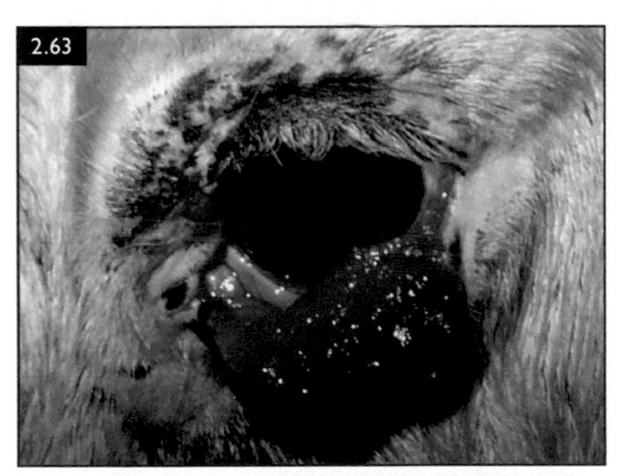

**FIGS. 2.61–2.63** SCC of the eyelid. (2.61) Early, superficial lesions can be categorised as plaques or SCC *in situ*. (2.62, 2.63) Advanced eyelid lesions have deeper involvement and are consequently more challenging to treat.

- other neoplasms such as sarcoids, papillomas, adenomas, adenocarcinomas, melanomas, mast cell tumours, basal cell tumours, fibromas.

## Diagnosis

- consider SCC with any persistent proliferative or ulcerative eyelid lesion.

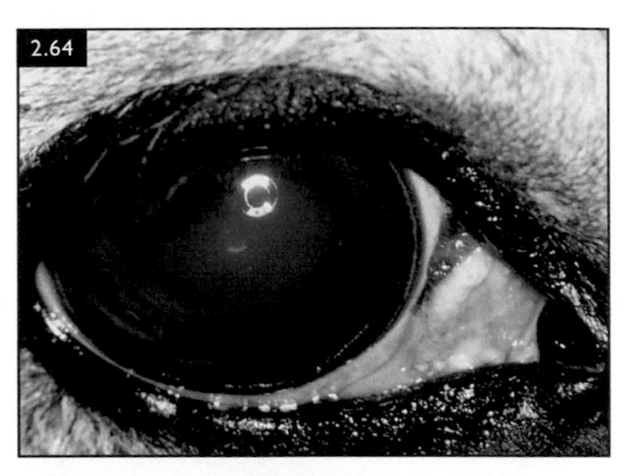

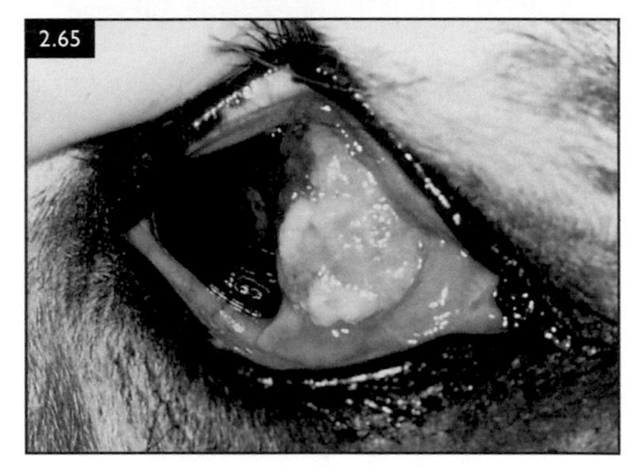

**FIGS. 2.64, 2.65** SCC of the nictitans. (2.64) Only the leading edge of the nictitans appears affected. (2.65) With retropulsion of the globe, the nictitans is elevated, demonstrating deeper tissue involvement.

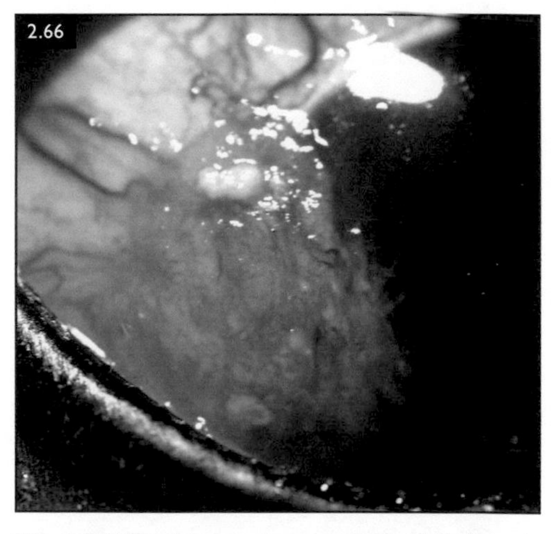

**FIG. 2.66** The temporal corneoscleral limbus is a predisposed location for SCC. A pink, fleshy mass is apparent laterally, invading the bulbal conjunctiva and adjacent cornea.

- clinical appearance may be suggestive.
- definitive diagnosis is based on:
  - cytology (fine-needle aspirate [FNA] or scrapings).
  - histology (incisional/wedge or excisional biopsy).

## Management

- varies with tumour size, location, extent of invasion, visual status, intended use of the animal, the equipment available and financial constraints.
- surgical debulking or excision followed by:
  - cryotherapy, radiofrequency hyperthermia and radiation therapy (Fig. 2.67).
  - immunomodulation (i.e. BCG) or intralesional chemotherapy (i.e. cisplatin).
- extensive lesions involving the eyelids, globe and/or orbit:
  - exenteration (surgical removal of all orbital contents) is recommended.
- beta radiation is most beneficial, following keratectomy or penetrating keratoplasty, in superficial SCC of the cornea and/or limbus.
- carbon dioxide laser ablation may also be used for limbal SCCs.

**FIG. 2.67** Strontium$^{90}$ treatment following conjunctival resection of a SCC on the dorsolateral bulbar conjunctiva.

## Prognosis

- varies according to the size and location of the tumour, and treatment modality selected.
- good overall prognosis for survival:
  - SCCs are usually locally aggressive and may invade orbital soft tissues and/or bone.
  - can metastasise to regional lymph nodes, salivary glands and the thorax.
- tumour size is inversely related to survival:
  - extensive tumours carry a poorer prognosis.
- tumour location influences survival:
  - eyelid and orbital SCCs carrying a poorer prognosis than those located on the third eyelid or limbus.
- recurrence rate can be high (>30%):
  - especially if eyelid or nictitating membrane is affected.
  - one or more recurrences of SCC following therapy markedly decreases the survival time:
    - early follow-up is recommended, with repeated treatment as necessary.
- incidence and recurrence of ocular SCC can be reduced by decreasing exposure to UV radiation:
  - using sports sunscreen, hoods or protective fly masks.
  - tattooing non-pigmented eyelids and margins of the third eyelid is not effective in preventing or decreasing the incidence of SCC.

# Melanoma

## Definition/overview

- primary neoplasms of the equine eye and adnexa that involve neoplastic melanocytes.
- single or multiple, usually darkly pigmented, dense masses.
- most commonly affect the ocular adnexa:
  - may affect the orbit, epibulbar tissue, conjunctiva, cornea and uveal tissue.
- usually slow-growing and benign:
  - may, over years, transform into malignant tumours:
    - grow rapidly and metastasise to regional lymph nodes, lungs, spleen and liver.

- ♦ haematogenous spread may also occur.

## Aetiology/pathophysiology

- cause is uncertain.
- incidence increases with age (rare in horses less than 6 years old).
- usually occur in grey or white horses:
  - Arabians, Lipizzaners and Percherons exhibit an increased risk.
  - rare in horses with coloured hair coats.
  - pathophysiology related to lightly pigmented or non-pigmented ocular and periocular tissues.
  - disturbance in melanin metabolism, possibly associated with greying, may lead to tumour formation.
  - aberrant immune response or cytotoxic reaction with subsequent local destruction of normal melanocytes has been implicated.

## Clinical presentation

- firm or soft, dome-shaped, hairless, grey or black pigmented masses on the eyelid (Fig. 2.68), conjunctiva and cornea or, rarely, intraocularly:
  - occasionally can be unpigmented.
- solitary or multiple masses may be found, and they may be ulcerated and infected.
- most frequently reported as locally expansive and destructive:
  - slow or rapid expansion.
- anterior uveal melanomas may cause secondary pupil distortion or obliterate the anterior chamber (Fig. 2.69).
- uveitis, keratitis (Fig. 2.70), secondary cataract formation and/or secondary glaucoma may develop due to their interference with the ICA.

## Differential diagnosis

- extraocular melanomas:
  - sarcoids, haemangiomas, haemangiosarcomas, dermoids, SCCs, granulomas and abscesses.
- orbital melanomas:
  - other orbital neoplasms and other causes of retrobulbar disease.
- includes uveal cysts for intraocular melanomas.

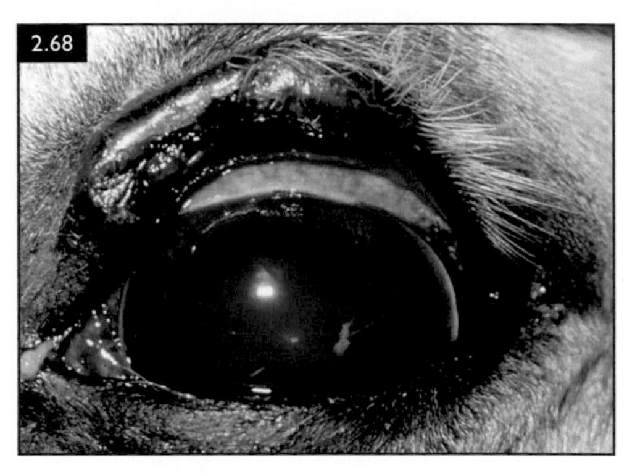


**FIG. 2.68** Melanoma involving the upper eyelid in a grey horse. There are several coalescing nodules. (Photo courtesy R Morreale.)

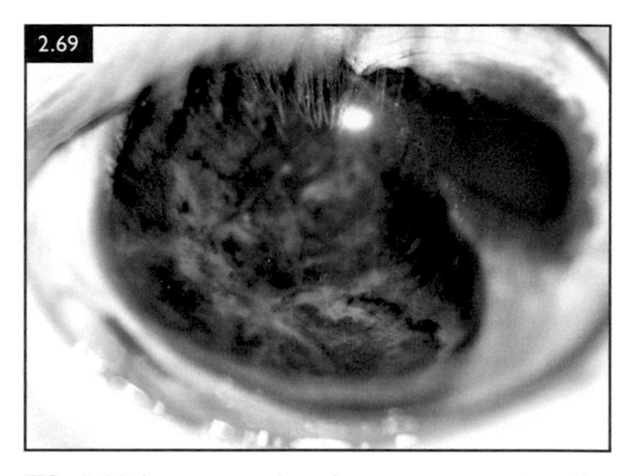


**FIG. 2.69** Large uveal melanoma occupying the lateral iris and pressing up against the corneal endothelium of the right eye of a horse. The pupil is distorted and miotic owing to a secondary uveitis. Anterior chamber haemorrhage can also be visualised on the medioventral border of the melanoma.

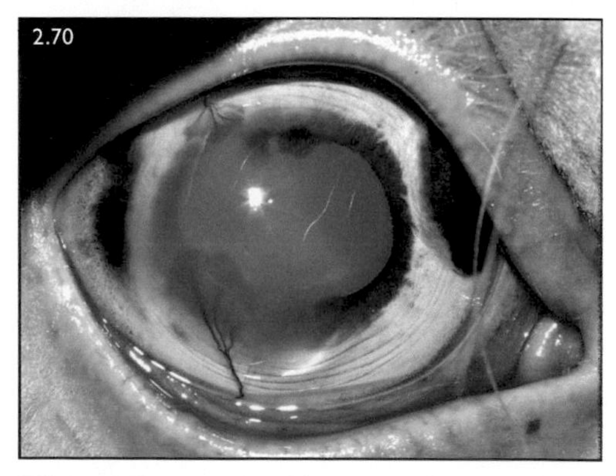


**FIG. 2.70** Connemara with two uveal melanomas (medially and laterally) and a mid-stromal vascular keratitis in the right eye.

## Diagnosis

- history and ophthalmic examination:
  - intraocular melanomas differentiated from anterior uveal cysts by using a focused beam of light (e.g. transilluminator) and/or high-frequency ultrasonography.
- cytology (from scrapings or FNA).
- histopathology:
  - categorised as benign or malignant, using the mitotic index.

## Management

- surgical excision may be used alone or in combination with adjunctive therapies in extraocular tumours:
  - depends on tumour size/location, equipment available and financial constraints.
  - cryosurgery, radiofrequency hyperthermia and immunotherapy.
  - radiation, diode laser photocoagulation and intralesional chemotherapy.
- long-term cimetidine (2.5 mg/kg p/o q8 h) has been used, alone and in combination, in the treatment of cutaneous melanomas to limit or stop the progression of the tumour:
  - beneficial in some horses by decreasing the size or number of melanomas.
  - stop if no clinical improvement within 3 months.
- intraocular melanomas:
  - may be monitored for progression or surgically excised via iridocyclectomy.
  - enucleation or exenteration is recommended in cases where the tumour is extensive or secondary intraocular changes cause blindness or chronic pain.

## Prognosis

- depends on the location, size/extent of melanoma and treatment modality chosen.
- generally, epibulbar melanomas have a good prognosis as typically they are slow-growing and benign:
  - surgical excision can be curative for small extraocular melanomas.
  - occasionally melanomas can be locally aggressive and metastasise:
    - ♦ especially conjunctival melanomas.
- larger the tumour, the poorer the prognosis.
- longer the interval between surgery and adjunctive therapies, e.g. intralesional cisplatin chemotherapy treatment:
  - more opportunity for tumour regrowth and poorer the prognosis.

# Lymphosarcoma (LSA)

## Definition/overview

- most common secondary neoplasm affecting the equine eye or adnexa:
  - approximately 25% of horses with systemic LSA develop lesions of the eye or ocular adnexa.
- life-threatening neoplastic disease of the lymphoreticular tissue:
  - capable of involving any system or body organ, alone or in combination.

## Aetiology/pathophysiology

- metastatic neoplastic disease of the adnexa and eye.
- unknown cause.
- systemic involvement typically may precede or accompany ocular LSA.

## Clinical presentation

- ocular manifestations of LSA include:
  - serous or mucopurulent discharge.
  - diffuse retrobulbar infiltrates leading to exophthalmos, lagophthalmos and exposure keratitis.
  - third-eyelid masses.
  - neoplastic infiltrate of the palpebral conjunctiva (Figs. 2.71, 2.72).
  - conjunctivitis
  - chemosis.
  - conjunctival haemorrhage.
  - corneoscleral masses.
  - corneal neovascularisation.
  - oedema and/or ulceration.
  - anterior uveitis
  - hyphaema.
  - hypopyon
  - secondary glaucoma.
  - chorioretinitis
  - retinal detachment.
- lymph node enlargement and signs of visceral involvement may be present.

## Differential diagnosis

- other causes of conjunctivitis, corneal ulceration, anterior uveitis, hyphaema, retinal detachment, chorioretinitis, glaucoma and orbital disease.

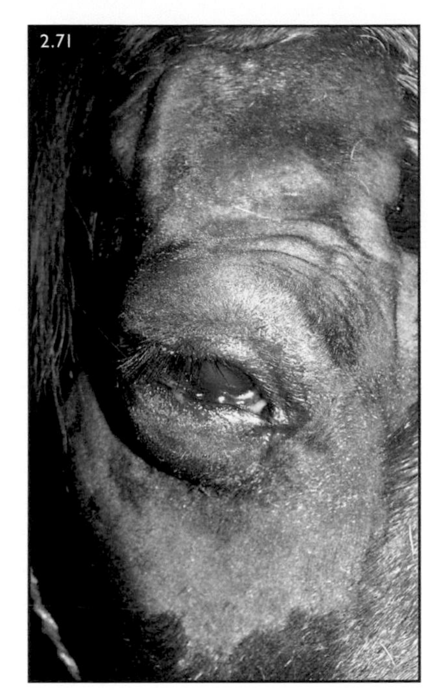


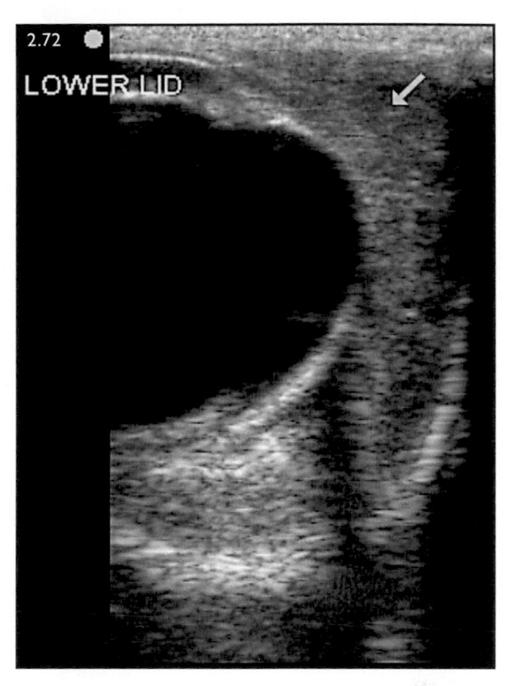


**FIGS. 2.71, 2.72** A 14-year-old black Cob with a periocular swelling of the right eye, which was confirmed on ultrasonography as a homogeneous mass infiltrating the upper and lower palpebral tissues (2.72). Biopsy confirmed this as an LSA.

### Diagnosis

- history and clinical presentation suggestive.
- cytological samples provided via FNA.
- histopathology of biopsy samples of affected periocular tissues, regional or enlarged lymph nodes and/or bone marrow can provide a definitive diagnosis.

### Management

- no specific treatment for LSA.
- supportive care.
- enucleation or exenteration.

### Prognosis

- poor for survival:
  - majority of affected horses die within 6–12 months.

## Orbital neoplasia

### Definition/overview

- may be primary, secondary or metastatic.
- usually, unilateral.

### Aetiology/pathophysiology

- cause unknown.
- most are primary and highly malignant (Table 2.5).
- may also result from:
  - secondary invasion by neoplasms of the nasal or paranasal sinuses.
  - extension from other adjacent structures.
  - metastases from distant sites.

### Clinical presentation

- typically, manifest as slowly progressive, often painless, unilateral exophthalmos:
  - varying amounts of globe displacement (strabismus), lagophthalmos and secondary exposure keratitis (Figs. 2.73, 2.74).
  - conjunctival hyperaemia, blepharoedema, elevated third eyelid, mydriatic pupil and resistance to/absent globe retropulsion may be present.

**TABLE 2.5** Types of orbital neoplasia

- Sarcoid
- SCC
- Adenocarcinoma
- Multi-lobular osteoma
- Lymphosarcoma
- Fibroma/fibrosarcoma
- Haemangioma/haemangiosarcoma
- Melanoma
- Lipoma
- Angiosarcoma
- Granulocytic carcinoma
- Neuroendocrine tumour
- Microglioma
- Medulloepithelioma
- Neuroepithelial carcinoma
- Osteoclastoma
- Extra-adrenal paraganglioma
- Neurofibromas (schwannoma, neurilemmoma)
- Undifferentiated carcinomas
- Mast cell tumour

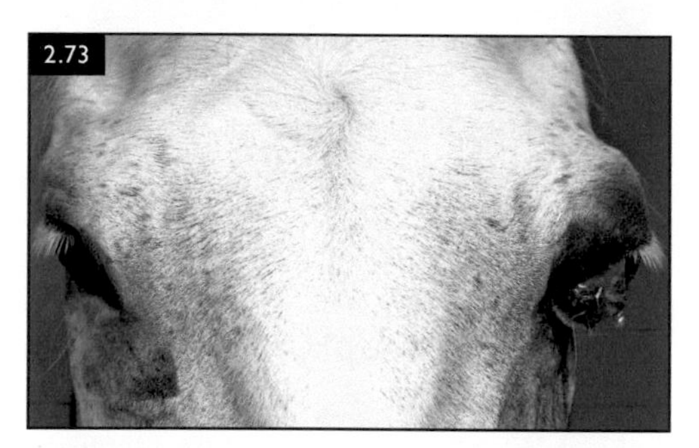


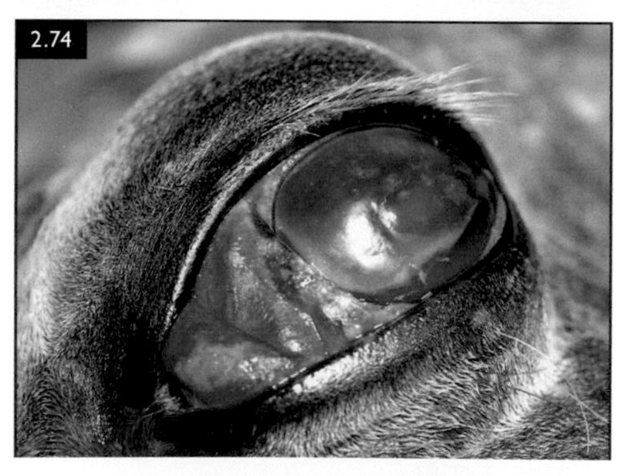


**FIGS. 2.73, 2.74** Exophthalmos affecting the left eye. The affected eye has gross swelling of the periocular tissues (including reduction in depth of the supraorbital fossa on that side), globe protrusion, exposure of the sclera, conjunctiva and cornea with adherent wood chippings (bedding) (2.73). The same eye prepared for exenteration of the orbit. Note the lateral strabismus, swollen and thickened conjunctiva and exposure keratitis (2.74).

  - soft-tissue mass may be detected in the retrobulbar space when the orbit is palpated over the supraorbital fossa.
  - occasionally, scleral indentation visible on ophthalmoscopy.
- facial and/or periorbital swelling, decreased air passage through the nostril(s), serosanguineous nasal discharge and vision impairment may be seen:
  - menace response and/or PLR may be normal, decreased or absent.

## Differential diagnosis

- orbital abscess
- cellulitis.
- foreign body.
- chronic sinusitis
- facial trauma.
- retrobulbar haemorrhage/haematoma.
- salivary gland mucocoele.
- guttural pouch mycosis or empyema.
- retrobulbar hydatid cystfungal granuloma EIA.

## Diagnosis

- based on history and clinical presentation.
- cytological examination and culture and sensitivity of discharge or FNA from the orbit.
- biopsy of masses.
- survey radiography, orbital angiography and dacryocystorhinography.
- ultrasonography of the eye and orbit.
- MRI and CT are invaluable when evaluating orbital disease.

## Management

- early surgical exenteration is generally the therapy of choice unless the orbital tumour is well circumscribed.

## Prognosis

- generally guarded to poor because most orbital tumours are primary and highly malignant.

# INFECTIOUS/INFLAMMATORY DISORDERS OF THE EYE

## Conjunctivitis

### Definition/overview

- inflammation of the conjunctiva.
- caused by a variety of primary aetiologies or secondary to systemic disease.

### Aetiology/pathophysiology

- causes are wide and varied (Table 2.6).
- normal microflora of the equine conjunctiva is variable depending on:
  - season of the year.
  - geographical location.
  - bacterial and fungal organisms such as:
    - *Staphylococcus aureus*, *Moraxella equi*, *Streptococcus zooepidemicus*, *Corynebacterium* and *Bacillus*.
    - *Aspergillus*, *Penicillium*, *Alternaria* and *Cladosporium* spp.
  - conjunctivitis can occur when there is a change in the normal conjunctival flora that allows either opportunistic commensal or pathological organisms to cause disease.
- **Note that not all causes are infectious.**

### Clinical presentation (Fig. 2.75)

- blepharospasm.
- conjunctival hyperaemia, oedema (chemosis) and thickening.

**TABLE 2.6** Common aetiologies of conjunctivitis

- Bacterial
  - *Streptococcus equi, Actinobacillus* spp., *Chlamydia, Rhodococcus* spp., *Moraxella equi, Leptospira* spp.
- Viral
  - Adenovirus, EHV-1 (rhinopneumonitis), EHV-2 (cytomegalovirus), EHV-4?, EIA, equine viral arteritis, equine influenza type A2, parainfluenza, AHS (reovirus)
- Mycotic
  - *Histoplasma capsulatum* var. *farciminosum* (also called *H. farciminosum*), *H. capsulatum*, sporotrichosis, blastomycosis
- Parasitic
  - *Habronema muscae, H. microstoma, Draschia megastoma, Onchocerca cervicalis*
- Allergic
- Follicular
- Systemic causes
  - Pneumonia, EPM, polyneuritis equi, vestibular disease syndrome, epizootic lymphangitis
- Eosinophilic keratoconjunctivitis
- Trauma
- Foreign body
- Entropion
- Dacryocystitis
- Environmental irritants/chemical irritation
- Neonatal maladjustment syndrome, neonatal sepsis, immune-mediated haemolytic anaemia
- Keratoconjunctivitis sicca
- Lymphosarcoma
- Idiopathic/immune-mediated

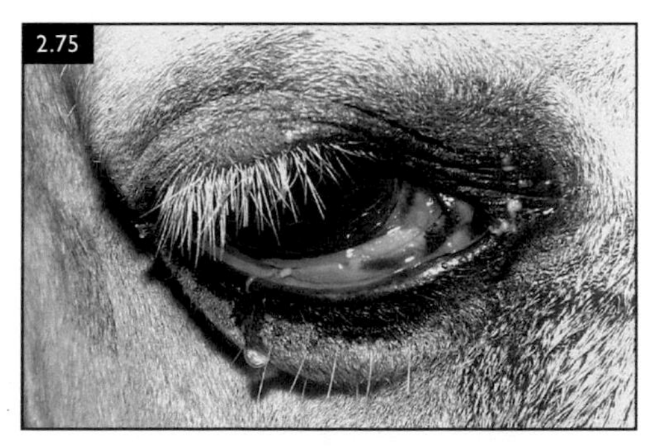


**FIG. 2.75** This right eye had a chronic foreign body in the lower conjunctival fornix, which led to chronic conjunctivitis with lower eyelid swelling, generalised conjunctival hyperaemia and oedema (chemosis), and a mucopurulent ocular discharge.

- ocular discharge (serous, mucoid or purulent) or epiphora.
- follicles on the palpebral and/or bulbar conjunctiva or third eyelid may be present.

### Differential diagnosis

- other causes of conjunctival hyperaemia or 'red eye':
  - orbital disease, corneal ulceration, uveitis or glaucoma.

### Diagnosis

- full ophthalmic and neuro-ophthalmic examination including:

- STT to check adequate tear production and distribution.
- examine eyelids and ocular surfaces for any causes of mechanical irritation:
  - entropion
  - distichia.
  - ectopic cilia.
  - foreign bodies (conjunctival fornices or behind the third eyelid):
    - patient's temperament and level of discomfort determine whether topical anaesthesia, regional nerve blocks and sedation may be needed.
    - sweep the conjunctival sac with a cotton bud, blunt instrument or digital palpation for evidence of foreign material.

## Management

- specific treatment depends on the underlying cause of the conjunctivitis.
- topical corticosteroids may be used if no corneal ulcer is present.
- topical antimicrobials may be used alone or in combination with corticosteroids.
- NLS should be flushed to ensure normal patency:
  - dacryocystitis commonly concurrent and can exacerbate the conjunctivitis.

## Prognosis

- excellent to poor, depending on the cause.

# Nasolacrimal system obstruction

## Definition/overview

- acquired obstruction of the NLS is uncommon in horses.

## Aetiology/pathophysiology

- causes include:
  - chronic dacryocystitis.
  - neoplasia (e.g. cutaneous SCC, nasal and paranasal sinus neoplasia).
  - Habronema blepharoconjunctivitis.
  - foreign bodies.
- blocked NLS can lead to dacryocystitis and associated ocular signs.

## Clinical presentation

- non-painful.
- epiphora or mucopurulent ocular discharge (if dacryocystitis is present) (see Fig. 2.43).
- facial dermatitis (chronic overflow of tears or discharge).
- +/– conjunctivitis.

## Differential diagnosis

- congenital NLS anomalies.
- other causes of chronic epiphora.

## Diagnosis

- failure of fluorescein dye to exit the nostril after application to the eye.
- inability to flush or cannulate the nasolacrimal duct:
  - any flushings subjected to culture and sensitivity testing.
- dacryocystorhinography will confirm the diagnosis.

## Management

- nasolacrimal duct should be catheterised, and the system flushed.
- additional surgical procedures may be required depending on the cause and location of the obstruction (see congenital NLS obstruction, page 85).
- topical and systemic antimicrobials and anti-inflammatories are helpful in preventing/treating infection and inflammation associated with the obstruction.

## Prognosis

- excellent to poor depending on the underlying cause:
  - obstruction secondary to neoplasia typically has a poor prognosis.
  - foreign body material obstruction: usually excellent prognosis following removal.

# Keratoconjunctivitis sicca (KCS)

## Definition/overview

- uncommon in the horse.

## Aetiology/pathophysiology

- cause in most cases is unknown.

- trauma to the head/orbital region leading to facial nerve paralysis is the most common.
- may result from lacrimal gland dysfunction:
  - toxic effects on glandular tissue due to locoweed poisoning or chemical exposure.
  - eosinophilic dacryoadenitis has been reported.
- topical atropine and other medications can temporarily decrease lacrimation.
- KCS reduces tear production, leading to superficial corneal damage, delayed epithelial healing and exposure of the corneal stroma:
  - secondary infections are common.

## Clinical presentation

- blepharospasm, conjunctivitis and mucopurulent ocular discharge.
- keratitis and corneal ulcers.
- chronic cases: corneal perforation and resultant panophthalmitis.

## Differential diagnosis

- any cause of chronic keratitis and/or conjunctivitis should be considered.

## Diagnosis

- history, clinical signs and STT should be diagnostic.

## Management

- treat the underlying cause if possible.
- any sequelae (e.g. corneal ulceration) must be treated appropriately.
- antimicrobial/steroid combination topical medication may be helpful to treat the inflammation and secondary bacterial invaders in some cases q4–12 h:
  - **rule out corneal ulceration prior to using corticosteroids.**
  - neomycin sulphate/polymyxin B sulphate/dexamethasone 0.1%.
  - triple antibiotic/steroid combination solution or ointment.
- topically applied tear lubricant used frequently to prevent conjunctiva and cornea from drying out.
- topical cyclosporine is a lacrimomimetic in horses.
- parotid duct transpositions have been attempted in the horse.

## Prognosis

- acute cases with no secondary disease processes have a favourable prognosis:
  - most of these cases are transient and of unknown aetiology.
  - appropriate management to maintain corneal health is paramount:
    - problems with owner and/or patient non-compliance.
- poor prognosis for those horses with KCS secondary to facial paralysis.
- panophthalmitis after perforation of the globe has a poor prognosis.

# Corneal ulcers/ ulcerative keratitis

## Definition/overview

- breaks in the corneal epithelium (corneal ulcers) are common in the horse.
- equines are predisposed to traumatic corneal injury and secondary infection:
  - large, prominent, laterally placed equine eye.
  - behavioural tendencies of the horse.
  - environment rich in bacterial and fungal organisms.
- corneal ulcers are typically classified based on:
  - aetiology.
  - depth.
  - presence of complicating factors, such as infection or collagenase activity.
  - rate of progression:
    - melting ulcers are rapidly deepening ulcers caused by collagenase enzymes that destroy the corneal stroma.
    - ulcers that progress to the level of Descemet's membrane (verge of perforating) are called descemetocoeles.
- **corneal ulcers can be sight-threatening:**
  - **require early diagnosis and appropriate, prompt medical and surgical management to avoid serious complications.**

## Aetiology/pathophysiology

- most ulcers in horses are initiated by trauma and then become secondarily infected.
- variety of other causes may be involved (Table 2.7).
- most common pathogens vary, based on geographical location and season:
  - *Streptococcus zooepidemicus*, alpha-haemolytic streptococci and *Staphylococcus aureus* are common Gram-positive organisms.
  - *Pseudomonas* spp. and *Actinobacillus* spp. are common Gram-negative organisms.

**TABLE 2.7** Aetiologies of corneal ulcers

- Trauma
  - Secondary anaerobic and aerobic bacterial infection: *Pseudomonas* spp., *Enterobacter* spp., *Actinobacter* spp., *Streptococcus zooepidemicus, S. equi, Staphylococcus aureus, Corynebacterium* spp., *Bacillus cereus, Klebsiella* spp., *Moraxella equi, Bacillus* spp., *Streptomyces* spp., *Neisseria* spp.
  - Fungal infection: *Aspergillus* spp., *Fusarium* spp., *Penicillium* spp., *Alternaria* spp., *Cladosporium* spp., *Pseudoallescheria* spp., *Geotrichum* spp., *Candida* spp., *Mucor* spp., *Exophiala* spp., *Torulopsis glabrata* (also known as *Candida glabrata*), *Hanseniaspora uvarum, Drechslera* spp., *Cylindrocarpon destructans, Curvularia* spp., *Trichosporin cutaneum*, *Phycomyces* spp., *Paecilomyces* spp.
- EHV-2, EHV-3 and EHV-5 infections
- Eosinophilic keratoconjunctivitis
- Indolent-like ulcers
- Immune-mediated keratitis/limbal keratopathy
- Conformational – entropion; lagophthalmos
- Abnormal hair growth – trichiasis, distichia, ectopic cilia
- Keratoconjunctivitis sicca
- Ocular foreign body
- Corneal degeneration/calcium deposition/band keratopathy
- Corneal sequestration
- Exposure to caustic substances/chemical irritants (e.g. alcohol, soaps, insect repellent), heat or radiation
- Exposure keratopathy due to cranial nerve deficits (i.e. facial paralysis or corneal hypoaesthesia), inadequate eyelid function, buphthalmos or exophthalmos

- intact corneal epithelium acts as a protective barrier to invasion by normal microbial inhabitants of the equine environment and corneal/conjunctival microflora.
- partial- or full-thickness defect in this barrier, usually result of trauma:
  - allows opportunistic and pathogenic bacterial and fungal organisms:
    - adhere to, invade and replicate in injured or diseased corneal surface.
    - initiating infection.
- fungi predilection for invading the deep stroma or Descemet's membrane.
- activation and/or production of excessive proteinolytic enzymes:
  - corneal epithelial cells.
  - leucocytes.
  - certain microbial organisms:
    - especially *Pseudomonas* and beta-haemolytic streptococci.
  - results in sudden, rapid degeneration of collagen and other stromal components.
  - induces corneal liquefaction or keratomalacia:
    - **leads to globe rupture in less than 12 hours if not controlled.**
- anterior uveitis secondary to corneal disease is common and can lead to:
  - scarring and/or blockage of the ICA.
  - and/or uveoscleral outflow pathway.
  - cause an elevation in the IOP or glaucoma.
- prolonged topical antimicrobial, corticosteroid or combinations of the two:
  - may inhibit growth of normal bacteria.
  - predispose to mycotic infections in corneal ulcer cases.

## Clinical presentation

- appearance varies:
  - simple, superficial breaks or abrasions in corneal epithelium not visible to naked eye.
  - deep stromal ulcers.
  - full-thickness corneal perforations with iris prolapse (Fig. 2.76).
- associated ocular signs vary considerably and include:
  - ocular pain with blepharospasm, photophobia and epiphora.
  - serous to mucopurulent ocular discharge.

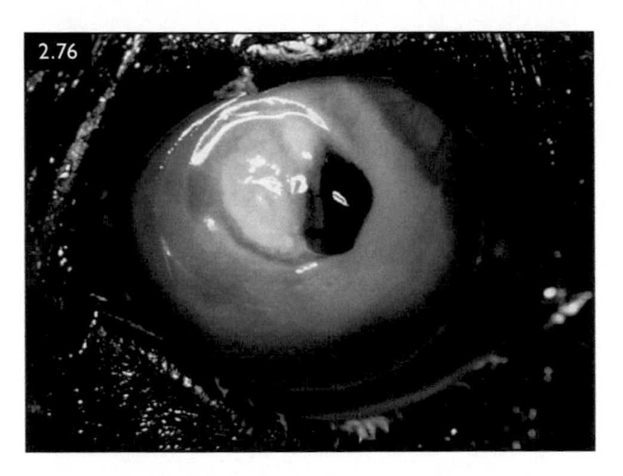


**FIG. 2.76** Iris prolapse. This horse has a descemetocoele with dark iris prolapsing through a perforation at the temporal aspect of the lesion. Focal perforations in the equine eye commonly become 'plugged' with iris.

- conjunctival hyperaemia, chemosis and conjunctivitis.
- corneal oedema.
- variable corneal neovascularisation (superficial and/or deep).
- white to grey to brown plaque adhered to the corneal surface (fungal infections) (Figs. 2.77, 2.78).
- interstitial keratitis.
- white-yellow or grey gelatinous corneal opacity or exudates (stromal necrosis/liquefaction or keratomalacia).
- quite subtle clinical signs in sick or hospitalised foals:
  - corneas significantly less sensitive than those of normal foals or adult horses.
- signs of secondary anterior uveitis, ranging in severity, are commonly seen:
  - miosis
  - aqueous flare.
  - hypopyon.
- other more chronic signs include:
  - corneal scarring and pigmentation.
  - anterior and posterior synechiae.
  - cataract formation.
  - endophthalmitis, phthisis bulbi and blindness.

## Differential diagnosis

- corneal facet:
  - re-epithelialised ulcer.
- stromal abscess.
- uveitis, glaucoma and other causes of a red or cloudy eye.

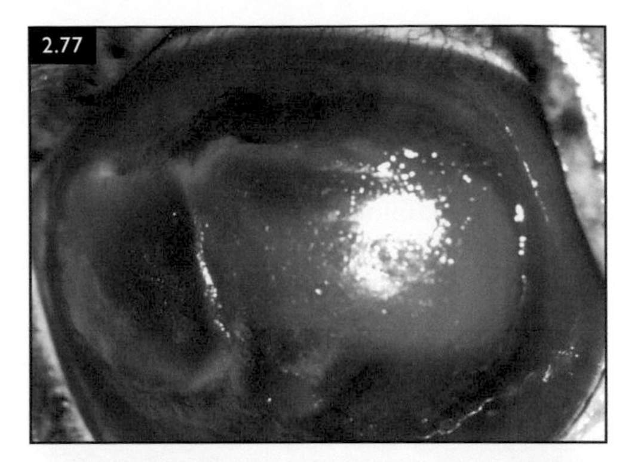


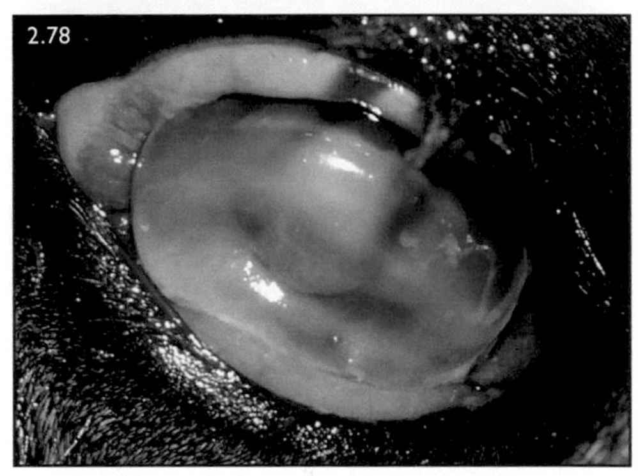


**FIGS. 2.77, 2.78** Fungal keratitis. (2.77) Chronic superficial fungal keratitis produced this necrotic plaque of corneal tissue, referred to as a corneal sequestrum. When the plaque was removed, the fungal organisms were recovered cytologically from the underlying corneal stroma. (2.78) This foal developed severe, progressive keratitis and secondary uveitis, determined to be associated with colonisation by *Fusarium* spp. Note the reactive conjunctival chemosis, diffuse corneal oedema, undulating corneal topography indicating stromal loss, and intraocular hypopyon.

## Diagnosis

- ophthalmic examination and fluorescein staining can identify corneal ulceration (see Fig. 2.14).
- corneal swabs collected from central and peripheral aspects of the ulcer for culture and sensitivity testing if secondary infection is suspected.
- followed by corneal scrapings for cytology unless perforation is imminent.
- mixed bacterial and fungal infections can occur:
  - fungal isolates have a predilection for Descemet's membrane.

  - aggressive and repeated scrapings are often required to collect relevant samples.
  - specialised stains, such as modified Wright–Giemsa, Gomori's methenamine silver and PAS help detection.
  - corneal samples for histopathology can be collected, usually at the time of surgery.
  - PCR tests may be helpful in detecting fungal and viral agents.

## Management

- more severe or complex cases:
  - early consultation with a knowledgeable clinician recommended to establish appropriate diagnostic procedures, therapy and criteria for referral.
- treatment costs in more complicated cases can be substantial.
- aggressive medical therapy may reduce the likelihood of surgical intervention.
- administration of medications may be as often as hourly:
  - hospitalisation may be helpful for successful management.
  - SPL systems are useful to facilitate administration of topical medications.
- therapy is based on the underlying aetiology, depth of the ulcer, presence of complicating factors, rate of progression and response to treatment:
  - identify and remove or treat the cause.
  - all ulcers should be treated initially and aggressively with topical and/or subconjunctival broad-spectrum antimicrobials to prevent or control infection:
    - preferably bactericidal, based initially on cytology staining.
    - pending culture and sensitivity results.
    - triple antibiotic (neomycin–polymyxin–bacitracin/gramicidin) or chloramphenicol q4–8 h good choice for initial treatment of simple ulcers.
    - if cytology reveals Gram-positive organisms, then ciprofloxacin, ofloxacin and erythromycin can be considered.
    - Gram-negative organisms: gentamicin, tobramycin (0.3%) and amikacin may be selected.
    - deep, complicated or melting ulcers:
      - cefazolin (5.5%) and ciprofloxacin.
      - triple antibiotic and ciprofloxacin.
  - topical atropine to help control ciliary body spasm and decrease pain:
    - **note that mydriasis may occur before ciliary body spasm is relieved.**
    - **further atropine administration required despite pupillary dilation.**
  - antifungals instituted in:
    - endemic areas.
    - cases where mycotic infection is suspected, pending laboratory results:
      - history of corneal injury with vegetative material.
      - corneal ulcer received prolonged antimicrobial and/or corticosteroid therapy with minimal or no improvement.
    - miconazole (1%), natamycin (3.33%), itraconazole and voriconazole are possible topical antifungal medications:
      - applied frequently (q1–8 hr).
      - severe inflammation may be produced by dead and dying fungal organisms so q4 hr or greater is often recommended.
- **where perforation or surgery is imminent, only ophthalmic solutions should be used, as ophthalmic ointments can induce severe endophthalmitis if they enter the globe.**
- systemic drugs are indicated in certain cases:
  - corneal perforation or surgery is present or impending.
  - disease process extends to the eyelids or orbit.
  - antibiotics:
    - trimethoprim/sulfadiazine (24–30 mg/kg p/o or i/v q12 h).
    - penicillin (sodium/potassium penicillin, 20,000 IU/kg i/v q6 h.
    - procaine penicillin (20,000 IU/kg i/m q12 h).

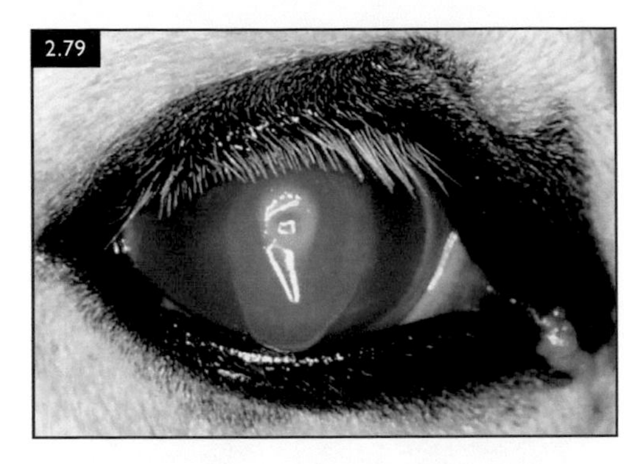


**FIG. 2.79** Keratomalacia, or 'corneal melting', is associated with the establishment of collagenase activity and can result in rapid deterioration of the cornea. Such wounds tend to progress in a circumferential pattern from the initial stimulus.

- gentamicin (6.6 mg/kg i/v q24 h).
- antifungals:
  - itraconazole (3 mg/kg p/o q12 h).
  - ketoconazole and fluconazole (1 mg/kg p/o q12 h).
- keratomalacia (Fig. 2.79) may lead to perforation within 12 hours:
  - topical antiproteinases used to help inhibit stromal necrosis (melting):
    - acetylcysteine.
    - sodium or calcium EDTA.
    - and/or autogenous serum:
      - administered topically as often as possible.
      - requires refrigeration.
      - short shelf life (replace every 72 hours).
      - strict attention to reducing any contamination.
  - tetracycline family of drugs, especially doxycycline, have strong anti-matrix metalloproteinase (MMP) effects and may be selected in cases of keratomalacia.
- uveitis is present in all types of corneal ulceration:
  - must be controlled to prevent associated complications and preserve vision.
  - topically applied atropine sulphate is effective in:
    - stabilising the blood–aqueous barrier (BAB).
    - reducing vascular protein leakage.
    - minimising pain from ciliary muscle spasm.
    - reducing the chance of synechia formation by causing mydriasis.
    - used to effect (as often as every 4 hours):
      - frequency of administration reduced as pupil dilates.
      - pupil dilation a good indicator of clinical improvement or control of intraocular inflammation.
    - **note that topical atropine may prolong intestinal transit time and decrease or eliminate intestinal sounds:**
      - monitor for signs of colic and changes in GI sounds.
      - limited exercise (walking in-hand) and grazing several times daily.
  - systemic NSAIDs:
    - reduce ocular inflammation and relieve ocular discomfort.
    - prevent further injury and limit complications.
    - flunixin meglumine (1.1 mg/kg i/v or p/o q12–24 h) or phenylbutazone (2.2–4.4 mg/kg i/v or p/o).
    - high doses of systemic NSAIDs used for extended periods of time:
      - systemic gastroprotectants may be administered:
      - omeprazole (4 mg/kg p/o q24 h)
      - sucralfate (20 mg/kg p/o q12 h).
- **topical or systemic corticosteroid therapy is contraindicated in the treatment of corneal ulcerations.**
- soft contact lenses or collagen shields can be used as corneal bandages to help encourage ulcers to heal (see Figs. 2.32, 2.33).
- cyanoacrylate adhesive (tissue glue) has been used in some cases of corneal ulceration:
  - can be expensive, infrequently indicated and application can be difficult.
  - not to be used to seal leaking corneal perforations.
  - not recommended for descemetocoeles.
- **well-fitting hood with a hard eyecup, cross-tying or constant supervision:**
  - prevent self-trauma to a painful eye.

  - perforation of an eye with a severely compromised cornea.
- frequent re-evaluations are required:
  - sequential photographs or detailed drawings can document changes.
    - how rapidly the ulcer is changing.
  - treatment modifications in response to therapy.
  - depth of the corneal ulcer and evidence of melting.
- **surgical intervention and referral may be necessary in some cases:**
  - may decrease the dose, length of time and frequency of medical treatment.
  - recommended where:
    - ulcer is greater than 50% of the corneal depth.
    - stromal necrosis (melting) is present.
    - ulcer is progressing despite appropriate medical management.
  - enucleation may be considered, based on:
    - financial constraints.
    - inability to heal or the progression of corneal ulceration.
    - animal's inability to tolerate long-term frequent application of topical medication.

## Prognosis

- variable (good to guarded) depending on:
  - underlying aetiology and the therapy selected.
  - corneal depth and rate of ulcer progression.
  - presence of stromal necrosis (melting).
  - mycotic infections can be difficult to heal medically.
  - keratomalacia can be difficult to control.

# Iris prolapse

## Definition/overview

- sequela to traumatic insult to globe/orbit or corneal perforation following ulcerative disease.

## Aetiology/pathophysiology

- blunt trauma can cause globe rupture at the limbus or equator (sclera is thinnest).

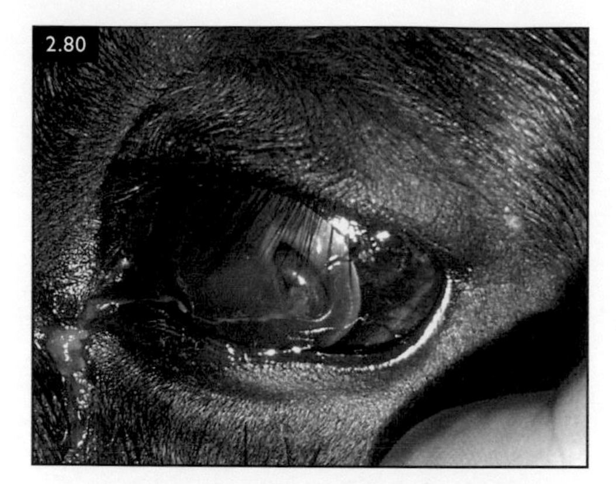


**FIG. 2.80** Prolapse of iris through a defect in the centrolateral cornea following corneal laceration. Note the dark-brown mass (staphyloma) with surrounding corneal oedema and a distorted pupil.

- sharp trauma can cause rupture of the cornea, limbus and/or sclera.
- corneal perforation secondary to enzymatic degradation by infectious and non-infectious ulcerative keratitis.
- deep stromal abscessation can progress to full-thickness corneal rupture in rare cases.
- **globe rupture may occur during examination of deep corneal ulcers or descemetocoeles if the horse is not amenable to examination.**
- globe rupture results in hyphaema, fibrin formation and uveal prolapse.
- perforations may initially seal but are unstable and may leak intermittently.

## Clinical presentation

- iris prolapse:
  - focal red to brown or tan corneal mass bulging from the surface:
  - associated corneal oedema and fibrin formation (Fig. 2.80).
- soft globe may be noted if the prolapse has not resealed and the globe re-inflated.
- fluid leakage from the globe may be present.
- other clinical signs may include:
  - lacrimation.
  - red-tinged serous or mucopurulent ocular discharge.
  - blepharospasm
  - photophobia.
  - enophthalmos.
  - blepharoedema
  - chemosis.
  - keratomalacia.

- miosis
- dyscoria.
- anterior synechiae.
- shallow or absent anterior chamber.
- hyphaema and/or intraocular haemorrhage.

## Differential diagnosis

- phthisis bulbi, ulcerative keratitis and corneal/conjunctival masses.

## Diagnosis

- history and clinical appearance.
- fluid leakage may be confirmed by a positive Seidel test.
- cases following corneal ulceration:
  - samples for cytology, culture and sensitivity, and histopathology.
  - collected at time of surgery to determine the underlying aetiology.
- transpalpebral ultrasonography:
  - evaluate the posterior segment and identify possible intraocular foreign bodies.
  - **gentle handling is required to prevent worsening of the injuries.**
- radiography is helpful for orbital disease (e.g. fracture) or when a radiopaque foreign body is suspected.

## Management

- **immediate referral for surgical repair is recommended in corneal perforation:**
  - presence of an indirect PLR in the other eye implies that retinal function persists in the damaged eye and a positive sign that vision may be saved.
  - absent PLR may reflect severe intraocular disease, intense miosis of the pupil or opacity of the intraocular fluids.
  - all criteria should be examined critically to determine whether therapy should be aimed at:
    - saving vision.
    - establishing a cosmetic globe.
    - removing a blind and chronically painful eye.

## Prognosis

- depends on the prolapse duration, size, location and the intraocular structures involved.
- generally guarded for vision:
  - approximately one third of eyes are blind at the time of discharge and over half of these eventually develop phthisis bulbi.
  - visual outcome is slightly better if the prolapse is the result of ulcerative keratitis rather than laceration (40% versus 33%).
  - poor visual outcome and usually require enucleation:
    - corneal perforations or lacerations present for more than 15 days.
    - perforations measuring 15 mm or more in length.
    - perforations extending to, along or beyond the limbus.
- overall rate of ocular survival is better in cases of laceration than in cases of ulcerative keratitis (80% versus 67%).
- perforation by blunt trauma has a worse prognosis than that due to sharp trauma, because greater accompanying damage to tissues.
- keratomalacia can lead to endophthalmitis and a worse overall prognosis.
- common sequelae following iris prolapse may include:
  - endophthalmitis, persistent intraocular inflammation, anterior and posterior synechiae, cataract formation, phthisis bulbi, blindness and enucleation.

# Corneal stromal abscess

## Definition/overview

- intrastromal accumulation of fungi, bacteria and/or inflammatory cell debris (sterile abscess) beneath an intact epithelium.
- may be a history of previous ulceration or clinical signs of ocular discomfort.
- difficult to treat:
  - may require weeks to months of medical therapy and/or surgical intervention.

## Aetiology/pathophysiology

- may be bacterial, fungal or sterile in origin (see Table 2.7).
- develop following focal trauma to the cornea that allows opportunistic pathogens and debris into the stroma beneath the corneal epithelium.

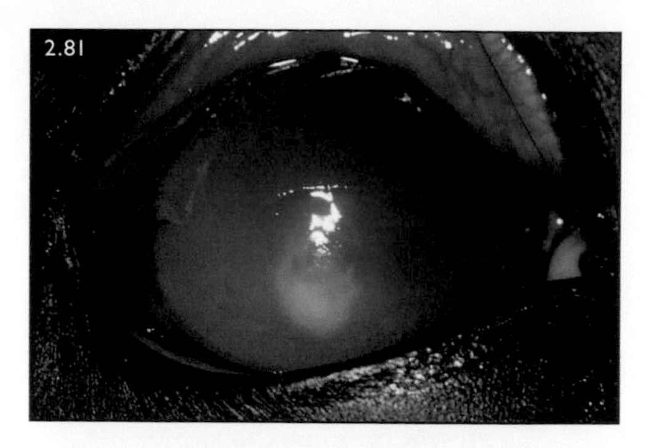

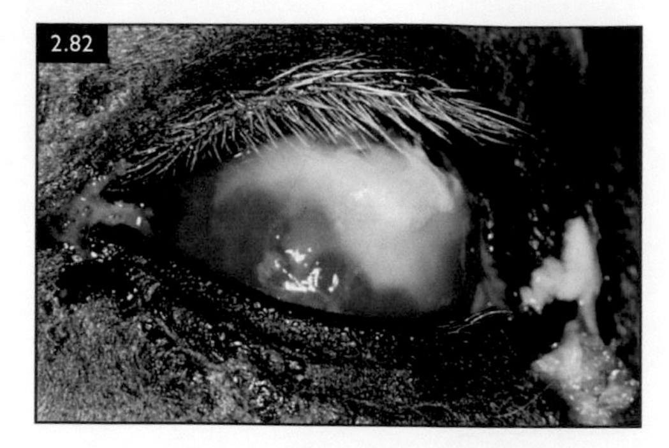

**FIGS. 2.81, 2.82** Stromal abscesses (2.81) appear as creamy white to yellow focal areas in the cornea and are usually accompanied by corneal oedema, corneal vascularisation and varying degrees of reflex uveitis. (2.82) If left untreated, or if treated inappropriately, a focal abscess can progress to involve the entire cornea, as in this eye treated for a suspected inflammatory problem with topical steroids.

- subsequent healing or re-epithelialisation of ulcer or epithelial micropuncture forms a barrier:
  - protects bacteria or fungi from topical antimicrobial medications.
  - seals in the microorganisms and allows ongoing infection.
- some cases, the initial treatment may kill the microorganisms:
  - subsequent release of toxins by dying bacteria/fungi and degenerating leucocytes.
  - continues stimulus for abscessation.
- topical antimicrobial/corticosteroid combination therapy can predispose to corneal abscessation.
- concurrent anterior uveitis is common.
- equine cornea vascularises extremely slowly with stromal abscesses:
  - vascularisation is essential for abscesses to heal.
  - contributes to the long recovery period.

## Clinical presentation (Figs. 2.81, 2.82)

- single or multiple, focal, white to yellow, stromal infiltrates or opacities:
  - occur axially, paraxially or peripherally.
  - very small to quite large in diameter.
- associated corneal oedema and variable corneal neovascularisation.
- occur at all depths of the cornea from superficial to deep:
  - may rupture into the anterior chamber.
- associated clinical signs can also include:
  - lacrimation
  - blepharospasm.
  - photophobia.
  - enophthalmos.
  - third-eyelid elevation.
  - conjunctival hyperaemia.
  - miosis and signs associated with anterior uveitis.

## Differential diagnosis

- ulcerative keratitis.
- corneal degeneration.
- calcific band keratopathy.
- corneal neoplasia (e.g. SCC, haemangioma, angiosarcoma).
- corneal foreign body or granulation tissue.
- parasitic infestation (e.g. *Onchocerca*).
- eosinophilic keratitis/keratoconjunctivitis.
- non-ulcerative keratouveitis.
- anterior segment dysgenesis.

## Diagnosis

- history of previous trauma and/or evidence of ulceration.
- clinical appearance of a yellow-white corneal stromal infiltrate.
- negative fluorescein staining over site of the abscess.
- following epithelial debridement:
  - collect samples for cytology, culture/sensitivity and histopathology:
    - can be difficult prior to surgery:
      - presence of an intact epithelium.
      - deep location of most abscesses.

- fungal isolates have a predilection for Descemet's membrane:
  - aggressive/repeated scrapings often required to obtain diagnostic samples.
  - specialised stains may be required to detect fungal organisms.
  - fungal PCR of corneal specimens or cytology may assist diagnosis.

## Management

- medical therapy similar to that for corneal ulceration:
  - aggressive use of topical and systemic antibiotics and antimycotics:
    - selection of topical medication important as only certain drugs can penetrate an intact epithelium satisfactorily:
      - ciprofloxacin.
      - chloramphenicol.
    - empirical therapy targeting both bacterial and fungal agents is often recommended as the aetiological agent/s are often unidentified.
    - SPL system facilitates treatment.
    - subconjunctival injections also used.
    - antimycotic drugs include:
      - voriconazole.
      - fluconazole.
      - itraconazole.
      - miconazole.
      - ketoconazole.
      - amphotericin B.
      - natamycin.
      - silver sulfadiazine.
    - voriconazole (1%) q4 h commonly selected first-line therapy.
  - topical atropine.
  - systemic NSAIDs.
- superficial abscesses may require periodic corneal debridement:
  - removes epithelium to enhance penetration of topical medications.
- corneal neovascularisation or vascularisation from a conjunctival graft is required for stromal abscesses to heal:
  - NSAIDs significantly inhibit corneal vascularisation so must be used with care.
  - **topical steroids contraindicated with corneal ulceration or abscessation.**
- significant improvement does not occur:
  - within a few days of intensive medical therapy.
  - or there is deterioration following an initial improvement.
  - **referral and surgery considered.**

## Prognosis

- variable, depending on the depth, rate of progression and therapy chosen:
  - can heal by vascularisation within 1 week or as much as 2–3 months.
- generally good with surgical intervention.
- progression, imminent or pre-existing rupture into the anterior chamber, and endophthalmitis are indications for a poor visual outcome.

# Immune-mediated keratitis (IMMK)

## Definition/overview

- term used to describe a group of non-ulcerative, non-infectious eye diseases:
  - believed to result from dysregulated immune responses in the normally immunologically privileged cornea.
- variability in clinical presentation suggests IMMK represents a syndrome with multiple disease processes, rather than a single entity.

## Aetiology/pathophysiology

- precise aetiopathogenesis has not been elucidated:
  - some dysfunction of the normal immune privilege of the cornea is suspected.
  - corneal trauma with subsequent neovascularisation may disrupt the normal corneal immunological tolerance in some cases.
  - other factors are likely to be important, such as exposure to antigens which cross-react with corneal proteins and the immunophenotype of the patient.

## Clinical presentation

- three categories of IMMK characterised by the depth of the corneal inflammatory response.

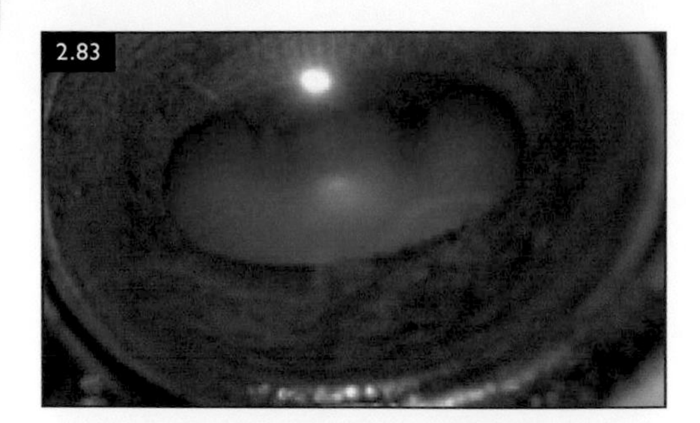

**FIG. 2.83** Superficial IMMK in a Thoroughbred gelding. Multifocal corneal epithelial and superficial stromal opacities are affecting a geographical region of the ventral cornea.

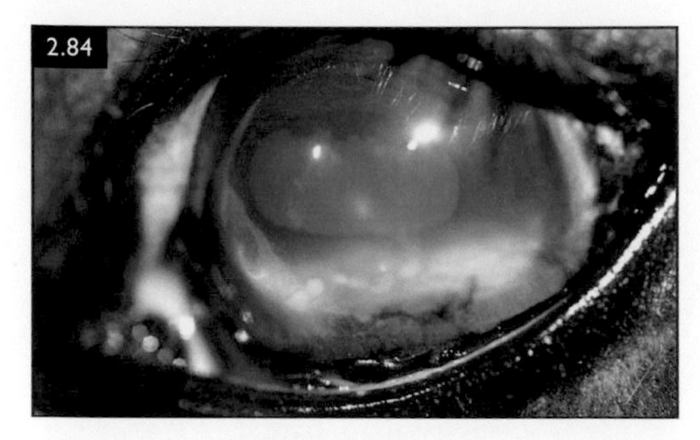

**FIG. 2.84** Mid-stromal IMMK in a Warmblood mare. There is extensive deep corneal stromal vascularisation, a haze to the central ventral cornea with dense multifocal yellowish infiltrates and some superficial pigment deposition.

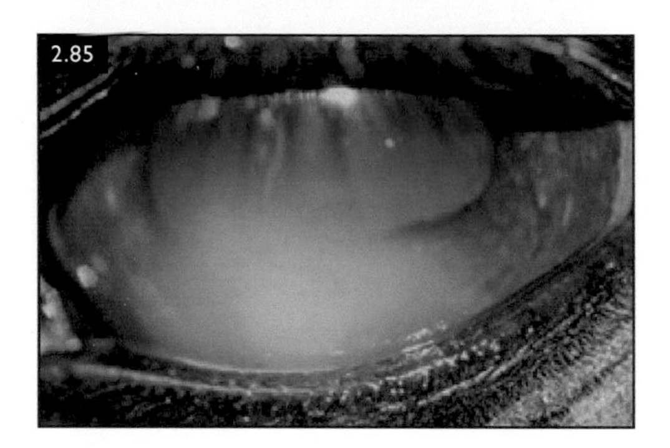

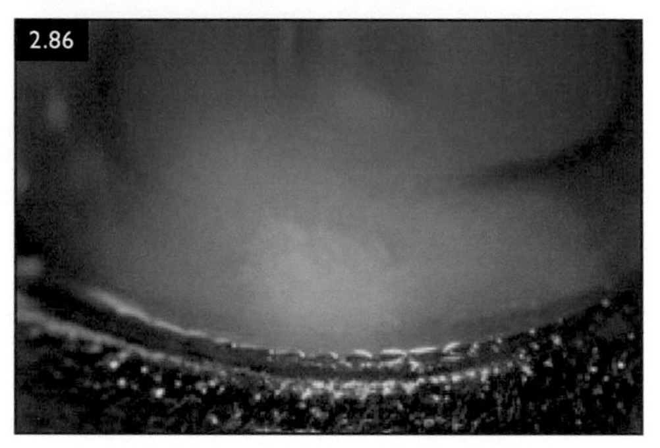

**FIGS. 2.85, 2.86** Endothelial IMMK. (2.85) Diffuse corneal oedema in the ventral cornea associated with a dense area of opacity. (2.86) A closer view of the lesion demonstrates a plaque of endothelial fibrosis and cellular deposition that is the cause of this corneal oedema.

- usually unilateral although both eyes may be affected, sometimes sequentially.

### Superficial IMMK

- often insidious and signs of discomfort may be mild.
- irregularity of the corneal surface due to thickening of the epithelial layer.
- sometimes cellular infiltrate into the superficial stroma (Fig. 2.83).
- superficial branching corneal neovascularisation with corneal oedema is often present.

### Deep/mid-stromal IMMK

- associated with extensive mid- to deep stromal vascularisation, oedema, cellular infiltration and scarring (Fig. 2.84).
- often minimal ocular pain.
- disease may cycle through relatively quiescent periods followed by recurrence of active disease episodes.

### Endotheliitis

- corneal endothelium is affected with disruption of its pump mechanism:
  - leads to varying degrees of corneal oedema:
    - may vary from mild to severe.
    - corneal decompensation (corneal hydrops) occurs in some cases.
- deep corneal vascularisation and cellular infiltration, including clumps of cells on the endothelium, may be present (Figs. 2.85, 2.86).
- concurrent anterior uveitis and pain may be present in some cases.

## Differential diagnosis

- no specific test for definitive diagnosis of IMMK:
  - rule out other causes of keratitis:
    - viral infection (EHV-2 and -5).
    - mycoses
    - corneal abscess.

- eosinophilic keratoconjunctivitis.
- foreign bodies, physical irritation associated with distichiasis, ectopic cilia and entropion.
- neurotrophic keratitis, and exposure keratopathy due to facial nerve dysfunction.
- lagophthalmos associated with exophthalmia or buphthalmia.

## Diagnosis

- based on identifying the characteristic corneal pathology associated with the three subtypes of IMMK.
- rule out other potential causes of corneal pathology.
- subsequent response to anti-inflammatory treatment helps strengthen the diagnosis.

## Management

- early diagnosis and initiation of treatment is important to minimise the recruitment of inflammatory cells to the cornea:
  - causes IMMK to become increasingly refractive to medical management with time.
- medical treatment includes:
  - topical corticosteroids
    - 1% prednisolone acetate.
    - 0.1% dexamethasone.
  - systemic NSAIDs (flunixin meglumine, phenylbutazone).
  - topical immunosuppressive therapy (cyclosporine A [CsA] and tacrolimus).
  - response to anti-inflammatory treatment is not predictable:
    - various combinations of anti-inflammatory and immunosuppressive therapy may need to be trialled.
    - upon remission, the anti-inflammatory medication weaned down to establish the minimum required to maintain remission.
    - some cases are poorly responsive and show a progression of signs or require long-term (potentially lifelong) treatment:
      - more superficial cases generally respond better to topical therapy.
      - implantation of an episcleral cyclosporine sustained-release device has been beneficial in superficial and endothelial cases.
- surgical intervention with a keratectomy to remove any residual diseased cornea:
  - beneficial in some cases of superficial and mid-/deep stromal IMMK.

## Prognosis

- depends on the response to treatment:
  - initially may be fair to guarded, especially in superficial cases.
  - chronicity, recurrence and deep pathology worsen the prognosis.
  - superficial cases amenable to treatment by superficial keratectomy can have a fair to good prognosis if the lesion can be excised completely.

# Uveitis

## Definition/overview

- inflammation of the iris and/or ciliary body is termed anterior uveitis.
- inflammation of the choroid is termed posterior uveitis.
- inflammation of all structures of the uvea is termed panuveitis.
- reported worldwide.
- wide variety of underlying causes which are often not identified in individual cases:
  - usually associated with systemic disease in the foal.
  - most often immune-mediated or due to direct invasion of microorganisms into the eye in adults.
- cases of equine uveitis share common clinical features and are often treated symptomatically.
- potential sequelae capable of causing blindness from uveitis include:
  - cataracts and pupillary seclusion.
  - retinal detachment or degeneration.
  - glaucoma.
  - phthisis bulbi.
- some breeds appear predisposed, e.g. Appaloosas, whereas others may have a reduced risk, e.g. Standardbreds.
- may occur unilaterally or bilaterally.

## Aetiology/pathophysiology

- proposed causes include trauma, neoplasia, infectious disease and other systemic diseases (Table 2.8).

- underlying initial insult in any case is tissue damage and breakdown of the BAB.
- Leptospirosis is the most commonly implicated infectious cause of equine uveitis:
  - horses that are seropositive to *Leptospira* serovar Pomona are 13 times more likely than seronegative horses to have signs of uveitis.
  - clinical uveitis does not occur until months or years after primary infection with leptospirosis.
  - antigen mimicry between *Leptospira* spp. and equine ocular tissues has been demonstrated:
    - possible immune-mediated process in the pathophysiology of equine recurrent uveitis (ERU).
- aberrant ocular migration of *Onchocerca cervicalis* microfilariae is the most commonly implicated parasitic cause of recurrent uveitis, although with modern deworming products this now occurs rarely.
- decreased IOP typically occurs owing to a decrease in production of aqueous humour resulting from ciliary body inflammation.

**TABLE 2.8** Main causes of uveitis

- Trauma – blunt or penetrating
- Iatrogenic from surgical trauma (i.e. intraocular surgery)
- Systemic infections:
  - Bacteria: *Leptospira* spp., *Brucella abortus, B. melitensis, Borrelia burgdorferi, Salmonella* spp., *Streptococcus equi, Escherichia coli, Rhodococcus (Corynebacterium) equi, Actinobacillus equuli,* leishmaniasis, tuberculosis
  - Parasites: *Onchocerca cervicalis, Halicephalobus deletrix,* intestinal strongyles, *Dirofilaria immitis*
  - Viruses: equine arteritis virus, EIA virus, EHV types 1 and 4, equine influenza virus
  - Protozoal: *Toxoplasma gondii, Neospora* spp.
  - Fungal: *Cryptococcus*
- Keratitis-associated axonal reflex uveitis
- Immune-mediated/idiopathic, including Equine recurrent uveitis
- Neoplasia
- Lens-induced uveitis

- aqueous flare is caused by exudation of protein and inflammatory cells from the uveal tissue:
  - may accumulate in large amounts as hypopyon and cause synechiae.
- posterior synechiae and the presence of products of inflammation interfere with lens metabolism:
  - secondary cataract formation.
- secondary lens luxation may result from:
  - inflammation on the lens zonules.
  - traction by fibrous adhesions.
  - globe enlargement and zonule rupture seen with secondary glaucoma.
- vitreal haze can occur secondary to inflammatory cell infiltrate and exudation of protein from the uveal tract:
  - may organise to form bands of tissue that adhere to the retina.
  - may contract to produce retinal detachment.
- chronic cases: corneal degeneration can occur secondary to pathological changes within the cornea and disruption of the epithelium, resulting in ulceration.
- anterior uveitis may also lead to pre-iridial fibrovascular membrane (PIFM) formation:
  - may limit aqueous absorption by the iris.
  - lead to physical and functional obstruction of the ICA.
- lens luxation, anterior synechiae, iris bombé and pupillary seclusion can cause:
  - glaucoma with subsequent buphthalmos.
  - interference with the flow of aqueous humour.
- in end-stage uveitis, fibrosis of the uveal tract results in phthisis bulbi.

## Clinical presentation (Figs. 2.87–2.90)

- varies depending on the severity of inflammation, the area(s) of the uvea involved and the duration of the problem, but may include:
  - variable vision deficits.
  - lacrimation
  - blepharospasm.
  - photophobia
  - enophthalmos.
  - conjunctival/episcleral congestion.
  - corneal oedema.

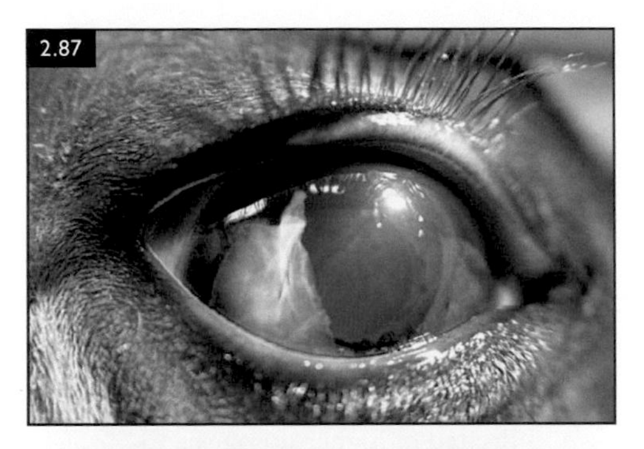


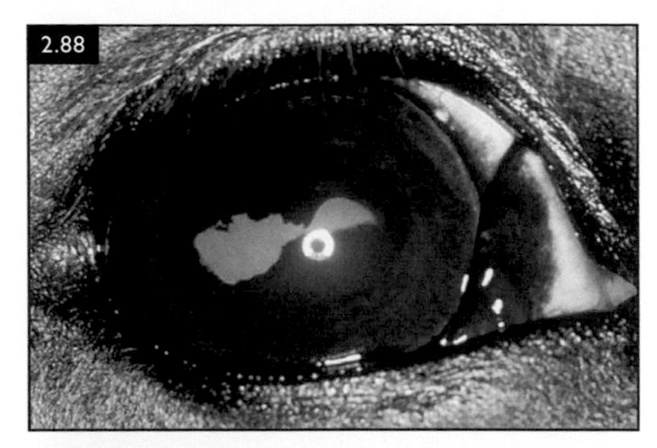


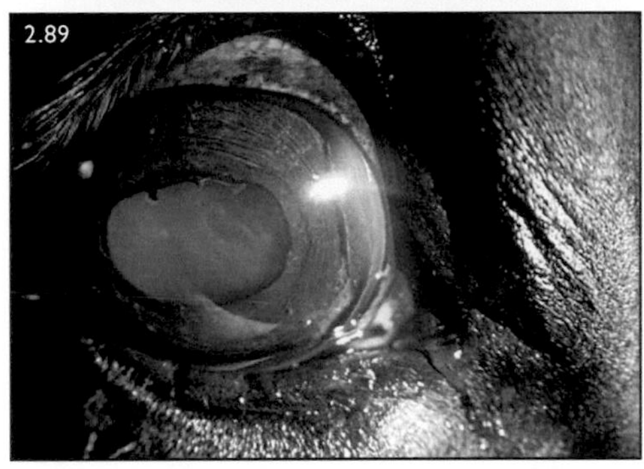


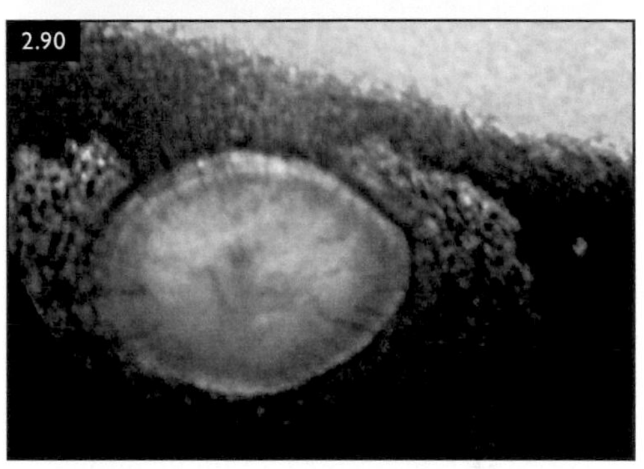


**FIGS. 2.87–2.90** Uveitis can present acutely, in which case the anterior chamber may contain fibrin (as seen in this horse), inflammatory cells or blood (2.87). With recurrent bouts of inflammation, sequelae include yellow staining in the vitreous due to protein accumulation, and posterior synechiae (2.88). Cataract and atrophy of the corpora nigra are common sequelae of uveitis (2.89). 'Butterfly lesions' are associated with chorioretinal scarring around the peripapillary retinal vasculature (2.90). (Figs. 2.87 and 2.89 courtesy I Jurk.)

  - corneal neovascularisation/ciliary flush.
  - keratic precipitates.
  - aqueous flare.
  - fibrin in anterior chamber.
  - hypopyon
  - hyphaema.
  - miosis.
  - cellular debris and/or pigment on the anterior lens capsule.
  - oedematous and/or hyperaemic iris.
  - yellow-green discolouration/vitreal haze.
  - peripapillary chorioretinitis and retinal detachment.
- possible sequelae include:
  - band keratopathy (white chalky spiculated opacities in the cornea).
  - dyscoria.
  - anterior and posterior synechiae.
  - iris bombé.
  - iris colour or pigment change.
  - corpora nigra atrophy.
  - cataract formation.
  - lens subluxation/luxation.
  - secondary glaucoma.
  - vitreal liquefaction/floaters.
  - vitreal traction bands.
  - retinal detachment.
  - depigmentation of the peripapillary region:
    - focal or alar pattern/butterfly lesions.
  - phthisis bulbi.
- posterior segment lesions cannot be directly visualised in many cases because of the severe anterior segment abnormalities.

## Differential diagnosis

- other causes of red eye or cloudy eye, including:
  - glaucoma, lens luxation, keratitis and conjunctivitis.

## Diagnosis

- definitive diagnosis of the cause of uveitis is often elusive.
- attempt to determine the underlying cause of the uveitis:
  - complete blood count and biochemistry panel, including inflammatory markers.
  - urinalysis and faecal float test.
  - urine, faecal and vitreal cultures.
  - serum *Brucella* test.
  - thoracic radiographs or ultrasonography (e.g. *Rhodococcus* pneumonia).
  - leptospirosis and toxoplasmosis titres.
  - conjunctival biopsy for *Onchocerca*.

## Management

- treat the underlying cause if one is identified.
- uveitis requires aggressive treatment to reduce or control ocular inflammation:
  - decrease pain.
  - minimise the progression of ocular lesions.
  - decrease the incidence of post-inflammatory sequelae.
  - preserve vision.
- topical, subconjunctival and/or systemic anti-inflammatory therapy is vital:
  - inhibit BAB breakdown.
  - topical prednisolone acetate (1%) is the anti-inflammatory of choice:
    - topical 0.1% dexamethasone also acceptable.
    - applied every 2–6 hours depending on the severity of inflammation.
- subconjunctival injections of corticosteroids may be helpful:
  - methylprednisolone acetate (20–40 mg), dexamethasone (0.2–1 mg) or triamcinolone (2 mg).
  - complications, including granuloma formation, reported following use of methylprednisolone.
- **steroids should not be used in the presence of corneal ulceration:**
  - use topical NSAIDs such as:
    - diclofenac, flurbiprofen or suprofen q6–12 h as needed.
  - anti-inflammatories and facilitate mydriasis.
- frequency of topical anti-inflammatory medication gradually reduced once clinical improvement of the uveitis occurs.
- systemic NSAIDs such as flunixin meglumine (1.1 mg/kg i/v or p/o q12–24 h) or phenylbutazone (2.2–4.4 mg/kg p/o or i/v q12–24 h) may be used:
  - provide analgesia and inhibit prostaglandin production associated with uveitis.
  - initiated at higher dosages and gradually reduced once inflammation subsides.
- topical atropine sulphate (1%) administration is essential for pupillary dilation:
  - prevent or decrease the risk of posterior synechia formation.
  - provide analgesia by eliminating ciliary body spasm (cycloplegia).
  - stabilise the BAB.
  - lasts a few hours in an inflamed eye:
    - administered topically to effect from 2 to 6 times per day.
  - GI motility should be monitored.
  - topical tropicamide may be used for short-term mydriasis to reduce the risk of inducing GI ileus or colic with frequent atropine administration.
  - topical or subconjunctival phenylephrine (2.5–10% or 5 mg per eye, respectively) may be used to help dilate the pupil.
- cases of band keratopathy:
  - superficial keratectomy may be performed.
  - and/or calcium-chelating drugs (e.g. 0.05% sodium or potassium EDTA).
- topical or systemic antimicrobials are not usually indicated unless:
  - corneal ulceration is present.
  - uveitis appears to be septic.
  - horse is pyrexic.
  - systemic disease responsive to antimicrobials (e.g. leptospirosis, Lyme disease) identified as cause.
- uveitis induced by *Onchocerca cervicalis*:
  - larvicidal medication indicated once active inflammation controlled.
  - ivermectin 0.2 mg/kg once, or diethylcarbamazine 4.4–6.6 mg/kg p/o for 21 days.

- eyes that are blind and chronically painful:
  - enucleation or evisceration.
- immunisation of horses against leptospirosis when *L. pomona* has been implicated as the cause of endemic outbreaks is controversial.

## Prognosis

- guarded:
  - likelihood of recurrence.
  - possible vision-threatening sequelae.
- peracute cases where therapy is prompt, intensive and prolonged:
  - prognosis for preserving vision is fair to good if recurrence does not occur.

# Equine recurrent uveitis (ERU)

## Definition/overview

- painful, chronic ocular condition characterised by:
  - recurrent and increasingly severe episodes of active inflammation of the uveal tract (iris, ciliary body and/or choroid).
  - separated by variable lengths of quiescence.
- initial ocular injury or infection associated with ocular inflammation leads to the establishment of immunologically sensitised cells.
- systemic re-exposure to:
  - similar circulating antigens that enter the eye through a destroyed BAB.
  - or native ocular antigens.
  - causes a non-specific immune-mediated delayed hypersensitivity reaction and recurrent bouts of uveitis.
- recurrent inflammation causes progressive ocular destruction with potential loss of vision:
  - leading cause of vision impairment and blindness in adult horses.
  - major cause of economic loss worldwide.
- recurrence can be frequent and long-term medication is often required.
- ultimately bilateral in approximately two-thirds of cases.
- Appaloosas appear predisposed (more than eight times more likely than other breeds), suggesting a possible genetic link.
- treatment is expensive and time-consuming:
  - enucleation/evisceration may be required in a chronically painful non-visual eye.

## Aetiology/pathophysiology

- pathogenesis appears immune-mediated but the specific causes of ERU are unknown.
- proposed causes have included:
  - trauma.
  - bacterial, viral, fungal, parasitic and other systemic diseases.
  - in most cases, an aetiological agent cannot be identified.
- theories on the pathogenesis include:
  - periodic episodes of inflammation can be directly induced and maintained by the persistence of a specific antigen in the ocular tissues:
    - *Leptospira interrogens* considered the most important infectious agent associated with ERU.
  - immune-mediated, delayed-type hypersensitivity reaction to self- or sequestered antigen (antigen mimicry) in the uveal tract.
- intraocular inflammation follows BAB breakdown with resultant infiltration of inflammatory cells and protein, and the clinical signs of anterior uveitis.
- active episodes of inflammation may last days to weeks:
  - gradually resolving to a relatively comfortable quiescent period.
- recurrent episodes of uveitis associated with progression of irreversible ocular damage:
  - cataracts and pupillary seclusion are common causes of blindness.
  - occasionally, glaucoma with buphthalmos develops:
    - secondary to anterior synechiae, lens luxation or pupillary seclusion.
  - vitreous may appear yellow green in colour:
    - due to the presence of fibrin and porphyrin metabolites.
  - vitreal fibrin may form vitreoretinal traction bands, which can lead to retinal detachment.

  - fibrosis can lead to phthisis bulbi in end-stage disease.

## Differential diagnosis

- previous ocular trauma and inflammation.
- other causes of equine uveitis.
- other causes of a cloudy, red or painful eye must be ruled out.

## Clinical presentation

- varies depending on the severity of inflammation, the area(s) of uvea involved and the duration of the problem (Fig. 2.91):
  - may include variable vision.
  - lacrimation
  - blepharospasm.
  - photophobia
  - enophthalmos.
  - conjunctival hyperaemia and conjunctivitis.
  - corneal oedema.
  - corneal neovascularisation.
  - keratic precipitates
  - aqueous flare.
  - hypopyon
  - hyphaema.
  - miosis
  - dyscoria.
  - peripheral and/or posterior synechiae.
  - iris bombé
  - corpora nigra atrophy.
  - debris and/or pigment on the anterior lens capsule.
  - oedematous and/or hyperaemic iris.
  - cataract formation
  - lens subluxation/luxation.
  - vitreal haze, degeneration/liquefaction and membrane/cellular infiltrates.
  - peripapillary chorioretinitis, butterfly lesions/retinal scarring and/or retinal detachment.
- associated lesions may also include:
  - corneal degeneration and ulceration.
  - secondary glaucoma.
  - retinal degeneration/atrophy.
  - phthisis bulbi.
- in many cases, vitreal or retinal lesions cannot be appreciated because of severe inflammation of the anterior segment or its sequelae.
- lesions may be unilateral or bilateral.
- in some cases of insidious ERU, acute inflammation and discomfort may not be seen:
  - signs of chronic ERU are identified:
    - peripheral and/or posterior synechiae.
    - corpora nigra atrophy and iris hyperpigmentation.

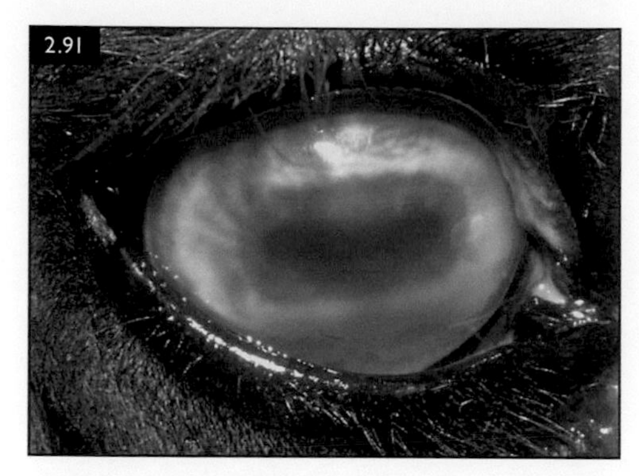


**FIG. 2.91** The right eye of a horse with acute anterior uveitis showing increased lacrimation, miosis, aqueous flare, swelling and discolouration of the iris, circum-limbal corneal vascularisation, mild corneal oedema and hypopyon.

    - pigment on the anterior lens capsule.
    - cataract formation.
    - lens subluxation/luxation.
    - vitreal degeneration and chorioretinal scarring.

## Diagnosis

- ophthalmic signs are sufficient in most cases.
- presumptive diagnosis of ERU:
  - history of previous, recurring episodes of inflammation:
    - responded to anti-inflammatory agents.
  - examination findings consistent with chronic uveitis.
  - lack of significant laboratory findings and no signs of systemic disease.

## Management

- major goals of treatment for each inflammatory episode:
  - preserve vision.
  - decrease pain.
  - minimise extent of ocular tissue damage:
    - depends on the severity and duration of the acute uveitis attack.
    - promptness and effectiveness of therapy.
  - prevent or decrease the recurrence of attacks of uveitis.
- specific prevention and therapy are difficult as the aetiology is rarely identified.

- **treatment is nearly always symptomatic:**
  - involving intense use of anti-inflammatory and mydriatic/cycloplegic drugs.
- anti-inflammatory medications to control the intraocular inflammation:
  - usually corticosteroids are used topically and/or subconjunctivally.
  - NSAIDs are given systemically.
  - suprachoroidal injections of triamcinolone used in patients with medically resistant ERU in referral settings.
  - systemic corticosteroids may be beneficial in severe, refractory cases:
    - **use with caution owing to side effects and potential complications.**
  - frequent recurrence treated by long-term, low-dose corticosteroid prophylactic therapy:
    - may predispose to corneal infection or degeneration.
    - requires client and patient compliance.
    - not always effective.
    - contraindicated in the presence of a corneal ulcer.
  - NSAIDs may be used and are effective at reducing intraocular inflammation when a corneal ulcer is present:
    - flunixin meglumine and phenylbutazone are frequently used systemically.
    - some horses become refractory to these medications:
      - substituting another NSAID may be helpful.
- topical atropine sulphate is used to:
  - minimise synechia formation by inducing mydriasis:
    - can increase the IOP.
    - tropicamide may be safer for use in uveitis with secondary glaucoma.
  - alleviate some of the pain by relieving spasm of the ciliary body.
- medications should be slowly reduced in frequency once clinical signs abate:
  - therapy can last for weeks or months.
  - **should not be stopped abruptly or recurrence may occur.**
- medical treatment can be ineffective, time-consuming and expensive:
  - lifelong therapy is often required.
- surgical treatments are available at referral centres:
  - suprachoroidal sustained delivery CsA implant has been developed:
    - significant decrease in the severity and frequency of recurrence.
    - very low rate of complications.
    - implant lasts, on average, 3.2 years and can be repeated if required.
    - retention of vision was achieved in 78% of ERU cases.
  - Pars plana vitrectomy has been used to treat ERU in European Warmbloods.
    - high success rates reported in Europe, but not always elsewhere.
    - stable vision in the majority of cases.
    - high complication rates have been reported in North America.
- management practices aimed at reducing exposure to potential antigens include:
  - parasite-control programmes.
  - eliminating environmental contact with cattle and wildlife.
  - excluding horses from ponds and swampy areas.
  - limiting rodent access.
  - decreasing incidence of respiratory and systemic infections.
  - maintaining good-quality feed.

### Prognosis

- long-term prognosis for vision in horses with recurrent uveitis is generally poor:
  - sequelae are inevitable.
  - vision loss occurs in one or both eyes in 44% of cases.
- affected by breed and leptospiral seroreactivity:
  - Appaloosas are approximately four times more likely to become blind in one or both eyes than other breeds.
  - leptospiral seropositivity horses are over four times more likely to lose vision than uveitis attributable to other causes.

## Acquired cataracts

### Definition/overview

- lenticular opacities:
  - focal or diffuse,
  - unilateral or bilateral.

- symmetrical or asymmetrical.
- stationary or progressive.

- approximately 5–7% of horses have cataracts.
- can involve the lens capsule, cortex and/or nucleus.
- present at birth (congenital) (see page 90), or acquired in early neonatal stage or later adult life.
- acquired cataracts may be primary/inherited or secondary to another ocular disease process.

## Aetiology/pathophysiology

- most often acquired with chronic inflammation of the anterior uvea, especially ERU (see Table 2.4):
  - diffusion of harmful inflammatory mediators across the lens capsule.
  - subsequent alterations in lens metabolism causing cataractous changes.
  - higher incidence in Appaloosas.
- any alteration in aqueous humour production, composition or flow can have adverse effects on lens metabolism and result in cataract formation.

## Clinical presentation

- opacity in the lens (see Figs. 2.56, 2.57).
- variable effects on the menace response and vision:
  - depends on extent of the cataract and underlying aetiology/sequelae.
  - small incipient lens opacities are common and not associated with blindness.
  - cataracts mature and become opaque, the degree of blindness increases.
  - horses with cataracts causing visual impairment are prone to traumatic injury.
- other ocular lesions that may be associated with cataract formation include:
  - conjunctival hyperaemia.
  - corneal ulceration.
  - uveitis (Fig. 2.92)
  - synechiae.
  - glaucoma.
  - lens luxation/subluxation (see Fig. 2.54).
  - retinal disease or detachment.

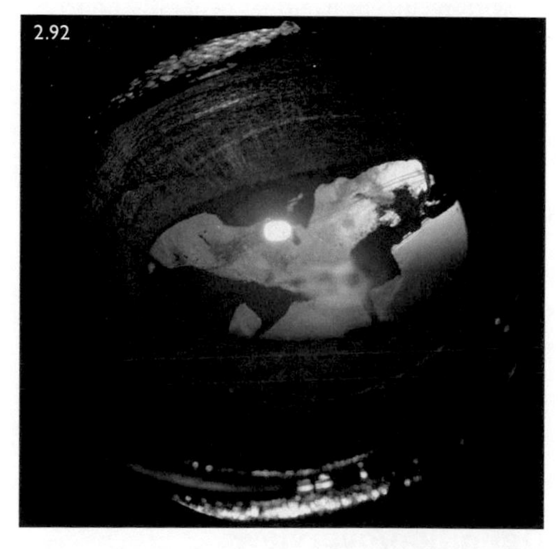

FIG. 2.92 Chronic uveitis with multiple posterior synechiae and iris rests, leading to a number of anterior capsular and cortical cataracts of varying density.

## Differential diagnosis

- any other cause of vision deficits.

## Diagnosis

- identification of a unilateral or bilateral opacity in the lens.
- careful assessment for evidence of other ocular pathology.
- ocular ultrasonography and ERG useful in diagnosing posterior segment abnormalities.

## Management

- many cases of acquired cataracts are not significantly visually impaired:
  - ridden and managed normally.
  - monitoring of the eyes on an annual basis:
    - will confirm any progression of lens or other ocular pathology.
- some horses with cataracts may become visually impaired:
  - cannot be ridden or used for their intended purpose.
  - occasional horses may be dangerous and prone to self-injury.
- horses with unilateral or bilateral immature or mature cataracts that interfere with vision should be referred to a veterinary ophthalmologist promptly for evaluation:
  - confirm the diagnosis and discuss treatment options.

- surgical cataract removal by phaecoemulsification may be recommended to restore functional vision.
- cataracts secondary to uveitis are poor candidates for surgery.
- enucleation may be performed in horses with painful, non-visual eyes when cataract surgery is contraindicated.

### Prognosis

- where the cataract is non-progressive and not affecting vision, the prognosis is good.
- cataracts that are progressive or affecting vision carry a guarded to poor prognosis.
- age of the patient influences the success rate of surgery:
  - older animals (65% in non-uveitic patients).
  - generally, success is less likely in ERU patients (25%).

## Glaucoma

### Definition/overview

- elevation in the IOP.
- caused by a decrease or reduction in the outflow of aqueous humour.
- eventually results in optic nerve damage and blindness.

### Aetiology/pathophysiology

- aqueous humour is produced constantly and flows from the posterior chamber into the anterior chamber between the lens and iris.
- drains from the eye through the ICA and uveoscleral outflow pathways.
- categorised as congenital or acquired.
- **congenital** glaucoma occurs rarely in horses:
  - reported in Thoroughbred, Arabian and Standardbred foals (see page 92).
  - often associated with multiple congenital ocular anomalies.
- **acquired** glaucoma may be categorised as:
  - **primary** due to an abnormal aqueous drainage pathway:
    - uncommon and often bilateral.
  - **secondary** due to other ocular diseases that cause either:
    - mechanical blockage of the pupil and/or ICA.
    - functional obstruction of the ICA and/or uveoscleral pathway:
      - damage from scarring, vascularisation or accumulation of inflammatory cells and debris.
    - most frequent as a sequela to anterior uveitis.
    - occasionally following trauma, intraocular neoplasia or lens luxation/subluxation.
    - increased risk of developing glaucoma in horses with:
      - previous or concurrent ERU.
      - animals older than 15 years.
      - Appaloosas.
    - secondary glaucoma may be bilateral in some cases of bilateral ERU.
- obstruction leads to retention of aqueous humour and subsequent increase in the IOP within the eye:
  - decreases retinal ganglion cell function, causing optic nerve axon degeneration and progressive visual deterioration.

### Clinical presentation (Fig. 2.93)

- vision impairment or blindness.
- buphthalmos.
- blepharospasm • mydriasis • epiphora.
- conjunctival hyperaemia.
- episcleral congestion.
- corneal oedema.
- Haab's striae (breaks in Descemet's membrane).
- uveitis • lens subluxation/luxation.
- tapetal hyperreflectivity, retinal degeneration/atrophy and optic nerve cupping/atrophy.
- secondary glaucoma may have a history and clinical signs of other ocular pathology:
  - chronic or recurrent uveitis.
  - posterior synechiae (adhesions), a miotic pupil and cataract formation.
  - ulcerative exposure keratitis and lens subluxation/luxation can also occur late in the disease (Fig. 2.94).
  - may or may not be painful.

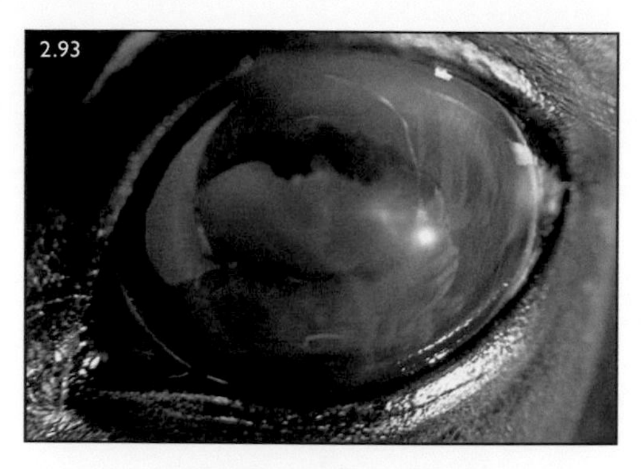


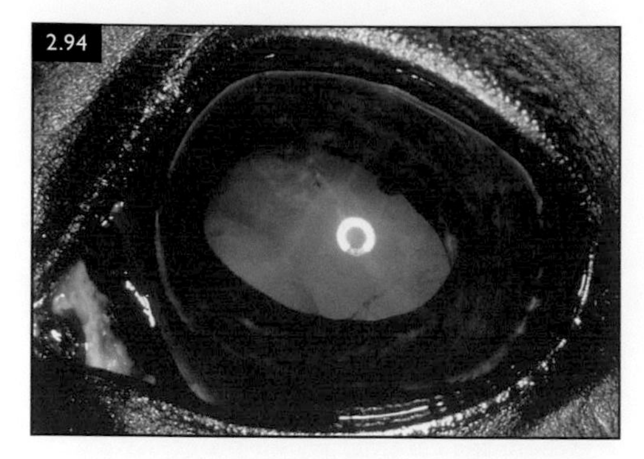


**FIGS. 2.93, 2.94** Glaucoma. A raised IOP results in globe-stretching and creates corneal striae (Haab's striae), linear track-like lesions associated with oedema (2.93). Posterior luxation with the dorsal edge of the lens visible in the ventromedial aspect of the pupil in a horse with glaucoma (2.94).

## Differential diagnosis

- consider in any case of unexplained corneal oedema, vision impairment or severe unrelenting ocular inflammation.

## Diagnosis

- based on the history, clinical appearance and applanation tonometry (elevation in IOP):
  - multiple episodes of intraocular inflammation.
  - followed by a severe unrelenting bout of ocular cloudiness and discomfort:
    - does not respond to traditional uveitis therapy.
- thorough and complete ophthalmic examination to rule out other causes and determine whether the glaucoma is primary or secondary.
- IOP averages 24 mmHg in the equine eye (normal range 15–30 mmHg):
  - Tono-Pen measurement greater than 35 mmHg is consistent with glaucoma.
- examination of the ICA may show abnormalities.
- ocular ultrasonography helps rule out other intraocular diseases (e.g. intraocular tumour).

## Management

- essential to determine the cause of the glaucoma:
  - therapy will vary according to aetiology.
  - most common cause in horses is ERU.
- treatment is centred on decreasing production of aqueous humour or increasing outflow.
- inconsistent response to antiglaucoma medications.
- medical treatment may include:
  - topical beta-adrenergic blockers (e.g. 0.5% timolol maleate q12 h).
  - topical carbonic anhydrase inhibitors:
    - 2% dorzolamide or 1% brinzolamide q8–12 h.
  - timolol/dorzolamide combination medication is available for use in the horse:
    - help decrease the number of medications necessary.
  - systemic carbonic anhydrase inhibitors are also available:
    - acetazolamide 2–3 mg/kg p/o q6–12 h.
    - dichlorphenamide 1 mg/kg p/o q12 h.
    - methazolamide 0.25 mg/kg p/o q24 h.
    - additional potassium supplementation is required.
  - topical and systemic anti-inflammatory medications should be used initially to help control intraocular inflammation and increase patient comfort.
- surgical treatments available on referral that do not respond to medications include:
  - laser cyclophotoablation.

- cyclocryoablation.
- surgical techniques to increase aqueous outflow:
  - gonioimplants, sclerostomies and iridectomies.
- horses should have their IOP measured regularly to monitor the response to therapy.
- affected eyes will often become blind and chronically painful:
  - enucleation or evisceration is the treatment of choice in these cases.
- primary glaucoma is suspected:
  - repeated measurements of the IOP should be taken in the fellow (predisposed) eye 3–4 times per year for life or until the eye becomes glaucomatous.

### Prognosis

- guarded for vision.
- most effective long-term therapy is cyclophotoablation in combination with topical antiglaucoma medications.

# NEUROLOGIC DISORDERS OF THE EYE

## Horner's syndrome

### Definition/overview

- not a specific disease, but a syndrome that involves the loss or disruption of sympathetic innervation to the eye and adnexa.
- characterised by ptosis of the upper eyelid, ipsilateral facial sweating, mild miosis, enophthalmos and regional hyperthermia.
- unilateral or bilateral and may or may not be permanent.

### Aetiology/pathophysiology

- possible causes of Horner's syndrome in the horse (Table 2.9).
- sympathetic innervation to the eye and adnexa can be divided into three neuroanatomical sections:
  - **central component** consists of fibres descending from the brainstem, down the spinal tract to synapse at spinal cord segments T1–T3.
  - **pre-ganglionic** sympathetic axons leave the spinal cord and enter the sympathetic trunk in the dorsal thorax, then travel through the cervicothoracic and middle cervical ganglia and up the neck in the vagosympathetic trunk, to synapse in the cranial cervical ganglion.
  - **post-ganglionic** sympathetic axons then continue through the middle ear, join the ophthalmic branch of the trigeminal nerve and distribute to the sweat glands of the head, smooth muscles of the periorbita and eyelids, and the iris dilator muscle.
- lesions causing denervation and Horner's syndrome can occur anywhere along this pathway.
- loss of sympathetic innervation to Muller's muscle of the upper eyelid and tissue of the lower eyelid results in:
  - narrowing of the palpebral fissure and ptosis (drooping of the upper eyelid).
- loss of sympathetic innervation to the sweat glands of the head:
  - ipsilateral facial sweating and regional hyperthermia.
  - caused by vasodilation and increased cutaneous blood flow.

| **TABLE 2.9** Causes of Horner's syndrome |
|---|
| • Severe head, neck and chest trauma |
| • Cranial thoracic neoplasia/space-occupying masses |
| • Otitis media/interna |
| • Cervical neoplasia or abscesses |
| • Drug injection into the carotid artery or jugular vein |
| • Guttural pouch disease or surgery (e.g. carotid artery ligation for facial surgery or guttural pouch epistaxis) |
| • Oesophageal rupture, obstruction or surgery |
| • Periorbital abscesses or tumours |
| • Post-anaesthetic myopathy |
| • Equine protozoal meningoencephalitis |
| • Cauda equina neuritis/polyneuritis equi |
| • Systemic aspergillosis |
| • CNS infection or neoplasia |

- lack of tone in the orbital smooth muscle causes the eye to retract slightly, leading to enophthalmos.
- loss of normal sympathetic tone to the iris dilator muscle results in ipsilateral miosis and anisocoria.

## Clinical presentation (Fig. 2.95)

- variable and often subtle, but can include:
  - increased lacrimation.
  - hyperaemia of nasal and conjunctival mucosa.
  - ipsilateral sweating at the base of the ear, face and neck.
  - increased cutaneous temperature on the affected side.
  - ptosis
  - miosis.
  - anisocoria
  - enophthalmos.
  - inspiratory stridor.
  - dermatitis due to chronic sweating.
  - prolapse of the third eyelid, less common in horses than in other species.

## Differential diagnosis

- anterior uveitis, corneal ulceration and other causes of anisocoria.

## Diagnosis

- based on history and complete physical, neurological and ophthalmological examinations.
- dilation of the pupil of the affected side will occur with dim lighting, but not as extensive as in the normal eye.
- pharmacological testing using topical phenylephrine (direct-acting sympathomimetic agent):
  - may help to determine whether the lesion is pre- or post-ganglionic.
  - treat both eyes for comparison.
  - dilute (0.1%) topical phenylephrine will cause more rapid and extensive pupil dilation on the affected side owing to denervation hypersensitivity.
  - with post-ganglionic lesions, mydriasis will occur within 20 minutes of administration.
  - the onset of dilation is at 30–50 minutes in animals with pre-ganglionic lesions.
- examine endoscopically the guttural pouches and the pharynx of all patients.

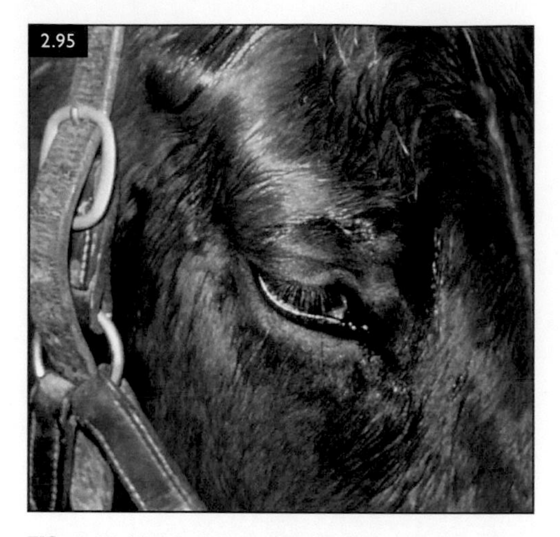


**FIG. 2.95** Horner's syndrome in horses is characterised by ipsilateral sweating of the face and neck, as well as enophthalmos, miosis and ptosis.

- jugular furrows should be palpated for swellings:
  - history of recent intravenous or intramuscular injections in the neck.
- radiographs of the cervical vertebrae or thorax may also be indicated.

## Management

- treatment depends on the underlying cause.
- no specific treatment for Horner's syndrome:
  - topical phenylephrine may be used therapeutically for temporary alleviation of the associated clinical signs.

## Prognosis

- depends on the underlying cause.
- generally guarded.
- neurological signs are often irreversible even when the primary cause has been treated and eliminated.

# Photic headshaking

## Definition/overview

- Trigeminal-mediated headshaking (TMH) is a poorly understood acquired neuropathic facial pain condition of the horse.

- Photic headshaking, where the same clinical signs appear to be induced in response to exposure to light, is thought most likely to be a small subset of this condition.
- aetiopathogenesis of TMH is still unclear.
- may be more common in geldings.
- clinical signs are usually characteristic, although vary in severity, and occur most commonly at exercise.

## Aetiology/pathophysiology

- unclear.
- infraorbital nerve of TMH-affected horses may be sensitised:
  - result in neuropathic pain.
  - sensitisation threshold may be normal in seasonally affected cases when they are not showing clinical signs out of season.
- pathogenesis for photic headshakers is unknown:
  - an untested theory is that exposure to sunlight stimulates parasympathetic activity in the infraorbital nerve or sensory branch of the trigeminal nerve.
  - causes an irritating nasal sensation to the horse, leading to clinical signs.

## Clinical presentation

- TMH may affect around 1% of the UK equine population.
- proportion of these which are photic headshakers is unknown.
- horses are usually more severely affected at exercise.
- onset usually in the young mature horse (4–9 years of age, although with a wide range).
- up to 60% of those affected are more severely, or even only, affected in the spring and summer.
- photic cases appear to show clinical signs following exposure to light, with relief being obtained by change to a darkened environment, eye mask, tinted contact lenses or even blindfolding.
- clinical signs are predominantly of vertical headshaking, which may be violent:
  - often accompanied by sharp vertical flicks.
  - signs of nasal irritation which may include snorting, sneezing, rubbing the nose on the legs, ground or other surfaces and striking at the nose.
  - sufficient severity to appear to cause distress to the horse and make handling or riding impossible and even dangerous.

## Differential diagnosis

- conditions that can cause horses to shake their head may include:
  - ear mite infestation
  - otitis interna.
  - cranial nerve dysfunction.
  - cervical injury.
  - damage to trigeminal nerve.
  - guttural pouch mycosis
  - sinusitis.
  - ocular disease including iris cyst.
  - dental periapical osteitis.
  - protozoal myeloencephalitis.
  - orthopaedic pain.
  - behavioural or rider issues.

## Diagnosis

- diagnosis of TMH is made by exclusion.
- history, signalment and observation of clinical signs:
  - trigeminal-mediated headshakers usually show strongly characteristic clinical signs.
  - role of light in a photic headshaker would be determined at this stage, with signs being alleviated by use of masks or tinted contact lenses.
- diagnostic local anaesthesia at the posterior part of the infraorbital nerve:
  - may be performed for investigation where cases are:
    - demonstrating consistent clinical signs in a consistent situation.
    - amenable to the procedure.
  - if signs resolve, confirms they are due to facial pain.
  - does not determine the origin of that facial pain.
  - failure to respond to blocking does not rule out clinical signs being due to facial pain.
- further investigations to eliminate other causes of headshaking include clinical, oral and ophthalmic examinations, upper respiratory tract and guttural pouch endoscopy, and CT of the head.

### Management

- true photic headshaker should be manageable by:
  - eliminating exposure of the eyes to sunlight by maintaining the horse in a darkened environment.
  - use of an eye mask, tinted contact lenses or even blindfolding:
    - these may not be practical or sufficient.
- treatments used for TMH include:
    - nose nets.
    - gabapentin or carbemazepine alone or in combination with cyproheptadine.
    - EquiPens™ neuromodulation.
    - caudal ablation of the infraorbital nerve via coil compression.
    - all of these treatments have different success and complication rates.

### Prognosis

- guarded for all TMH cases as no treatment has proven consistently effective:
  - 5% of horses with TMH expected to recover spontaneously.
- unknown for photic headshaking.
- horses can become severely affected and cannot be ridden at intended level or at all.
- affected horses may be retired if clinical signs occur only at exercise.
- where signs occur even at rest, euthanasia on humane grounds should be considered.

## Equine motor neuron disease (EMND)

### Definition/overview

- oxidative neurodegenerative disorder of the somatic lower motor neurons in horses.
- deprived of adequate dietary vitamin E for an extended period.
- neurological disease is the primary complaint in most cases (see page 57):
  - ophthalmological disease can occur concurrently.

### Aetiology/pathophysiology

- vitamin E is a fat-soluble antioxidant that counteracts the harmful free radicals normally produced during metabolism in animals.
- deficiency in protective antioxidants may predispose animals to neurotoxic and/or oxidative injury.
- photoreceptor outer segments in the neurosensory retina are extremely susceptible to oxidative stress.
- EMND cases have accumulations of ceroid-lipofuscin in the retinal pigmented epithelial cells over the tapetal and non-tapetal fundus due to light-generated oxidative injury to the retina:
  - increased retinal pigmentation visible on fundoscopy in affected animals.
- remaining clinical findings of EMND are the result of dysfunction and/or death of somatic efferent motor neurons, which leads to axonal degeneration in the ventral roots and the peripheral and cranial nerves.

### Clinical presentation

- retinal lesions are common (50% of cases) and seen on fundoscopic examination:
  - yellow-brown to black pigmentation.
  - irregular mosaic (reticulated) pattern and/or horizontal band at the tapetal–non-tapetal junction or generalised throughout the fundus (Fig. 2.96).
  - effect on vision appears variable, with most cases not showing obvious deficits.
  - PLRs may be abnormal.

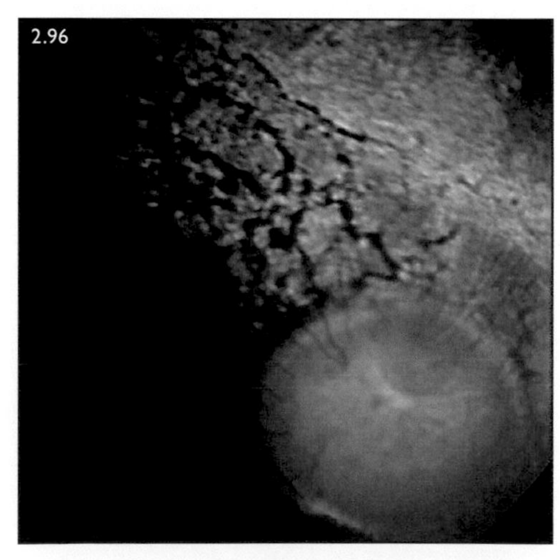


**FIG. 2.96** A pigmented mosaic pattern is apparent in the peripapillary region of this fundus and is typical of the lesions seen in EMND. (Photo courtesy A Gemensky-Metzler.)

## Differential diagnosis

- Equine protozoal meningitis.
- Equine grass sickness/dysautonomia.
- lead toxicosis • botulism.
- laminitis and other causes of lameness.
- rhabdomyolysis.
- PSSM and other chronic myopathies.
- colic • iliac thrombosis.
- senile retinopathy.

## Diagnosis

- based on history, clinical appearance (i.e. musculoskeletal signs), fundoscopy, muscle biopsies and laboratory results.
- fundic lesions alone can be suggestive of EMND.
- definitive diagnosis only made post-mortem with retinal histopathology.

## Management

- involves dietary supplementation with vitamin E and access to pasture or fresh forage.

## Prognosis

- variable:
  - 40% of cases have a marked improvement in clinical signs within 4–6 weeks after relocation to another stable and/or administration of dietary antioxidants.
  - 40% of horses are euthanased or die owing to continued deterioration (i.e. inability to stand or respiratory distress) within 4 weeks of the onset of clinical signs.
  - some horses survive and regain weight:
    - disease progression is arrested.
    - may never fully compensate for the irreversible loss of motor neurons and often suffer permanent chronic debilitation.

# Traumatic optic neuropathy

## Definition/overview

- occurs following severe blunt head trauma:
  - trauma to the poll of the skull is often listed in the history.
- concussive cranial injuries cause damage to the optic nerve(s) or chiasm.
- acute onset of unilateral or bilateral blindness immediately following or soon after injury.
- Optic nerve atrophy occurs within a few weeks:
  - manifests as a pale ONH.
- Peripapillary chorioretinitis may also occur with chronicity.

## Aetiology/pathophysiology

- severe blunt head trauma caused by rearing up or falling over backwards and striking the occipital region.
- allows posterior movement of the brain away from the fixed intracanalicular portion of the optic nerves.
- may cause stretching, shearing and/or avulsion of the retinal ganglion cell axons/optic nerve(s) or chiasm, resulting in optic nerve atrophy and sudden blindness.
- partial or complete visual loss occurs in the affected eye(s) within 24 hours of injury.

## Clinical presentation

- present with a history of sudden onset of blindness with or without a known history of trauma.
- pupil(s) is (are) fixed and dilated with sluggish to absent PLRs in the affected eye(s).
- ophthalmic lesions:
  - often not seen initially (retrobulbar nature of the injury in many cases).
  - when present within 24–48 hours of injury:
    - peripapillary and/or ONH oedema or haemorrhage (Fig. 2.97).
    - exudation into the vitreous.
  - with chronicity (Fig. 2.98):
    - lamina cribrosa becomes more prominent.
    - ONH will appear pale and atrophied.
    - peripapillary retinal vessels will appear diminished/attenuated.
    - focal grey patches medial, lateral and ventral to the ONH:
      - indicates choroidal degeneration.

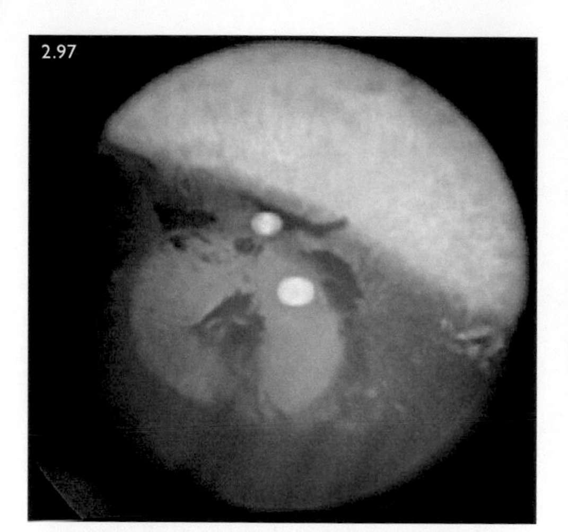

FIG. 2.97 Traumatic optic neuropathy. This horse presented blind and was suspected to have fallen backwards after rearing up. Note the peripapillary and ONH haemorrhages, as well as haemorrhagic streaming into the vitreous.

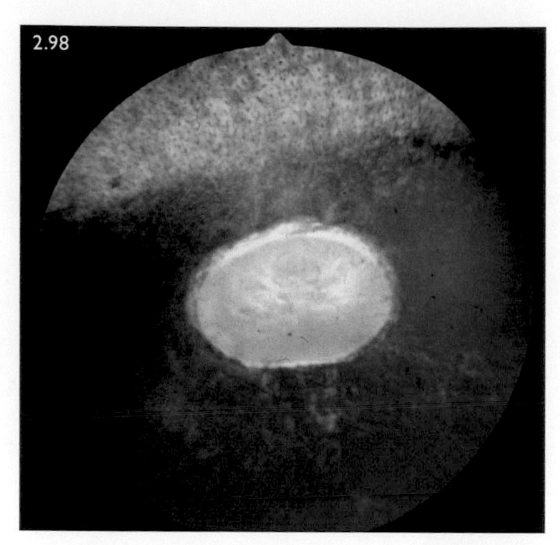

FIG. 2.98 Chronic traumatic optic neuropathy. This yearling fell over backwards 2 months previously and was immediately blind but without any fundic changes. The fundus of both eyes now shows classic changes of the condition in the optic disc and tapetal/non-tapetal fundus.

## Differential diagnosis

- optic nerve and retinal degeneration in the horse reported secondary to:
  - ERU
  - glaucoma.
  - hypovolaemia/blood loss.
  - exposure to toxins.
  - progressive retinal atrophy and carotid artery ligation.
  - other differentials include exudative optic neuropathy and optic neuritis.
  - brain injuries.

## Diagnosis

- based on history and findings on physical and ophthalmic examination.

## Management

- treatment when cases are presented acutely traditionally involves:
  - high doses of anti-inflammatories such as:
    - systemic corticosteroids (e.g. dexamethasone).
    - NSAIDs.
    - DMSO.
  - decrease optic nerve swelling and inflammation.
- no treatment is available for chronic cases.

## Prognosis

- restoration of vision is guarded to poor.

# Exudative optic neuritis/ neuropathy

## Definition/overview

- sudden onset of unilateral or bilateral blindness in middle-aged or older horses.
- most prominent finding is marked exudation with/without haemorrhages present over the surface of the ONH.

## Aetiology/pathophysiology

- cause is unknown.
- pathophysiology is unclear.

## Clinical presentation

- acute onset blindness.
- retinal lesions on fundoscopy may vary:
  - white to grey exudates that radiate from the ONH and raised into vitreous.
  - may obscure the ONH completely.
  - oedema of ONH and multiple small haemorrhages are often present if disc is visible.

  - ONH will appear atrophied with chronicity.

## Differential diagnosis

- sepsis and other causes of optic neuritis (Table 2.10).
- traumatic optic neuropathy.
- benign exudative/proliferative optic neuropathy.
- ONH tumours.

**TABLE 2.10** Causes of optic neuritis

- Idiopathic/immune-mediated
- Fungal
  - Systemic aspergillosis, *Cryptococcus neoformans*
- *Toxoplasma gondii*
- *Onchocerca cervicalis*
- *Leptospira* spp.
- Neoplasia
- Toxins
  - lead, arsenic, thallium, ethyl/methyl alcohol, chlorinated hydrocarbon?
- Sepsis
- Other bacterial
  - *Streptococcus equi, Actinomyces* spp., *Actinobacillus equuli, Rhodococcus equi*
- Vitamin A deficiency?
- Orbital inflammation
- Trauma
- Vascular compromise/ischaemia
- Borna disease
- Parasite migration
- Brain abscess/meningitis caused by *Pseudomonas mallei* (glanders), *Streptococcus equi*, cryptococcosis
- Equine protozoal encephalomyelitis
- Hepatoencephalopathy, leucoencephalomalacia, hydrocephalus, idiopathic epilepsy
- Verminous migration
- Profound blood loss
- Rabies

## Diagnosis

- based on history and clinical findings (Fig. 2.99).
- exclusion of optic neuritis as a differential requires:
  - neurological examination.
  - additional investigations that may include:
    - serology and toxicological testing.
    - ERG.
    - cerebrospinal fluid sampling.
    - CT/MRI.

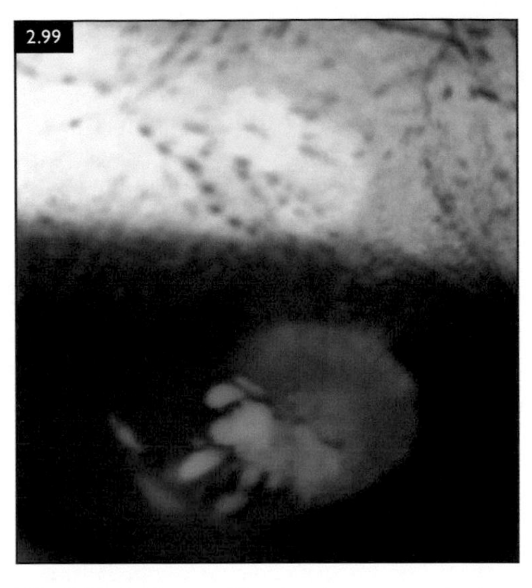


**FIG. 2.99** Optic neuritis of unknown cause in a young Welsh Section A pony. Note the exudation of white-cream material and haemorrhage into the vitreous from the ONH.

## Management

- no treatment.

## Prognosis

- poor because the disease typically progresses to ONH atrophy.

# Benign proliferative optic neuropathy

## Definition/overview

- white or grey material protruding anterior to the optic disc and into the vitreous in an otherwise normal fundus.
- exudate may slowly enlarge over months or years.
- generally considered to be a benign lesion.

## Aetiology/pathophysiology

- cause is unknown.
- pathophysiology is unclear.

## Clinical presentation

- no or minimal effect on vision:
  - unless it becomes large enough to obscure portions of the retina.

- white or grey masses on or near the optic disc and protruding into the vitreous humour:
  - typically attached at the periphery of the optic disc.
  - vascularised and can be pedunculated or multi-lobular.
- incidental finding seen unilaterally.
- primarily in middle-aged or older horses (>15 years old).

## Differential diagnosis

- differentiate from exudative optic neuritis, traumatic optic neuropathy and optic nerve neoplasia.

## Diagnosis

- based on history and clinical appearance (Fig. 2.100).

## Management

- no therapy is available or necessary.

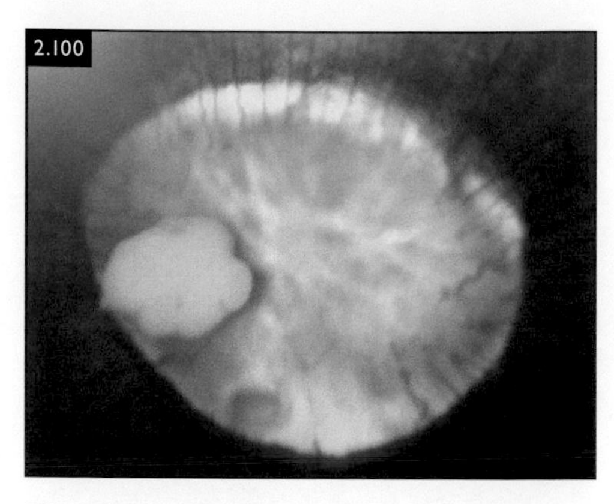

**FIG. 2.100** Proliferative optic neuropathy present at 8–9 o'clock was identified in this pony as an incidental finding. There is also a small optic nerve coloboma at the 6 o'clock position (typical) that was not associated with any identifiable visual deficits.

## Prognosis

- benign, incidental finding.
- non-progressive and the prognosis for vision is excellent.

# PARASITIC DISEASES OF THE EYE

## Onchocerciasis

### Definition/overview

- caused by aberrant migration of the microfilariae of the parasite *Onchocerca cervicalis*.
- non-seasonal and non-pruritic disease, with the incidence increasing with age.

### Aetiology/pathophysiology

- adult form of *O. cervicalis* resides harmlessly in the ligamentum nuchae.
- microfilariae migrate through the subcutaneous tissues to the dermis and become ingested by the intermediate host, a biting midge (*Culicoides* spp.).
- microfilariae are then transmitted to the horse by a bite and develop into adults.
- aberrant migration of the microfilariae may involve the eyelids, conjunctiva, cornea, sclera, anterior chamber, uvea and/or fundus, producing ocular signs of the disease.

### Clinical presentation

- ocular signs include:
  - lacrimation
  - blepharospasm.
  - conjunctival thickening (Fig. 2.101).
  - depigmentation of the temporal limbus (vitiligo) (Fig. 2.102).
  - conjunctivitis.
  - corneal oedema, vascularisation and stromal cellular infiltration (keratitis).
  - small nodules and corneal opacities.
  - anterior uveitis and peripapillary chorioretinitis.
- other clinical signs may include:
  - lesions of diffuse, patchy alopecia, erythema and scaling along the ventral midline, face, base of the mane and craniomedial forearm.
  - cranial pectoral 'bull's-eye' lesion in the centre of the forehead.

### Differential diagnosis

- SCC, habronemiasis, mycotic infection and other causes of keratitis, uveitis and chorioretinitis.

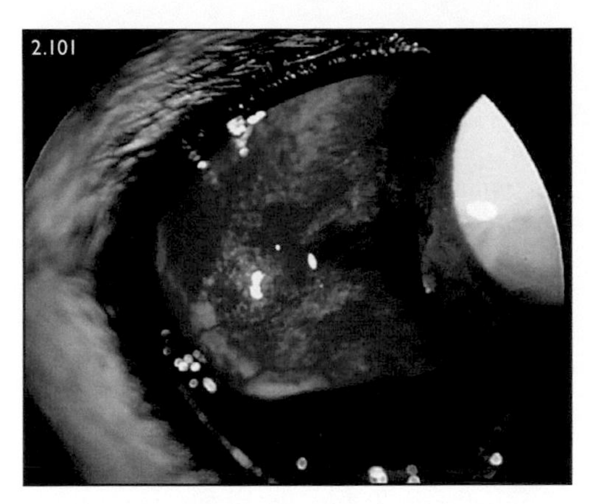


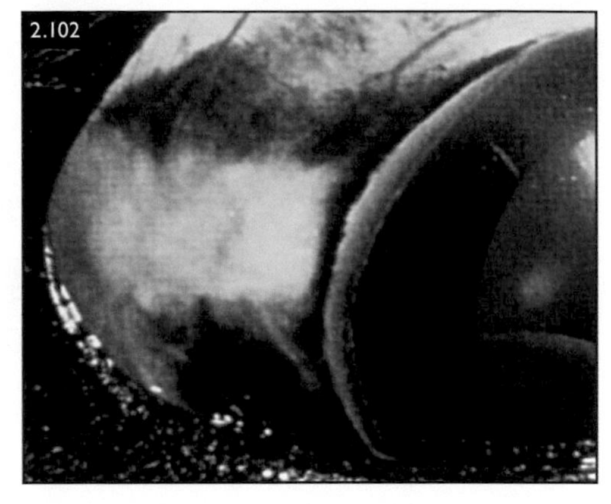


**FIGS. 2.101, 2.102** Onchocerciasis. (2.101) Active conjunctival hyperaemia and chemosis. (2.102) Vitiligo at the lateral limbus is a sequela to *Onchocerca*-related inflammation. (Photos courtesy American College of Veterinary Ophthalmologists.)

## Diagnosis

- history, clinical signs, exclusion of other differential diagnoses and response to therapy.
- conjunctival biopsy illustrating free microfilariae (Fig. 2.103), eosinophils and lymphocytes is diagnostic.

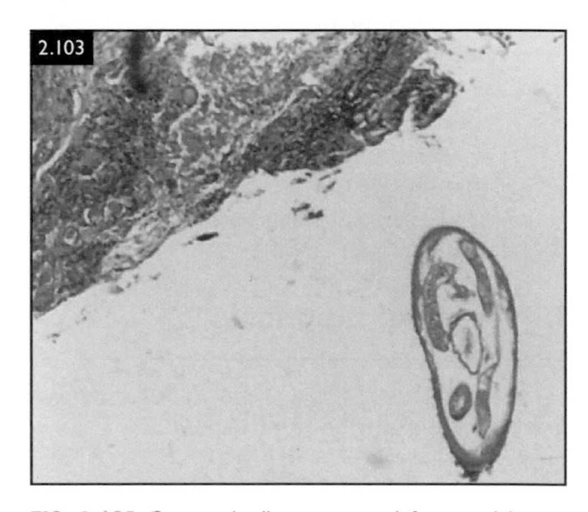


**FIG. 2.103** *O. cervicalis* recovered from a biopsy of the conjunctiva at the lateral limbus. (Photo courtesy American College of Veterinary Ophthalmologists).

## Management

- microfilariae may be eliminated using ivermectin (0.2 mg/kg p/o):
  - minor adverse reactions (fever and swelling) occur in 25% of horses.
- no treatment for the adult parasites:
  - recurrence is possible.
- topical and/or systemic anti-inflammatories may help control the inflammation incited by the dying microfilariae.
- ocular signs should be treated symptomatically.

## Prognosis

- most horses will improve within 2–3 weeks of treatment.
- disease may recur in 2 months.
- routine deworming practices are therefore recommended for all horses.

# Habronemiasis

## Definition/overview

- caused by the aberrant migration of nematode larvae.
- tends to occur in the warmer summer months ('summer sores').

## Aetiology/pathophysiology

- caused by the aberrant migration of nematode larvae of the species *Habronema muscae*, *H. microstoma* and *Draschia megastoma*:
  - adult parasite resides in the stomach of the horse.
  - eggs or larvae are passed in the faeces and ingested by the larvae of the intermediate host:

    - either house fly *Musca domestica* or the stable fly *Stomoxys calcitrans*.
  - horses are infected following:
    - ingestion of an infected adult fly.
    - infectious L3 larvae deposited on wounds around the eye or near the mouth.
    - larval migration through tissue incites a granulomatous inflammatory response, which can become walled off.
    - lesions of medial canthus skin result from inflammation and ulceration caused by larval migration into the NLS.

## Clinical presentation

- ocular signs include:
  - raised, irregular, proliferative wounds or nodular masses on the medial canthus:
    - may be ulcerated.
  - conjunctivitis.
  - pruritus and lesions associated with self-trauma.

## Differential diagnosis

- SCC, sarcoids, onchocerciasis, phycomycosis, foreign body reaction and exuberant granulation tissue.

## Diagnosis

- history and clinical signs are suggestive.
- larvae are easily missed on conjunctival scraping and/or faecal examination.
- conjunctival biopsy may reveal:
  - eosinophilic infiltrates, mast cells and granulation tissue.
  - and/or gritty caseated lesions (almost pathognomonic) 'sulphur granules'.
- gastroscopy may reveal eggs or larvae.

## Management

- systemic ivermectin (0.2 mg/kg p/o) or moxidectin (0.4 mg/kg p/o) is treatment of choice.
- topical (e.g. flurbiprofen) and/or systemic (e.g. flunixin meglumine or phenylbutazone) NSAIDS can help decrease the inflammatory reaction associated with treatment.
- surgical debulking or removal of lesions may be performed for large granulomas prior to ivermectin therapy.
- preventive measures should include fly control, regular removal of manure, an appropriate anthelmintic treatment regime and topical organophosphates.

## Prognosis

- good.

CHAPTER 3

# Cardiovascular Disorders

## Introduction

- cardiovascular disease is relatively rare in the horse.
- cardiac murmurs and arrhythmias are common:
  - often physiological or reflect disturbances in other systems.
  - resolve with treatment of the underlying condition.
- congenital anomalies are uncommon.

## Cardiological examination

### Clinical history and presentation

- cardiac conditions are usually discovered incidentally:
  - during routine or pre-purchase examinations.
- less often, they are suspected primarily based on compatible clinical signs:
  - range from subtle changes in performance to overt clinical signs of heart failure.
  - thorough history and performing a comprehensive physical examination are critical first steps.
- during history-taking, it is important to establish the following facts.
- **signalment of the animal:**
  - breed:
    - congenital heart disease is overrepresented in Arabians.
  - age:
    - older horses more likely to have severe and haemodynamically significant cardiac murmurs.
    - young, athletic horses commonly have innocent, functional cardiac murmurs.
    - congenital disease is most detected in neonates.
  - sex:
    - aged stallions are more likely to have aortic aneurysm.
- discipline and level of work horse is currently undertaking:
  - has this changed?
  - how has the horse tolerated this?
- length of time in the owner's/trainer's possession.
- change in body condition.
- has the horse recently had or been exposed to:
  - viral diseases.
  - known or suspected toxins.
  - episodes of collapse or syncope.
  - history of respiratory disease (infectious or non-infectious).
- management of the horse:
  - diet, turnout, training, recent competition, etc.
- any known cardiac condition (such as valvular regurgitation or atrial fibrillation):
  - duration of the condition.

### General physical examination

- general appearance, mentation and body condition score.
- respiratory rate, effort and presence/character of nasal discharge if applicable:
  - auscultate the lung fields.
  - rebreathing examination often indicated.
- obtain vital parameters:
  - heart and respiratory rate and rectal temperature.
  - assess hydration status and cardiovascular auscultation.
- detailed evaluation of cardiovascular parameters includes:
  - **jugular filling** on both sides:
    - careful attention to detect thrombosis, distension and pulsation.

DOI: 10.1201/9781003451921-3

- does the vein fill too quickly or, if occluded rostrally, does it empty very slowly:
  - signs of congestive heart failure or increased preload.
- vein distended without pulsation:
  - occlusion to venous return.
  - bilateral – thoracic inlet obstruction.
  - unilateral – complete thrombosis.
- vein fills very slowly:
  - sign of reduced preload.
- normal to see slight jugular pulsation extend to the lower one third of the jugular groove in non-sedated horses with a neutral head position (withers height).
- sedated horses with their head down will have jugular pulsation extending two thirds of the way or more up the jugular groove:
  - resolves once sedation wears off.

○ **mucous membrane colour and capillary refill time:**
- normal: <2 seconds.
- not specific to cardiovascular disease.
- helps assess general hydration, and circulatory and inflammatory disturbances.

○ **peripheral arterial pulse** quality:
- transverse facial artery or facial artery (Fig. 3.1).
- may be possible to palpate the pulse of the carotid artery in the jugular groove.
- assess rate, rhythm and strength of pulse.
- poor pulse quality could reflect poor pump function, reduced cardiac output or poor vascular tone.

○ **palpation of the femoral artery or aortic bifurcation** (via palpation per rectum):
- thromboembolic disease is suspected.

○ **palpate the extremities** for assessment of temperature.

○ **examine the superficial venous vasculature** for signs of:
- distension and prominent superficial facial veins.

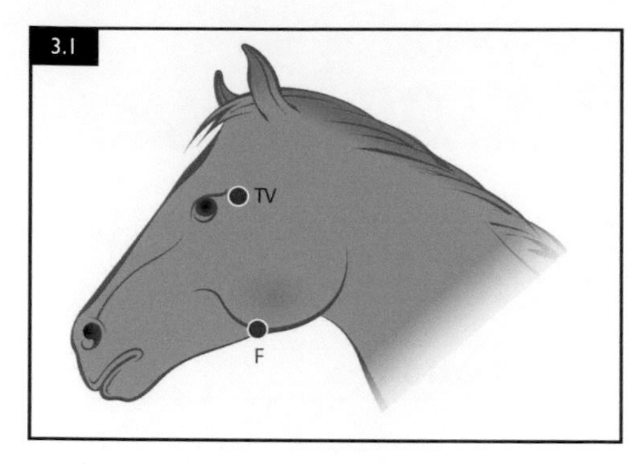


**FIG. 3.1** Areas on the head for palpation of the peripheral pulse. The transverse facial artery (TV) is palpable caudoventral to the eye (this vessel is also useful for arterial blood sampling in adults). The facial artery (F) is palpable over the mandible rostral to the masseter muscle.

- apparent in horses with thrombus formation in the jugular vein.

○ **thoracic percussion:**
- assess location and distribution of the apex beat.
- detect presence of cardiac thrill.
- cardiac impulse normally palpable:
  - left side at 5th or 6th intercostal space above the elbow.
  - right side is slightly further forward, lower and normally weaker.

○ **evaluate for evidence of peripheral oedema:**
- distal limbs, ventral abdomen or pectoral region.
- not pathognomonic for cardiac disease.

- following a comprehensive physical examination, the veterinarian should be able to have a general idea of whether there are signs of:
  - cardiac failure.
  - physiological response to underlying disease process (e.g. fever, hypovolaemia).
  - incidental cardiovascular finding.
- **Note:**
  - fever of unknown origin without a murmur does not preclude a diagnosis of endocarditis.
  - respiratory and cardiac disease are often intimately related.

## Cardiac auscultation

- close attention to timing, heart sounds, rate, rhythm and character of sound (Fig. 3.2 and Tables 3.1–3.3).
- cardiac dysrhythmias may be transient:
  - listen for at least 60–120 seconds on both sides of the chest.
  - dysrhythmias can be easily missed at high or low heart rates:
    - lower heart rates:
      - irregularly irregular rhythm of AF can be mistaken for the regularly irregular rhythm of a second-degree atrioventricular (AV) block.
    - electrocardiogram is required whenever a cardiac dysrhythmia is suspected or when there is persistent tachycardia at rest (heart rate [HR] >48 bpm).
  - other clues of a rhythm disturbance include:
    - varied intensity of heart beats.
    - pulse deficits when simultaneously auscultating the heartbeat and palpating the arterial pulse.
  - physiologic dysrhythmias such as second-degree AV block should be:
    - abolished by light exercise or excitement (increase in sympathetic tone).

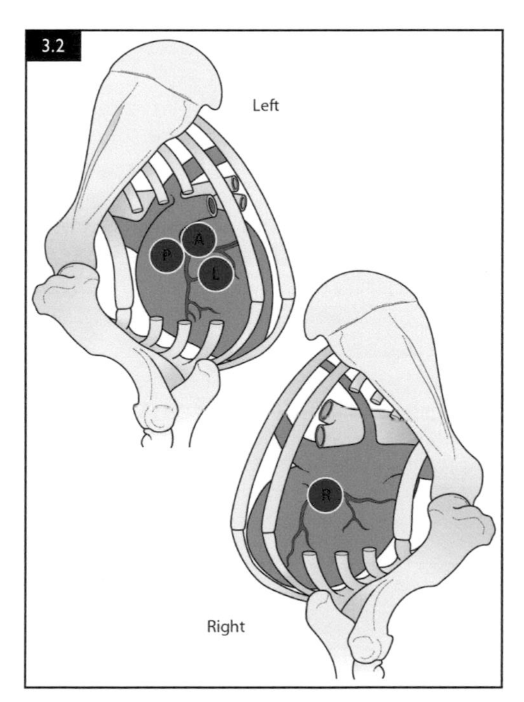


**FIG. 3.2** Location of heart valves and area of auscultation for each valve. The pulmonic valve (P) is auscultated well forward under the triceps muscle on the left side of the thorax in the area of the 3rd ICS. The aortic valve (A) is auscultated well under the triceps muscle on the left side of the thorax in the area of the 4th ICS just dorsal and caudal to the pulmonic valve. The left atrioventricular (AV) valve (L) is auscultated slightly under and just caudal to the triceps muscle on the left side of the thorax. The right AV valve (R) is auscultated under the triceps muscle on the right side of the thorax.

## Clinical signs of left-sided heart failure

- exercise intolerance.
- increased respiratory rate and effort.
- serous to mucoid to frothy nasal discharge (might just detect dried white discharge).
- reduced peripheral arterial pulse quality/ intensity (hypokinetic pulse).
- tachycardia with or without arrhythmia.
- cardiac murmur:
  - left-sided systolic murmur of variable intensity with point of maximal intensity over the heart base (just dorsal to mitral valve [MV]) most common.
  - lack of murmur possible if heart failure due to primary myocardial disease.

## Clinical signs of right-sided heart failure

- rapid jugular filling.
- jugular distension with pulsation.
- lethargy
- weight loss.

**TABLE 3.1** Location of heart valves

| VALVE | LOCATION | HEART SOUNDS |
|---|---|---|
| Pulmonic | Left 3rd ICS | S2 loudest |
| Aortic | Left 4th ICS | S2 loudest |
| Left atrioventricular (mitral) | Left 5th ICS | S1 loudest |
| Right atrioventricular (tricuspid) | Right 3rd–4th ICS | S1 loudest |

ICS = intercostal space.

- peripheral oedema (pectoral/sternum, distal limb, ventrum).
- pendulous appearance to abdomen (ascites).
- cardiac murmur:
  - variable (depends on the underlying aetiology).

## Diagnostic techniques

### Electrocardiography (ECG)

- indicated whenever there is a rhythm disturbance.
- single-lead electrocardiogram is used to assess cardiac rate and rhythm in horses.
- base–apex Y-lead (Fig. 3.3) is conventionally used:
  - generates a reliable electrocardiogram.
  - complexes are large and can be easily evaluated (Fig. 3.4).
  - some variation in exact lead placement is possible with alternative positions:
    - negative electrode (black) left side, at the apex of the heart (black heart).
    - positive electrode (red) is placed in the right jugular furrow or in front of the shoulder) (red neck).
    - ground electrode is placed on either side of the withers (white).
- 12-lead systems adapted for horses have been described:
  - potentially advantageous for detecting bundle branch blocks or ischaemia.

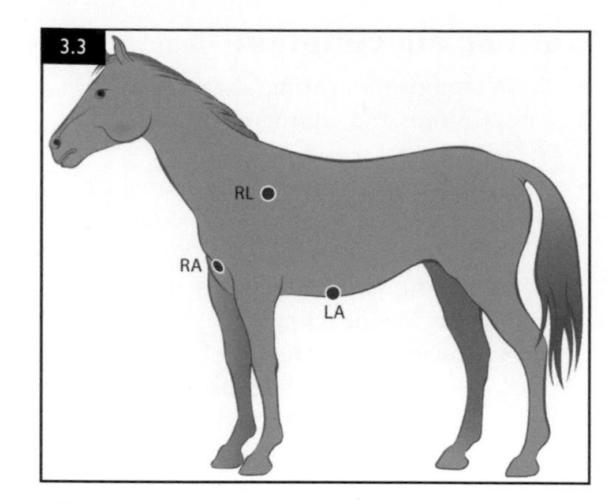


**FIG. 3.3** Locations for electrode placement for Y-lead ECG recording in the horse. The right leg (RL) is used as the ground lead and is here placed over the left shoulder. The right arm (RA) is attached over the manubrium and the left arm (LA) is attached over the xyphoid.

**TABLE 3.2** Murmur characterisation

| TERM | DEFINITION | DESCRIPTIONS |
|---|---|---|
| Timing | When it occurs in the cardiac cycle | • Systolic: between S1 and S2<br>○ Holosystolic: throughout systole, S1 and S2 audible<br>○ Pansystolic: throughout systole, S1 and S2 obscured<br>○ Early systolic<br>○ Mid-systolic<br>• Diastolic: occurs after S2<br>○ Holodiastolic<br>○ Early diastolic<br>• Continuous: present throughout all phases |
| Intensity | Ease of detection; graded from 1 to 6 | • Grade 1: not easily detected<br>• Grade 2: soft, but easily detected<br>• Grade 3: easily detected<br>• Grade 4: intense sound, louder than transients<br>• Grade 5: loud murmur with palpable thrill<br>• Grade 6: palpable thrill, murmur heard when stethoscope held away from body wall |
| Character | Shape of sound | • Crescendo–decrescendo: sound appears to intensify and then decrease in intensity<br>• Band/plateau: smooth/uniform intensity<br>• Decrescendo: sound appears to decrease in intensity<br>• Musical: variable high-pitch components<br>• Coarse: not smooth sounding, but intensity stable<br>• Complex: sounds vary or multiple sounds present |

**TABLE 3.3** Murmur types and causes

| PHASE OF CYCLE | TIMING | CHARACTER | INTENSITY (GRADE) | POINT OF MAXIMAL INTENSITY | RADIATION | LESION |
|---|---|---|---|---|---|---|
| Systolic | Early to mid-systolic | Crescendo–decrescendo or band/plateau | 1–3 | Heart base | No | Functional |
| | Holo-pansystolic | Band/plateau; complex if valve prolapse | 2–6 | LAVV | Yes; dorsal to valve, variable extent | LAVVR |
| | Holosystolic | Band/plateau; complex if valve prolapse | 2–6 | RAVV | Yes; dorsal to valve, variable extent | RAVVR |
| | Pansystolic | Band/plateau; may be coarse | 3–6 | Variable; right sternal border; left sternal border; less common | Yes; often widely and may radiate to the left | VSD |
| Diastolic | Early to mid-systolic | Whoop or decrescendo | 1–3 | Heart base | No | Functional |
| | Holodiastolic | Decrescendo or musical | 1–6 | Left heart base | Yes | AR |
| | Holodiastolic | Variable | 1–4 | Right heart base | Variable | Aortic root rupture |
| Continuous | Throughout; diastolic quieter | Variable | 1–4 | Left heart base | Variable | PDA |
| | Throughout | Variable, often complex | 1–4 | Heart base; often on right | Variable | Aortic root rupture |
| | Throughout | Variable | 1–6 | Right heart base | Variable | Arteriovenous fistula |

LAVV = left atrioventricular valve; RAVV = right atrioventricular valve; LAVVR = left atrioventricular valve regurgitation; RAVVR = right atrioventricular valve regurgitation; VSD = ventricular septal defect; AR = aortic regurgitation; PDA = patent ductus arteriosus.

- rhythm determination should occur at rest in a quiet environment:
  - muscle movement creates considerable artifact.
  - recorded at 25 mm/s and at an amplitude of 1 mV/cm for several minutes:
    - sufficient for persistent dysrhythmias or those present at examination.
  - 24-hour Holter monitor is required when:
    - complaint of poor performance, weakness or collapse.

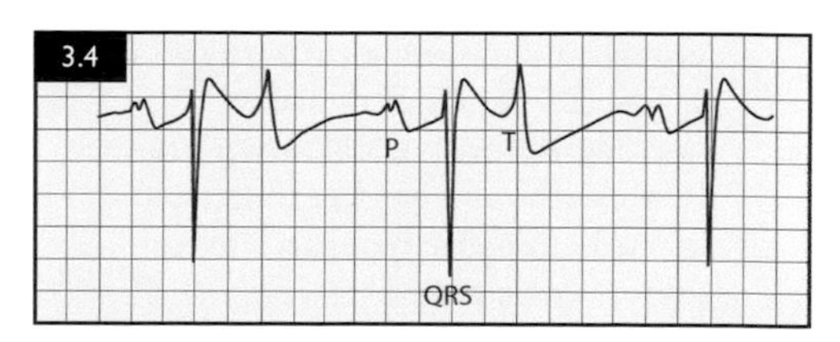


**FIG. 3.4** Base–apex (Y-lead) ECG recording from a horse in normal sinus rhythm. A bifid P wave is normal in the horse. A typical QRS complex with only an R (positive deflection) and an S (negative deflection) is present. The T wave is highly volatile in the horse and can be a positive or a negative (or both) waveform. The slight negative deflection following the P wave is due to atrial repolarisation (atrial T wave).

  - exercising ECG (telemetry) indicated for:
    - poor performance workup.
    - some horses with aortic and mitral regurgitation (MR).

### ECG interpretation

- determine quality of recording:
  - motion artifact.
  - ECG of adequate diagnostic quality.
- determine rate:
  - fast, normal or slow.
- examine each complex in turn:
  - P wave for every QRS complex.
  - complexes should have the same morphology.
- determine the relationship between P-to-P and R-to-R intervals:
  - rate should be regular but minor variation is expected:
    - represents normal variations in autonomic tone (heart rate variability).
  - Fixed P-to-P, R-to-R or P-to-R intervals can be abnormal.

## Echocardiography

- transthoracic cardiac ultrasonography is a non-invasive procedure to evaluate cardiac structure, dimensions and function.
- indicated in any situation where cardiac disease is suspected:
  - cardiac murmurs:
    - provide information about the haemodynamic impact of the murmur.
  - functional murmurs (low grade, transient) do not require echocardiography.
  - potentially pathological arrhythmia.
  - suspected pericardial disease.
  - horse with fever of unknown origin (with/without a cardiac murmur).
  - known exposure to a cardiotoxin (ionophores, plants).
  - cases of unexplained poor performance.
- assessment of chamber enlargement and valvular abnormalities.
- repeated echocardiography if cardiac pathology is detected:
  - useful to document progression.
  - predict cardiovascular effects on performance or lifespan prognosis over time.

### Technique

- quiet environment.
- clipping the hair is necessary in horses with a rough or long hair coat.
- contact best achieved with alcohol and coupling gel.
- position horse with forelimb on the side being examined pulled slightly forward.
- avoid sedation if possible:
  - if required, take effects of sedation into consideration.
  - alpha-2 agonists may:
    - worsen valvular regurgitation.
    - have negative influence on LV function.
  - acepromazine may:
    - improve indices of LV function.
    - reduce valvular regurgitation.
- low-frequency probe (2–4 MHz) with depth of 24–30 cm is required.
- simultaneous ECG displayed on screen to allow assessment of timing of cardiac events.
- systematic approach and ability to acquire the routine images is important for accurate echocardiographic interpretation:
  - measuring cardiac dimensions from non-routine images will distort interpretation.
  - most images are acquired from the right parasternal window:
    - general impression of heart size.
    - cardiac dimensions relative to normal structures.
    - crude assessment of cardiac pump function (fractional shortening).
    - basic 2-dimensional (2D) and motion mode (M-mode) images routinely obtained are demonstrated (Figs. 3.5–3.10) and used to determine:
      - measurements of cardiac chamber and great vessels size.
      - morphology of the valvular structures.
      - left ventricular function.
    - critical to use reference ranges published for the breed being examined:
      - Thoroughbreds, Standardbreds and Warmbloods available.

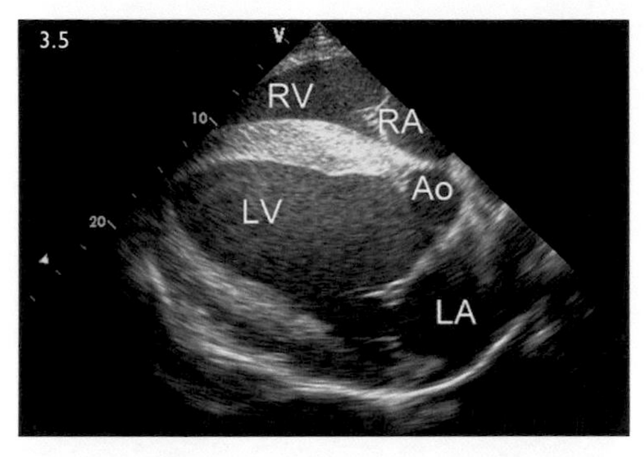


FIG. 3.5 Right-sided, long-axis, four-chamber view. Image taken from the right 4th ICS. The probe is angled slightly caudally until all four cardiac chambers are visible. RV = right ventricle; RA = right atrium; LV = left ventricle; LA = left atrium; Ao = aorta.

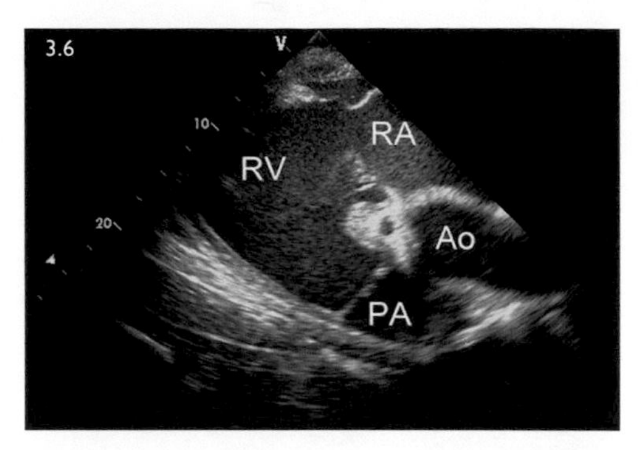


FIG. 3.6 Right-sided, long-axis, outflow tract view highlighting the pulmonic valve. Image taken from the right 4th ICS. The probe is angled cranially until the pulmonary artery is visualised. RV = right ventricle; RA = right atrium; Ao = aorta; PA = pulmonary artery.

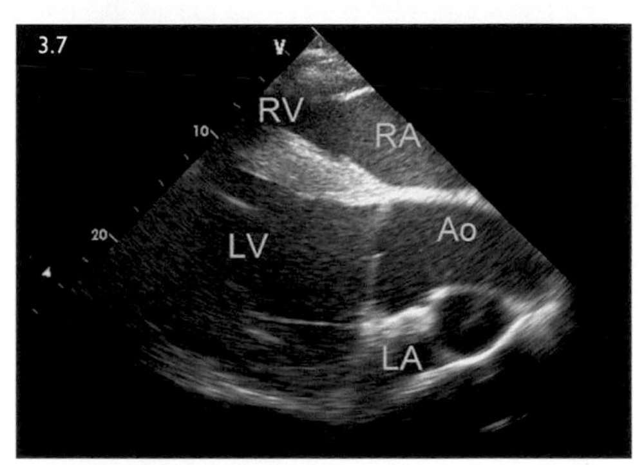


FIG. 3.7 Right-sided, long-axis view, highlighting the aorta. Image taken from the right 4th ICS aiming across the thorax. Slight rotation of the probe is necessary to have a true long-axis view of the aorta. RV = right ventricle; RA = right atrium; LV = left ventricle; Ao = aorta; LA = left atrium.

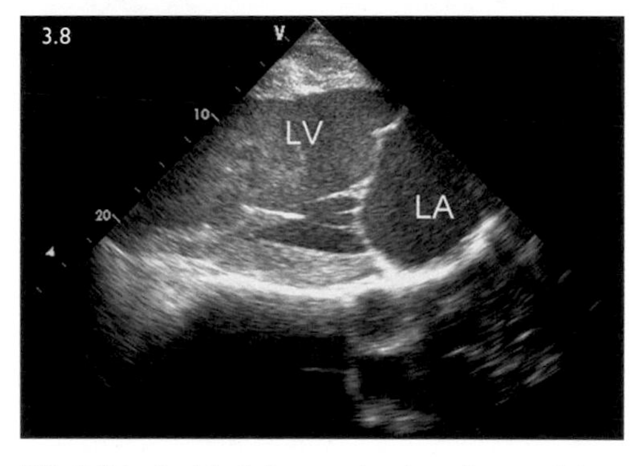


FIG. 3.8 Left-sided, long-axis view. Image taken from the left 5th ICS aiming across the thorax. LV = left ventricle; LA = left atrium.

    - allometric scaling of cardiac dimensions validated in Warmbloods but no other breed.
- Doppler imaging (colour flow and pulsed wave) is used to assess:
    - extent and direction of valvular regurgitations.
    - intracardiac and extracardiac shunts.
- Contrast echocardiography – 'bubble study':
    - intravenous administration of agitated saline and simultaneous echocardiography.
    - information regarding direction of blood flow across an intracardiac shunt.
    - beneficial when colour-flow Doppler is not available.

## Laboratory evaluation of cardiac disease

- complete blood count and fibrinogen concentration.
- serum biochemical analysis.
- urinalysis and fractional excretion of electrolytes:
    - investigation of potential triggers for rhythm disorders:
        - potassium depletion is a risk factor for AF.

### Cardiac specific

- Cardiac troponin I (cTnI):

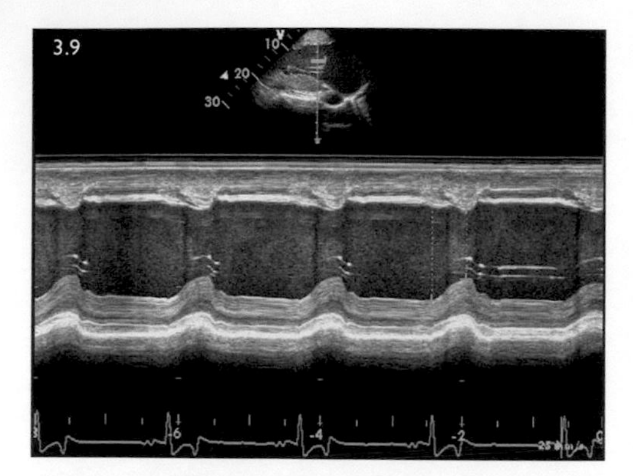

FIG. 3.9 M-mode image of a left ventricle. Image taken from the left 5th ICS in a normal horse. The small image shows the long-axis view from which the M-mode image was collected. In this case, M-mode of the left ventricle was collected for the evaluation of left ventricular function. Left ventricular internal dimensions in diastole and systole were collected and then ejection fraction and fractional shortening were calculated.

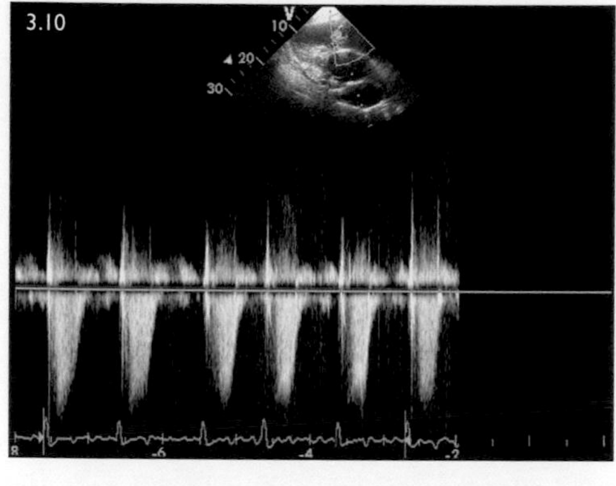

FIG. 3.10 Continuous-wave Doppler image of right AV valvular regurgitation (RAVVR). Image taken from the right 4th ICS of a 4-year-old Standardbred gelding with moderate RAVVR. Colour-flow Doppler is used to locate the most intense regurgitant jet and then continuous-wave Doppler to assess the flow velocity. This can then be used to calculate the pressure differential across the valve.

  - cardiac-specific cytosolic protein released from damaged (reversibly/irreversibly) or necrotic myocardial cells.
  - routinely on serum, stable when separated and kept refrigerated or frozen.
  - multiple assays are available:
    - stall side assays can be performed on anticoagulated whole blood.
    - sensitivity of the assay varies.
    - half-life of the protein in the serum is approximately 6 hours.
    - rapid decline is expected once myocardial insult has finished.
  - serial measurements have greater prognostic value.

### Blood pressure measurement

- invasive blood pressure measurements should be obtained in suspected primary hypertension cases.
- non-invasive blood pressure measurements are helpful for determining trends, but are generally inaccurate when compared to invasive pressures.

# ARRHYTHMIAS OR RHYTHM DISORDERS

## *Supraventricular rhythm disturbances*

## Atrial fibrillation (AF)

### Definition/overview

- most common clinically relevant dysrhythmia in the horse:
  - estimated prevalence ranges from 0.3–2.5% of the equine population.

### Aetiology/pathophysiology

- initiating event is unknown.
- possible trigger factors:
  - electrolyte abnormalities.
  - premature atrial contractions.
- possible genetic factor in the Standardbred.
- horses may be predisposed to the condition due to:
  - high vagal tone.

    - large atrial dimension.
- loss of coordinated electrical and mechanical function within the atria:
    - multiple pathways of depolarisation within the atria occur simultaneously.
    - may lead to increased numbers of electrical signals around a self-sustaining circuit maintaining the AF.
- duration of AF dictates treatment approach and informs the prognosis for successful conversion.
- AV node blocks most of the fibrillation wavelets such that the ventricular response rate in horses with AF (not from an underlying cardiac condition) is within a normal range.
- horses in heart failure often develop AF secondary to severe dilatation of the atria.

## Clinical presentation

- history:
    - incidental finding during routine examination (pre-purchase or preventative care).
    - poor performance in elite athletes (fading or tiring easily).
    - collapse/syncope.
    - agitation and colic-like episode.
    - exercise-induced pulmonary haemorrhage.
    - clinical signs compatible with congestive heart failure.
    - document approximate or exact time of onset whenever possible:
        - duration of 4 months or greater has more variable response to conversion.
- examination:
    - auscultation:
        - irregularly irregular heart rhythm with a normal or increased rate.
        - not abolished with exercise.
- AF may be persistent, paroxysmal or permanent:
    - persistent AF has irregular rhythms at rest and exercise.
    - paroxysmal AF spontaneously converts to sinus rhythm:
        - most cases present for 2–48 hours before spontaneous conversion.
    - permanent AF:
        - persistent AF where cardioversion was attempted unsuccessfully.
        - cases in heart failure.
        - some other co-morbidity which means cardioversion is contraindicated.

## Differential diagnosis

- Second-degree AV block, atrial flutter, atrial tachycardia with variable AV response, sinus arrhythmia, third-degree AV block and AV dissociation.

## Diagnosis

- physical examination:
    - detection of signs of heart failure is important.
- auscultation:
    - unpatterned or irregularly irregular rhythm.
    - not abolished with sympathetic stimulation (exercise/excitement).
- electrocardiographic diagnosis (Fig. 3.11):
    - absence of P waves.
    - irregular R-to-R interval.
    - irregular, undulating baseline waveforms (F waves):
        - represent unstructured atrial electrical activity.
    - contrast with atrial flutter – repeatable appearance to baseline.
- continuous ECG (Holter or telemetric ECG) required when paroxysmal AF is suspected.
- echocardiography is always indicated in AF:
    - abnormal echocardiogram impacts therapeutic approach and prognosis.
- Cardiac troponin (cTnI):
    - mild increases are possible in cases of acute onset < 7 days or paroxysmal AF.
    - typically, normal in horses with persistent AF.
- serum chemistry and urinary fractional excretion of electrolytes (potassium) indicated.

## Management

- indicated for most horses with persistent AF not related to underlying cardiac disease.
- pharmacologic cardioversion should only be attempted by veterinarians with:
    - ability to monitor physical examination.
    - close monitoring of continuous electrocardiograms (hourly or more).

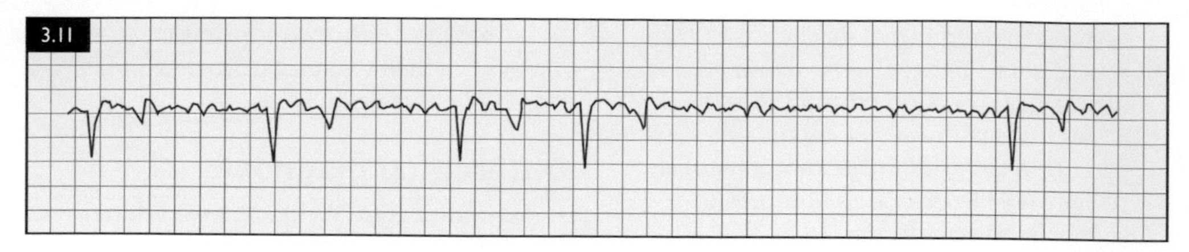


**FIG. 3.11** Atrial fibrillation. Base–apex lead recorded from an 8-year-old Standardbred gelding with sudden onset of poor performance during a race. Absence of P waves, presence of fibrillatory (F) waves, normal ventricular complexes and irregular ventricular rhythm are seen. The ventricular rate of 40/minute is within normal limits.

- traditional management involves quinidine sulphate (orally) or quinidine gluconate (i/v):
  - currently quinidine gluconate is not universally available.
- **adverse effects are common and range from:**
  - mild (depression, mild tachycardia, loose manure and colic).
  - severe and life-threatening:
    - tachyarrhythmia, neurological abnormalities, collapse and colitis.
  - limit adverse effects by careful monitoring and limiting total number of doses.
  - quinidine sulphate was administered intragastrically at 22 mg/kg (10 g/450 kg horse) every 2 hours:
    - until conversion to sinus rhythm was achieved.
    - clinical evidence of toxicity developed.
  - more conservative current recommendations are that no more than 4 doses be administered before extending the dosing interval to 6–8 hours.
- **with quinidine toxicity, the following medications should be available:**
  - magnesium sulphate – agent of choice for associated torsades de pointes.
  - sodium bicarbonate – decrease free quinidine levels by increasing protein binding.
  - digoxin – to treat tachycardia and improve cardiac output.
- transvenous electrical cardioversion (TVEC) is limited to referral centres and requires specialised equipment and training.
- **cardioversion (quinidine or TVEC) contraindicated in patients in heart failure.**
- management of cases should concentrate on avoiding trigger factors:
  - fluctuations in serum potassium.
  - oral or i/v sodium bicarbonate.
  - levothyroxine
- train owners/trainers to monitor for recurrence.

### Prognosis

- excellent for return to previous level of performance in horses with lone AF that are successfully converted.
- recurrence is common and increased with:
  - mitral valve insufficiency.
  - previous unsuccessful conversion attempts.
  - low active left atrial fractional area change (echocardiographic variable).
  - immediate recurrence is increased with a high frequency of atrial premature contractions.
- horses with AF due to underlying cardiac disease have a guarded prognosis for life.

## First-degree heart block

### Overview/aetiology/pathophysiology

- normal finding in horses.
- associated with high vagal tone:
  - may be more prevalent in fit horses.

### Diagnosis/clinical presentation

- normal sinus rhythm.
- electrocardiographic diagnosis:
  - P waves and QRS complexes have normal conformation (Fig. 3.12).
  - defined by prolongation of PR interval beyond 0.425 to 0.47 ms.

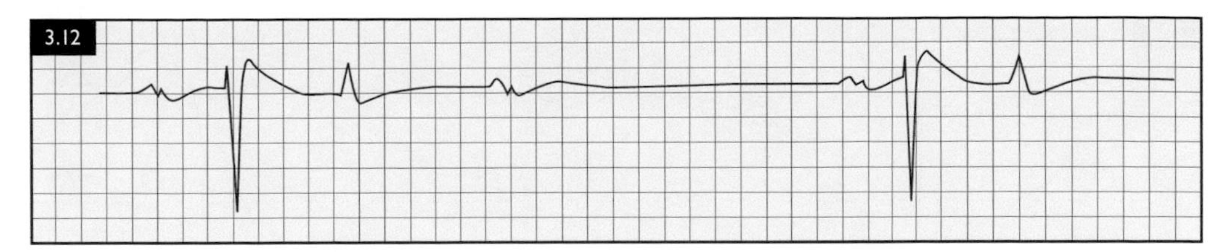


**FIG. 3.12** Base–apex lead recorded from a clinically normal Standardbred gelding at rest, with both first-degree and second-degree heart block. The PR interval is >0.42 seconds (first-degree block). The next P wave is not followed by a QRS complex (second-degree block). All complexes are normal in conformation. P waves followed by QRS and T are present either side of a blocked beat, where only the P wave is present. PR interval is variable and therefore this is a Mobitz type I second-degree block.

### Management

- no treatment is required.

## Second-degree heart block

### Overview/aetiology/pathophysiology

- typically, a normal finding in horses.
- associated with high vagal tone and decrease in conduction through the AV node:
  - may be more prevalent in fit horses.
- advanced block (multiple cardiac cycles in a row) may be a result of myocardial inflammation.

### Clinical presentation

- normal to slightly slow ventricular rate with regularly irregular rhythm.
- abolished with increase in sympathetic tone (light exercise, excitement).
- electrocardiographic diagnosis (see Fig. 3.12):
  - P waves occur without QRS complexes.
  - P-to-P interval is consistent.
  - QRS have normal conformation.
  - Mobitz type I (more common):
    - gradual prolongation of consecutive PR interval.
    - until a P wave occurs without a QRS (dropped beat).
  - Mobitz type II (less common):
    - constant PR interval with intermittent block.
- advanced block may be associated with pronounced bradycardia and collapse.

### Management/treatment

- none required.
- Mobitz type II may be associated with atrial myocardial disease.
- advanced blocking may require additional investigation:
  - minimum data base, cardiac troponin measurement and echocardiogram.
- suspect myocardial disease (not active viral or other infection) may benefit from corticosteroid therapy.

## Third-degree heart block

### Aetiology/pathophysiology

- variable.
- associated with myocarditis, pericarditis and aortic aneurysms.

### Clinical presentation

- bradycardia with or without syncope:
  - ventricular rate 10–20 bpm.
- electrocardiographic diagnosis:
  - no relationship between P and QRS complex.
  - P waves may be lost in QRS trace.
  - QRS may be wide or bizarre.

### Management/treatment

- rarely attempted.
- pacemaker implantation is possible.
- pharmacologic therapy is of limited value.
- if myocardial disease suspected (not active viral or other infection), may benefit from corticosteroid therapy.

## Wandering atrial pacemaker

### Aetiology/pathophysiology

- normal finding in horses.

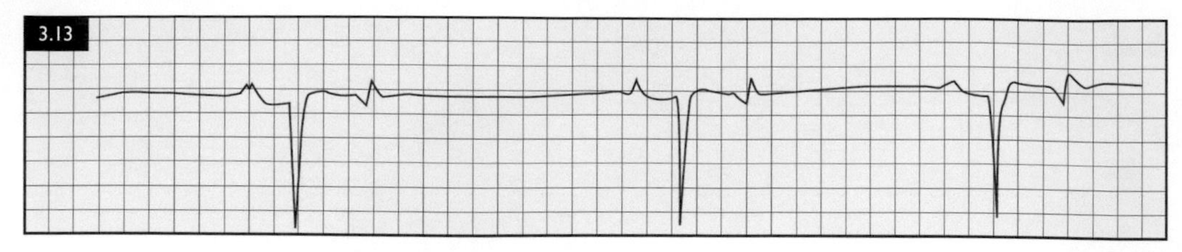


**FIG. 3.13** Wandering atrial pacemaker. Base–apex lead recorded from a clinically normal Standardbred gelding at rest. Two different P wave morphologies are present. One P wave has a single peak and two are bifid P waves.

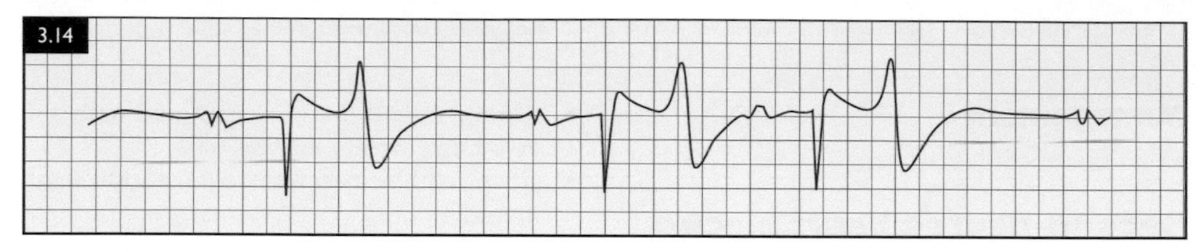


**FIG. 3.14** Atrial premature contraction (APC). Base–apex lead recorded from a clinically normal 10-year-old Standardbred gelding at rest. Two normal cycles are followed by an APC. The APC is characterised by early occurrence of altered-waveform P wave, followed by a normal ventricular complex.

### Diagnosis

- electrocardiographic diagnosis:
  - P wave has variable morphology.
  - normal QRS complex.
  - normal association between P waves and QRS complexes (Fig. 3.13).

### Management/treatment

- no treatment required.

## Atrial premature contractions (APCs)

### Aetiology/pathophysiology

- usually, a single event or infrequent finding <1 APC/hour.
- persistent or high-frequency APCs can be associated with myocardial disease or systemic inflammation.

### Clinical presentation/diagnosis

- auscultation of early beat with or without a compensatory pause.
- electrocardiographic diagnosis (Fig. 3.14):
  - shortened P-to-P interval with conformation change of the P wave.
  - QRS morphology is normal.
- comprehensive workup indicated for horses with atrial tachycardia:
  - minimum data base, cardiac troponin and echocardiogram.

### Management/treatment

- rest.
- rate control with digoxin if APCs are frequent.
- may predispose to development of AF.

## Atrioventricular (AV) dissociation or accelerated idioventricular rhythm (AIVR)

### Aetiology/pathophysiology

- variable.
- associated with myocardial irritation or inflammation.
- observed in some horses with systemic endotoxaemia or GI disease.
- ectopic focus develops in ventricular tissue with a higher intrinsic rate than the sinoatrial (SA) node.
- technically AV dissociation occurs in all cases of ventricular tachycardia:
  - AIVR is a better term and refers to a ventricular rate of <100 bpm.

### Clinical presentation

- tachycardia <100 bpm.
- electrocardiographic diagnosis (Fig. 3.15):

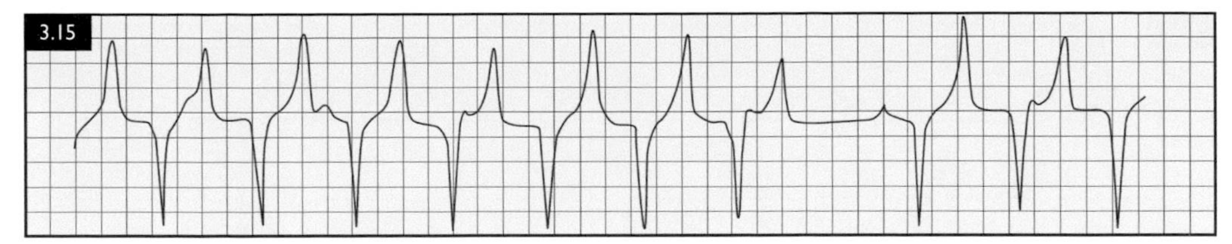


FIG. 3.15 Atrioventricular dissociation. Base–apex lead recorded from a 16-year-old Quarter horse gelding with colitis. A sudden increase in heart rate, which was not consistent with other clinical signs, precipitated the ECG recording. All waveform morphology is normal; however, there is no relationship between P waves and QRS complexes. The ventricular rate of 80 bpm is higher than the atrial rate of 60 bpm and some P waves are obscured by the ventricular waveforms. A capture beat is present near the end of the recording, where a relatively longer diastolic interval is followed by P, QRS and T complexes with a normal relationship.

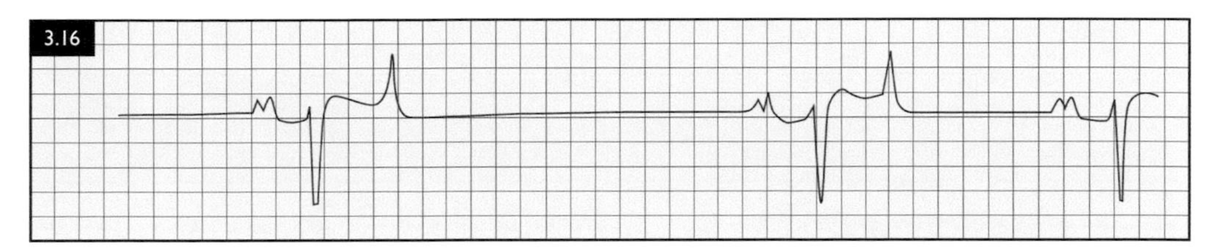


FIG. 3.16 Sinus arrhythmia. Base–apex lead recorded from a clinically normal Standardbred gelding presented for evaluation of arrhythmia detected at routine physical examination. Normal P, QRS and T complexes with normal temporal relationship (sinus rhythm) occurring at irregular intervals. An association with respiration was not detected and the arrhythmia abated with exercise.

 - no relationship between P and QRS complex.
 - ventricular rate is higher than the atrial rate.
 - ventricular and atrial complexes are often normal in conformation.

### Management/treatment

- typically resolves with treatment of the underlying condition.

## Atrial tachycardia

### Aetiology/pathophysiology

- uncommon.
- unknown underlying pathology:
  - suspected due to myocarditis or an electrolyte disturbance.

### Clinical presentation

- tachycardia with signs of perfusion deficits, anxiety and sweating.
- heart rate 100–200 bpm:
  - irregular rhythm if second-degree block present.
- electrocardiographic diagnosis:
  - multiple (4 or more) consecutive APCs.
  - QRS normal appearance and preceded by P waves.
  - P waves can have variable morphology.
  - superimposition on T wave can make it difficult to identify the P wave.

### Management/treatment

- identify and correct underlying inflammation or electrolyte disturbances.
- anti-arrhythmic medication is indicated.

## Sinus arrhythmia

### Aetiology/pathophysiology

- common in fit horses during exercise deceleration (exercise-associated arrhythmia).
- respiratory sinus arrhythmia (respiration-induced changes in vagal tone) is uncommon in horses.

### Clinical presentation

- respiratory sinus arrhythmia characterised by:
  - variations in RR interval associated with respiration (Fig. 3.16).
- exercise-associated arrhythmia:

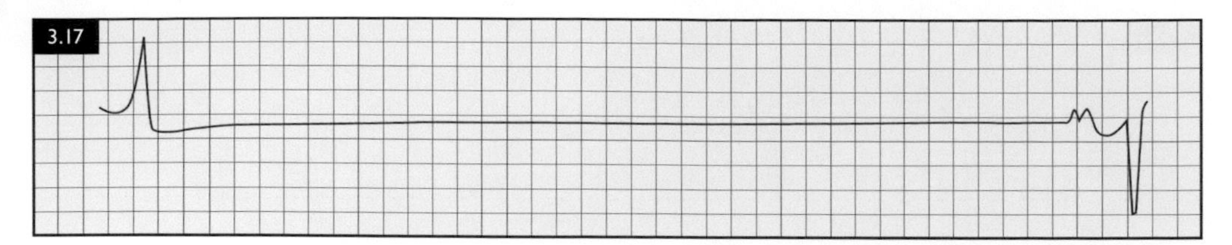


**FIG. 3.17** Sinus arrest/block. Base–apex lead recorded from the same horse as in Fig. 3.14. A 6-second interval between complexes is present in this recording and pauses of up to 8 seconds were detected. No clinical abnormalities were detected, the cardiac rhythm was normal at exercise and the horse was performing to expectations. The cause of the arrhythmia in this horse was unknown.

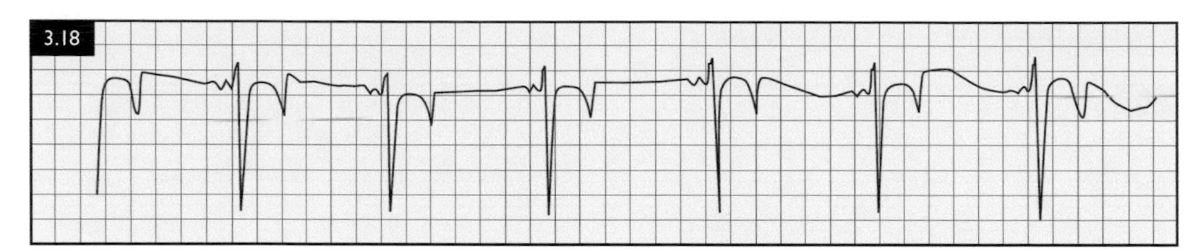


**FIG. 3.18** Pre-excitation (Wolff–Parkinson–White) syndrome. Base–apex lead recorded from a 7-year-old Standardbred gelding with a variable performance record. An extremely short PR interval and a positive delta wave are present at the beginning of the QRS complex, giving it a widened appearance. Variable conduction with some normal sinus complexes was present at elevated heart rates during treadmill exercise. The horse continued to race following this diagnosis, with variable results.

  - heart rate slows suddenly then gradually rises.
  - stepwise appearance of cardiac deceleration.

## Management

- none required.

# Sinus arrest/block

## Aetiology/pathophysiology

- normal variation in resting horses.
- associated with variation in vagal tone.
- persistence at increased heart rates may have a pathological basis.
- prolonged blocked intervals (4 seconds or more) or those associated with syncope are clinically relevant and likely have a pathological basis.

## Clinical presentation

- incidental finding.
- collapse/syncope.
- electrocardiographic diagnosis (Fig. 3.17):
  - normal cardiac rhythm with periods of absent P wave and QRS complex.
  - RR interval is usually twice the normal RR interval during the 'blocked beat'.
  - may see junctional escape beats:
    - QRS not preceded by P wave.
    - normal or abnormal QRS morphology.
- Sinus arrest: variable periods of electrical silence.

## Management/treatment

- benign form: no treatment required.
- pacemaker implantation is possible in advanced block.

# Pre-excitation syndrome (Wolff–Parkinson–White syndrome)

## Aetiology/pathophysiology

- rare in horses.
- accessory conduction pathway between the atria and ventricles or bundle of Hiss and surrounding ventricular myocardium.
- results in reduced cardiac output from ventricular dysynchrony, notably during exercise.
- predisposes to AF.

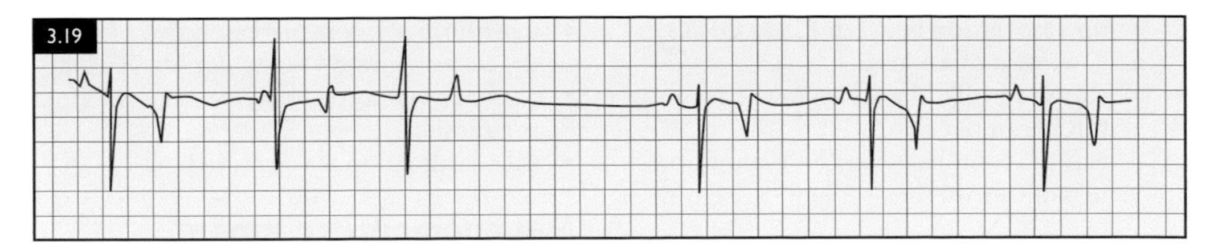


**FIG. 3.19** Premature ventricular complexes. Base–apex electrocardiogram recorded from a normal horse at rest. There is a normal PQRST (note negative deflection of QRS complex), followed by a P wave with an abnormal QRS almost obliterating the P wave. This premature ventricular complex is followed by another with the same conformation (note the absence of a P wave with this complex). A compensatory pause is then present, followed by normal complexes.

### Clinical presentation

- poor performance
- collapse/syncope.
- electrocardiographic diagnosis (Fig. 3.18):
  - abnormally short P-to-R interval with a prolonged QRS.
  - may be intermittent (a couple of cardiac cycles and then disappear).
  - QRS morphology can be variable:
    - delta wave (slurring of the R wave upstroke) at the beginning of the QRS.
- can become underlying rhythm following conversion from AF.

### Management/treatment

- no treatment options.
- **not safe to ride or drive.**

## *Ventricular rhythm disturbances*

## Premature ventricular complexes (VPCs)

### Aetiology/pathophysiology

- common and of variable clinical significance.
- isolated, singlet, infrequent VPCs are likely to be insignificant.
- long runs or frequent occurrence may indicate myocardial disease or inflammation.

### Clinical presentation

- early beat, often with a compensatory pause auscultated.
- incidental finding or present in horses with active systemic inflammation.
- electrocardiographic diagnosis (Fig. 3.19):
  - heart rate is normal.
  - QRS complex is often abnormal, wide and bizarre.
  - no associated P waves.
  - abnormal complex closely follows the sinus complex.
  - may occur in singlet or train of < 4.

### Management/treatment

- treat the underlying disease if present.
- no treatment required for short runs of 2 or 3 VPCs:
  - recommend 24-hour telemetry to determine the frequency.
- recommend exercising telemetry prior to return to work.

## Ventricular tachycardia (VT)

### Aetiology/pathophysiology

- typically indicates underlying myocardial disease or systemic inflammation.
- may reflect serum electrolyte abnormalities.

### Clinical presentation

- tachycardia, with loud (pounding) heart sounds on auscultation.
- jugular pulsation.
- syncope/collapse (rest or exercise).
- persistence can lead to heart failure due to myocardial fatigue/failure.
- electrocardiographic diagnosis (Fig. 3.20):
  - heart rate is often high (>100 bpm).
  - runs of abnormal ventricular beats with a bizarre QRS conformation.
  - P wave may be obscured (buried in T wave or QRS complex).

- no association between P and QRS.
- T wave is often long.
- capture beats (P wave followed by a QRS) can occur.
- ectopic complexes may be monoform or multiform.

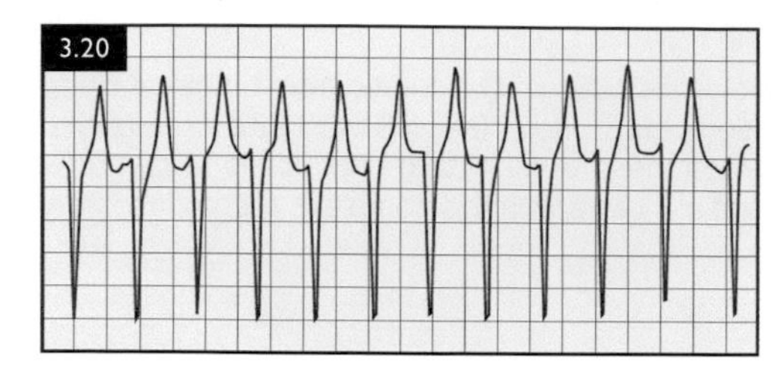

**FIG. 3.20** Ventricular tachycardia. Base–apex electrocardiogram recorded from a horse undergoing treatment for AF with the oral medication quinidine. Quinidine intoxication resulted in monomorphic ventricular tachycardia. The ventricular rate is 140 bpm. All the QRS complexes have a similar appearance.

- multiform VT is at increased risk of degenerating into ventricular fibrillation.
- cardiac troponin I – may be beneficial for prognosis.
- echocardiogram is warranted.

## Management/treatment

- correct underlying conditions and electrolyte imbalances.
- pulmonary oedema present:
  - intranasal oxygen insufflation and furosemide are indicated.
- intravenous lidocaine bolus may be corrective.
- other medications used to treat VT include:
  - magnesium sulphate (1.0–2.5 g/450 kg/min over 20–30 minutes (Table 3.4).
  - do not exceed 25 g total dose.

**TABLE 3.4** Common drugs used in the management of cardiovascular disease

| Digoxin |
|---|
| • **Main application.** Slow heart rate during heart failure. Increases cardiac output. Also used in the management of supraventricular tachyarrhythmia. Does not treat arrhythmia directly but slows rate through decreased AV nodal conduction.<br>• **Toxic effects.** Depression, anorexia and colic are common. Bradycardia, total AV block and ventricular tachyarrhythmias are all possible. Interaction with quinidine may increase potential for toxic effects. Predisposed by hypokalaemia, hypoproteinaemia, dehydration, and renal disease.<br>• **Dose.** 0.011 mg/kg p/o or 0.0022 mg/kg i/v q12–24 h. An initial i/v dose is often recommended. i/v administration is indicated if used to treat quinidine-associated tachycardia. Dose should be reduced by up to 50% if used concurrently with quinidine. Low therapeutic index; therefore, monitoring of blood levels is indicated. Therapeutic range 1–2 ng/ml. |
| **Quinidine** |
| • **Main application.** Most commonly used in the management of AF. Also used in treatment of other supraventricular and ventricular arrhythmias. Quinidine sulfate (oral) is administered via nasogastric tube and is used mainly in the management of AF. Quinidine gluconate is administered intravenously and used in the management of acute AF (<4 weeks) or ventricular arrhythmias.<br>• **Toxic effects.** Variable. Mild signs such as depression, nasal oedema and increased frequency of defecation are common and are often tolerated. More severe signs such as marked hypotension, ataxia, colic, diarrhoea, laminitis, sustained tachycardia, syncope and sudden death have been reported. Idiosyncratic responses may occur with first dose. Torsades de pointes may be more likely in hypokalaemic patients.<br>• **Dose.** Quinidinc gluconate: 0.5–2.2 mg/kg i/v bolus q5–10 minutes to effect, maximum dose 12 mg/kg. Often conversion of ventricular tachycardia occurs with one dose at 0.5 mg/kg. Quinidine sulfate: 22 mg/kg via nasogastric tube q2 h until conversion, toxic effects, therapeutic levels or six doses. Continue administration every 6 hours until conversion or adverse effects. Monitor ECG closely during treatment. Heart rate >80 bpm, widening of QRS complexes to 125% of the pre-treatment width and abnormal complexes are all indicators to cease medication. Therapeutic range 3–5 µg/ml. |
| **Lidocaine (without epinephrine)** |
| • **Main application.** Emergency treatment for ventricular arrhythmias. Does not have effects on supraventricular arrhythmias. Intravenous boluses are used in acute cases, while slow i/v administration is used in subacute cases. Short duration of action. |

- **Toxic effects.** Horses are very susceptible to lidocaine-induced CNS signs. Excitability, muscle fasciculations and convulsions may occur after i/v bolus. Ventricular tachycardia and sudden death have been reported.
- **Dose.** 0.25 mg/kg bolus. 0.5–1.0 mg/kg slowly to effect. Can repeat in 5–10 minutes. 20–50 mg/kg/minute continuous rate infusion (CRI). Therapeutic concentrations 1.5–5.0 μg/ml.

**Magnesium sulphate**

- **Main application.** Treatment of quinidine-induced torsades de pointes. Has been used in the management of ventricular arrhythmias that were not responsive to other antiarrhythmic medications.
- **Toxic effects.** Colic and syncope have been reported.
- **Dose.** 1.0–2.5 g/450 kg/minute over 20–30 minutes. Do not exceed 25 g total dose.

**Furosemide**

- **Main application.** Used in the management of heart failure to decrease volume overload, and therefore decrease vascular volume and cardiac workload. Will only have effect if cardiac output is sufficient for adequate renal perfusion.
- **Toxic effects.** Prolonged or aggressive use will cause dehydration, azotaemia, electrolyte abnormalities and metabolic alkalosis.
- **Dose.** 0.5–1.0 mg/kg i/v, i/m or p/o q12 h.

**Procainamide**

- **Main application.** Has been used in the management of AF but is considerably less effective than quinidine. Has been used in the management of ventricular tachycardia.
- **Toxic effects.** Similar to quinidine. Death due to ventricular arrhythmia has occurred during treatment for AF.
- **Dose.** 1 mg/kg/minute i/v, not exceeding a 20 mg/kg total dose; 25–35 mg/kg p/o q8 h.

**Propranolol**

- **Main application.** Has been used in the management of ventricular tachycardia. Decreases ventricular rate.
- **Toxic effects.** Bradycardia, AV block and arrhythmias may occur. It is a negative inotrope and may cause hypotension. Use with caution in animals with airway disease, as it may exacerbate bronchospasm.
- **Dose.** 0.03 mg/kg i/v; 0.38–0.78 mg/kg p/o q8 h.

**Atropine and glycopyrrolate**

- **Main application.** Used for the treatment of bradycardia, most commonly during anaesthesia. Have been used as diagnostic tools in heart block and sinus arrhythmia (removal of parasympathetic tone should remove these arrhythmias).
- **Toxic effects.** Mydriasis and decreased intestinal secretions and motility are possible even at therapeutic levels. Ileus and colic may develop. Tachycardia without increased contractility and arrhythmias may also occur.
- **Dose (therapeutic).** 0.005–0.01 mg/kg for bradycardia (both agents).

**Enalapril**

- **Main application.** Inhibits angiotensin-converting enzyme (ACE), decreasing afterload. May be beneficial in management of horses with aortic regurgitation or aortic root rupture/aneurism. Considered potentially to slow progression of disease in aortic regurgitation. Must be withdrawn prior to competition.
- **Toxic effects.** Hypotension, diarrhoea and anorexia may develop.
- **Dose.** 0.25–0.5 mg/kg p/o q24 h.

**Other agents**

- A number of other medications have reportedly been used in the management of equine cardiac arrhythmias, but much of the information available on these is anecdotal.
- **Hydralazine** (0.5 mg/kg i/v q12 h) has been used as a vasodilator to reduce afterload.
- The use of **flecainide** in the management of limited cases of equine AF has been reported. Anecdotal reports and one published report on the use of this medication in clinical cases have not been favourable.
- The use of amiodarone in equine AF has been limited both in scope and efficacy.

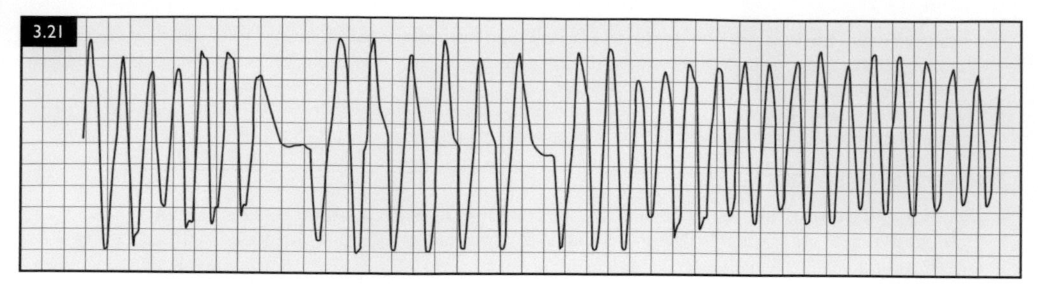

**FIG. 3.21** Torsades de pointes. Base–apex lead recorded from a 2-year-old Standardbred colt undergoing quinidine therapy for AF. Polymorphic ventricular tachycardia with twisting around the baseline is present in this recording. A ventricular rate of 240 bpm is present in the last third of the recording.

## Torsades de pointes ('twisting of the points')

### Aetiology/pathophysiology

- uncommonly encountered during quinidine treatment of AF:
  - predisposed by concurrent hypokalaemia.
  - sudden death during therapy may be due to development of torsades de pointes:
    - with rapid deterioration into ventricular fibrillation.

### Clinical presentation

- sudden death.
- electrocardiographic diagnosis (Fig. 3.21):
  - wide polymorphic VT with twisting of complexes around the baseline.
  - may progress to ventricular fibrillation and death.

### Management/treatment

- magnesium sulphate i/v.
- lidocaine (2%) bolus followed by constant-rate infusion.
- if associated with quinidine therapy:
  - do not give additional doses of quinidine.
  - administer i/v sodium bicarbonate:
    - reduces free, active quinidine through enhanced protein-binding.

## Ventricular fibrillation

### Aetiology/pathophysiology

- causes are variable.
- precedes death by a few seconds.
- peripheral pulse and heart sounds are absent.

### Clinical presentation/diagnosis

- collapse/sudden death.
- electrocardiographic diagnosis:
  - course, low-frequency undulations with no discernible atrial or ventricular activity.

### Management/treatment

- **practical, therapeutic options are limited.**
- electrical defibrillation may be attempted, but success is uncommon.
- horse is under general anaesthesia and biphasic defibrillation is available (TVEC case):
  - electrical defibrillation is indicated.

# CONGENITAL CARDIAC DISEASE

## Overview

- uncommon in the horse.
- no proven heritability, although more commonly reported in Arabians.
- possible contributory factors include:
  - maternal infection.
  - hypoxia due to placental insufficiency.
  - fetal infection or toxin exposure.
- single or complex anomalies.
- ventricular septal defect (VSD) is the most common:
  - small VSD may be an incidental finding in an adult horse.

- other defects reported include:
  - patent ductus arteriosus (PDA).
  - persistent truncus arteriosus.
  - tetralogy of Fallot.
  - atrial septal defects.
  - pulmonic stenosis.

## Clinical presentation

- murmurs are common in foals:
  - quiet, soft, systolic murmur unlikely to indicate congenital cardiac disease:
    - innocent, physiological flow murmur.
  - continuous holosystolic or holodiastolic murmur in a foal:
    - raises suspicion for congenital cardiac disease.
  - murmur that radiates widely or has a palpable thrill:
    - should prompt early echocardiographic investigation.
- cyanosis at rest may occur with complex cardiac disease:
  - when present raises the suspicion of right-to-left shunting of blood.
- clinical signs are variable and can be detected at various ages:
  - profound lethargy, with cyanosis, and/or congestive heart failure in a neonatal foal.
  - ill-thrift or poor growth in a foal.
  - poor performance in adults.
  - incidental finding on a post-mortem examination.

## Diagnosis

- complete physical examination with particular attention to:
  - mucous membranes
  - jugular veins.
  - arterial pulses.
  - cardiac and thoracic auscultation.
- arterial blood gas may help determine the presence of a right-to-left shunt.
- thoracic radiography may indicate evidence of pulmonary changes and cardiac enlargement.
- echocardiography with continuous ECG is indicated:
  - 2D echocardiography with Doppler and contrast echocardiography.
  - determine site of the lesion and haemodynamic impact.

# Ventricular septal defect (VSD)

## Definition/overview

- most reported congenital cardiac abnormality in the horse
- single defect (isolated) or as a part of complex abnormalities.
- any breed but Arabian, and possibly Welsh ponies, are overrepresented.

## Aetiology/pathophysiology

- incomplete formation of the interventricular septum during embryogenesis:
  - failure of any part of the normal formation process results in VSD.
  - location of defect reflects the point within the process when it failed (Fig. 3.22).
- Membranous septal defects are the most common:
  - just below the septal cusp of the tricuspid valve at the top of the septum.
  - due to failure of formation of the smooth septa.
- Infundibular defects:
  - involve the ridge of muscle that divides the inflow from the outflow tracts of the right ventricle (crista supraventricularis).

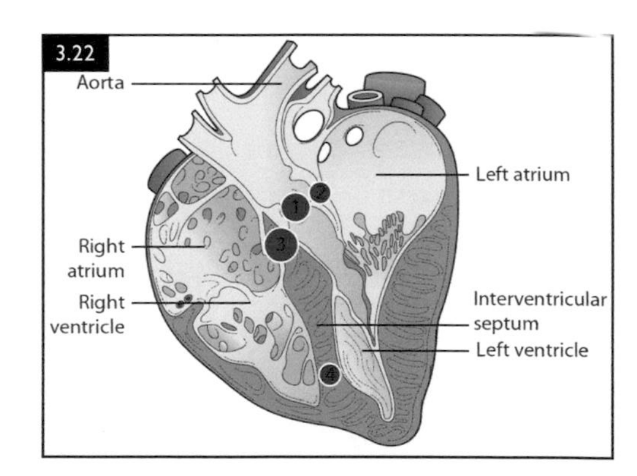


**FIG. 3.22** Representation of the equine heart opened to visualise the interventricular septum. Membranous defects (1) lie just below the septal cusp of the right AV valve. Infundibular defects (2) occur just below the outflow tracts. Smooth septum defects (3) occur in the muscular portion of the septum just below the membranous septum. Apical defects (4) are rare in the horse.

  - may disrupt the integrity of the base of the aortic and pulmonic valves.
- lower muscular septum defects are apical or trabecular and are rare in horses.
- higher pressure in the systemic circulation:
  - means blood flow across a VSD is from left to right in systole.
  - most cases (except apical defects), directly from left to right ventricular outflow tract.
  - no increase in right ventricular volume.
  - pulmonary hypertension and vasculature over-circulation.
  - left atrium and ventricle become volume overloaded.
- overtime, progressive pulmonary hypertension results in:
  - pressure overload of right ventricle with progressive right ventricular hypertrophy.
  - right-to-left shunting may occur:
    - increased right-sided pressure.
    - not reported with defects smaller than 4 cm or animals less than 2 years of age.
    - cyanosis is common with right-to-left shunting.
- infundibular defects:
  - progressive aortic valve instability can lead to aortic regurgitation (AR).
- thickening of tricuspid valve septal leaflets because of repetitive trauma of the shunted flow across membranous defects.

## Clinical presentation

- course or harsh holosystolic to pansystolic band-shaped murmur:
  - audible on both sides of the thorax (loudest on the right).
- palpable thrill may be present.
- membranous defect: point of maximum intensity (PMI) usually far forward on right side of the thorax.
- infundibular defects: PMI loudest on left side at the outflow tract.
- murmur grade does not correlate with extent of defect or haemodynamic significance:
  - small defects often generate louder murmurs but less haemodynamic effects.
  - pressure equalisation across VSD generates a soft murmur but is of greater haemodynamic significance.
- severity of clinical signs is related to the size and location of the VSD:
  - cyanosis at birth.
  - poor performance in young animals in training.
  - no haemodynamic effect.
  - large defects commonly have clinical signs resulting from volume overload and congestive heart failure.
  - pulmonary oedema may result in dyspnoea.
  - small defects may result in a loud murmur that is noted as an incidental finding in a horse that is performing well at a high level of training.
- cardiac arrhythmia (AF) may be present with atrial enlargement.

## Diagnosis

- thoracic radiography may be of value in determining the haemodynamic effects:
  - presence of pulmonary oedema.
  - cardiac silhouette changes are uncommon with VSDs in horses.
- 2D echocardiography with colour-flow Doppler:
  - small defects may be missed without careful assessment of the septum in both short- and long-axis views.
  - infundibular defects are particularly difficult to evaluate.
  - long-axis view of the left ventricular outflow tract from the right:
    - most helpful view to identify membranous or high-smooth septal defects.
  - peak velocity measurement across the defect helps to evaluate haemodynamic significance.
- contrast echocardiography with agitated saline is a viable method to confirm the defect when colour-flow Doppler echocardiography is not available.

## Management/treatment

- prognosis is variable:
  - echocardiographic evaluation to assess progression and long-term prognosis.
  - small defects:
    - only clinical abnormality might be the cardiac murmur.

        - performance and life expectancy considered normal.
    - large defects: poor, especially with disruption of the aortic valve.
    - evidence of congestive heart failure, and right-to-left shunt: grave.
- defects <25 mm and shunt velocity >4 m/s:
    - performance is good (including racing).
- VSD cases are at increased risk for bacterial endocarditis:
    - disturbed blood flow and increased probability of endocardial damage.

## Atrial septal defect (ASD)

### Definition/overview

- rare communication between the left and right atria.
- abnormal septation (ostium primum *vs* ostium secundum) – true ASD.
- failure of closure of foramen ovale at the time of birth (persistent foramen ovale).

### Aetiology/pathophysiology

- commonly occurs in combination with other defects in complex congenital heart disease.
- unknown cause.
- blood flows left to right due to normal pressure gradient across the atria:
    - smaller pressure differential than for the ventricles.
- small defects are of no clinical consequence.
- large defects may develop right-to-left shunting eventually and cause hypoxaemia.
- foramen ovale functionally closes after birth due to increases in left-sided cardiac pressures:
    - anatomic closure follows over the next several days.
    - patent foramen ovale expected in a foal that dies soon after birth from any cause.

### Clinical presentation

- clinical signs are dependent upon size of defect and presence of other congenital defects.
- murmurs do not typically accompany ASD:
    - holosystolic murmur over the pulmonic valve area has been described.
    - murmurs are more likely to be generated by accompanying congenital defects:
        - tricuspid dysplasia or pulmonic stenosis.
- small ASD may be an incidental finding on an echocardiogram.
- large defects result in hypoxaemia and heart failure.

### Diagnosis

- echocardiography:
    - careful examination of the interatrial septum with 2D echocardiography.
    - contrast or colour-flow Doppler:
        - determine the presence of shunting.
        - confirm true ASD (versus echocardiographic drop-out artifact).
    - careful investigation for other congenital defects indicated if ASD present.
- evidence of right-sided cardiac overload include:
    - pulmonary artery dilatation (echocardiography).
    - increased right heart or pulmonary artery pressure via cardiac catheterisation.

### Management/treatment

- no treatment for complex cardiac deformities and euthanasia may be indicated.
- septal closure with Amplatzer device is possible in theory, but difficult in practice.
- variable prognosis.
- small ASD as an incidental finding or as an isolated lesion:
    - routine echocardiographic evaluation to assess progression and long-term prognosis.
    - prognosis for performance and life expectancy are often normal.

## Patent ductus arteriosus (PDA)

### Definition/overview

- True PDA is rare in foals and almost always associated with complex congenital heart disease.

- ductus arteriosus does not close immediately after birth:
  - diagnosis of a true PDA should not be made in a foal <5 days of age.

### Aetiology/pathophysiology

- ductus arteriosus in the fetus allows blood to pass from the pulmonary artery to the aorta allowing oxygenated blood to supply the fetal circulation:
  - pressure gradient reverses at birth and ductus arteriosus closes over the first few days of life.
- complex congenital cardiac disease cases have abnormal pressure gradients resulting in a PDA.
- ductus failing to constrict may lead to left-to-right flow, resulting in:
  - pulmonary overcirculation and pulmonary hypertension.
  - right ventricular hypertrophy and heart failure.
  - sufficient increases in pulmonary arterial pressure may lead to right-to-left shunting.

### Clinical presentation

- range from none to severe signs:
  - determined by size of the PDA.
  - presence of other congenital cardiac disease.
- cyanosis occurs if there is reversal of the shunt (right to left).
- caudal cyanosis may occur if the PDA enters the aorta distal to brachiocephalic trunk.

### Differential diagnosis

- typical murmur is a continuous machinery murmur:
  - present in the normal foal and typically disappears over the first 96 hours.
- continuous murmurs are otherwise rare in the foal:
  - if detected at or after day 5, or in any foal with clinical signs of hypoxaemia:
    - complex congenital cardiac disease should be considered.

### Diagnosis

- continuous machinery murmur loudest over the left side:
  - diastolic component may be quiet to inaudible.
- thoracic radiography:
  - not specific for PDA.
  - enlargement of the cardiac silhouette.
  - evidence of pulmonary over-circulation and oedema.
- 2D echocardiography with colour-flow Doppler is required for diagnosis:
  - PDA may be visible from the left cardiac window (far forward – through triceps).
  - left atrial enlargement and left ventricular volume overload are common.
  - Doppler echocardiography may demonstrate:
    - disturbed blood flow within PDA and in the pulmonary artery.

### Management/treatment

- treatment for this condition has not been thoroughly evaluated in the horse:
  - referral is indicated.
  - surgical correction is possible.
- pharmacological closure with prostaglandin inhibitors is described in other species.
- prognosis is grave, particularly when other cardiac defects are present.

## Congenital valvular deformities

### Definition/overview

- uncommon, but described cases include:
  - pulmonic valve stenosis.
  - tricuspid atresia.
  - congenital mitral chordal rupture.
  - papillary muscle deformity.
  - mitral valve dysplasia, and parachute valve.

### Aetiology/pathophysiology

- all cases are developmental.
- tricuspid atresia and pulmonic valve stenosis are typical components of complex congenital deformities.
  - concurrent VSDs and ASDs critical for foals with tricuspid atresia to survive the fetal transition.
  - provide an alternative path for obstructed flow.

- ASD allows blood to flow right to left.
- shunted blood flows into left ventricle and right ventricle across the VSD.
- resulting in severe left-sided volume overload.

## Clinical signs

- sudden death in the neonatal period.
- weak, dyspnoeic foal.
- cyanosis is common.
- foals that survive neonatal period are typically tachycardic and may have episodes of collapse.
- grade 4–6/6 pansystolic murmur is present with PMI over the left thorax.
- clinical signs of left-sided heart failure are present.
- foals with pulmonic stenosis or tricuspid atresia are often severely stunted in growth.
- mitral valve deformities are variably stunted depending on severity of stenosis or regurgitation:
  - cardiac chamber failure may ensue.
  - loud systolic murmurs are common.

## Differential diagnosis

- acquired valvular disease, bacterial endocarditis and complex congenital cardiac disease.

## Diagnosis

- no characteristic radiographic changes, but chamber enlargement may be present.
- echocardiography diagnosis:
  - 2D echo may identify abnormal valvular morphology, post-stenotic dilatation and cardiac chamber abnormalities (dilatation, hypertrophy).
  - colour-flow Doppler echocardiography is essential for demonstrating the direction of blood flow and extent of the regurgitation.
  - pulsed-wave and continuous-wave Doppler required to identify a step up in pressure across stenoses.

## Treatment

- no specific treatments available.
- euthanasia often indicated because of the severity of clinical signs.

# Tetralogy of Fallot

## Definition/overview

- four congenital cardiac anomalies are present:
  - overriding aorta:
    - sits across the interventricular septum.
    - straddles both the left and right ventricular outflow tracts.
  - pulmonic stenosis.
  - VSD.
  - right ventricular hypertrophy.
- if ASD also present, 'pentalogy of Fallot'.

## Aetiology/pathophysiology

- conal septum, develops abnormally and is not completed:
  - narrowing of the right ventricular outflow tract and incomplete ventricular septum.
- aorta overrides both outflow tracts because of the pulmonic stenosis:
  - pulmonic stenosis results in right ventricular pressure overload and hypertrophy.
  - reduced pulmonary perfusion.
- systemic hypoxia from mixing of oxygenated with unoxygenated blood in the aorta.
- equalisation of pressure may occur across the VSD.
- cyanosis is more pronounced, the more severe the obstruction of the RVOT.

## Clinical presentation

- cyanosis with light exercise.
- loud pansystolic murmur, often with a thrill over the left 3rd to 4th ICS:
  - murmur may be complex due to the presence of multiple anomalies.
- severe exercise intolerance.
- cyanosis not responsive to intranasal oxygen therapy.

## Differential diagnosis

- other causes of cyanosis, including:
  - respiratory distress and cardiac failure with pulmonary oedema.
- other congenital cardiac anomalies such as:
  - right ventricular hypoplasia or persistent truncus arterious.

### Diagnosis

- clinical pathology: polycythaemia may occur with prolonged hypoxia.
- decreased pulmonary vascularity may be observed radiographically.
- echocardiographic diagnosis is required:
  - all four components are readily identified.
  - VSD is usually large, and the aorta sits above the VSD.
  - RV wall is sufficiently hypertrophied to detect with echocardiography.
  - Doppler or contrast echocardiography may characterise direction of blood flow.

### Management/prognosis

- no treatment.
- prognosis for survival is poor and euthanasia is often indicated.

# ACQUIRED CARDIAC DISEASE

## Introduction

- more common than congenital cardiac disease.
- majority of cases involve the heart valves and degenerative valve disease is common.
- myocardial, pericardial and peripheral vascular disease are less common.
- acquired arrhythmias have been addressed earlier.
- no breed predisposition.
- some conditions are more common in certain age groups.
- knowledge of age, breed, performance history and desired future use are important in:
  - determining the significance of the condition.
  - guiding recommendations.
  - determining the prognosis related to use.

## Valvular disease

- most cases of valvular heart disease are acquired.
- caused by:
  - degenerative changes (most common).
  - damage or rupture of chordae tendinae.
  - bacterial endocarditis.
- aortic and mitral (left atrioventricular) valves are the most affected.
- valvular insufficiency is a common cause of heart murmur in the horse:
  - may or may not be associated with degenerative valvular changes.
  - extent of regurgitation determines the haemodynamic impact:
    - moderate to severe insufficiency will influence performance.
    - could be of concern in horses undergoing general anaesthesia.

## Mitral valve disease (left atrioventricular valve)

### Definition/overview

- murmurs associated with this valve are relatively common (~3.5% of population).
- left-sided systolic murmurs most common cause for referral for cardiac evaluation:
  - grade 3/6 left-sided systolic murmur or greater should be referred for evaluation.
- commonly associated with performance limitations, compared to tricuspid valve insufficiency.
- most likely valvular insufficiency to lead to congestive heart failure.

### Aetiology/pathophysiology

- no breed predilection.
- structural changes in the valve leaflets may contribute to regurgitation:
  - valvular thickening, fenestration and cystic changes are described.
  - changes similar to endocardiosis in other species are also reported.
- valvular incompetency allows backward flow into the left atrium during systole:
  - valvular changes generally occur gradually.
  - overtime this increases the volume load on the left atrium.

   - subsequently atrial dilatation occurs and predisposes to the development of AF.
- left-sided heart failure occurs due to chronic left-sided volume overload:
   - may progress to pulmonary hypertension and right-sided pressure overload:
      - right-sided heart failure may then occur.
- cases of a major or minor chordae tendinae rupture are different:
   - sudden increase in volume allowed back into the left atrium.
   - major rupture can result in fatal left-sided volume overload and pulmonary oedema.
   - minor rupture may have an acute onset of lethargy and a new loud murmur.

## Clinical presentation

- often no clinical signs are apparent:
   - murmur discovered during routine examination.
- severe regurgitation may have respiratory signs such as:
   - laboured breathing, increased respiratory rate and effort.
   - exercise intolerance and prolonged recovery from exercise.
   - coughing.
   - left-sided signs are frequently interpreted as primary lower respiratory disease:
      - delays diagnosis and treatment until signs of right-sided heart failure occur.
- acute death from pulmonary oedema may occur with chordae tendinae rupture.

## Differential diagnosis

- innocent flow murmurs (systolic outflow tract murmur).
- endocarditis.
- primary non-infectious respiratory disease (equine asthma).

## Diagnosis

- usually, no clinical signs are apparent at the time of detection.
- left-sided systolic murmur:
   - band-shaped, soft, blowing, holosystolic basilar murmur (just above MV).
   - mid-systolic crescendo murmur may occur in cases with valve prolapse:
      - beat-to-beat variability in murmur quality may be present.
   - intensity/grade of murmur does not correlate with severity or haemodynamic significance.
- no characteristic radiographic or ECG changes.
- AF is present:
   - mild mitral insufficiency, cardioversion is often indicated.
   - moderate to severe mitral insufficiency:
      - prognosis for performance is negatively impacted.
      - risks/benefits of cardioversion should be carefully considered:
         - moderate MR is risk factor for AF recurrence.
   - severe MR and overt signs of congestive heart failure:
      - cardioversion is contraindicated.
      - control heart rate with treatment of clinical signs of heart failure.
- echocardiography is required for diagnosis and prognosis (Fig. 3.23):
   - 2D descriptions of mitral valve and apparatus (papillary muscles, chordae).
   - M-mode to assess left ventricular dimensions and functional indices.
   - atrial dimensions assessed from the left and often the right sides.

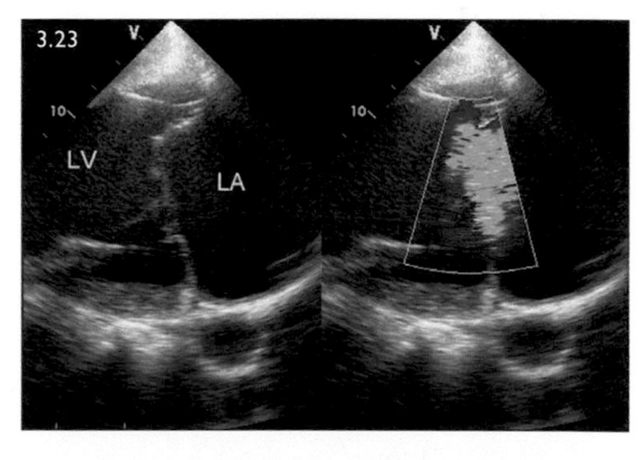


**FIG. 3.23** Left atrioventricular (AV) (mitral) valvular regurgitation. Left heart long-axis view taken from the left 5th ICS. Two-year-old Standardbred colt with poor performance and a grade 5/6 left-sided holosystolic murmur. Colour-flow Doppler echocardiography reveals high-velocity regurgitant flow at the left AV valve in green in the right-hand image. LV = left ventricle; LA = left atrium.

- colour-flow/pulsed-wave Doppler help determine the extent of regurgitation.
- onset of signs of congestive heart failure is a poor prognostic indicator.

## Management

- no treatment to correct mitral valve regurgitation.
- progression is generally slow and over years.
- serial echocardiographic examinations are indicated to monitor progression.
- exercising telemetry is indicated in moderate to severe disease (evidence of cardiac chamber enlargement).
- evidence of left-chamber volume overload might benefit from:
  - administration of an angiotensin-converting enzyme inhibitor (ACE-I):
    - benazepril or quinapril.
    - slow progression to left-sided failure, prolonging performance career.
    - effects of this treatment are not well documented.
- heart failure treatment is palliative and aimed at controlling clinical signs.

## Prognosis

- affected by:
  - underlying cause of the valvular disease (e.g. degeneration vs. chordal rupture).
  - age of onset (mild regurgitation in a 4 year old vs. in a 14 year old).
  - severity and haemodynamic impact of the regurgitation:
    - cardiac chamber dimensions, presence of arrhythmia and systolic function.
- major chordal rupture is rare:
  - usually results in sudden onset of heart failure.
  - tachycardia and respiratory distress from pulmonary oedema.
  - mortality rate is high.
- minor chordae ruptures have a more favourable prognosis:
  - typically present with a newly detected loud murmur.
- serial echocardiograms provide the most accurate prognosis for use and lifespan.

# Tricuspid valve disease (right atrioventricular valve)

## Definition/overview

- murmurs associated with the tricuspid valve (TV) are relatively common.
- pathology associated with the valve is seldom detected.
- right-sided regurgitation is uncommonly associated with any clinical signs or performance effects.
- up to 9% incidence reported (highest in Thoroughbred and Standardbred racehorses):
  - suspected to have tricuspid regurgitation (TR) associated with hypertrophic changes in response to extensive training rather than a hereditary component.
- grade 4/6 or greater right-sided systolic murmur should have echocardiography.

## Aetiology/pathophysiology

- common in elite athletes:
  - right ventricular hypertrophy in response to intense training.
  - results in altered coaptation of TV leaflets.
  - mild to moderate TR.
- deformities of the TV are uncommon.
- severe TR:
  - regurgitation into right atrium during systole results in right atrial enlargement overtime.
- tricuspid valve regurgitation may be:
  - primary.
  - secondary to left-sided heart failure or pulmonary hypertension from severe primary respiratory disease.
- chordal rupture is less common as compared to the left heart.

## Clinical presentation

- often no clinical signs are apparent, and murmur discovered during routine examination.
- murmur is classically holosystolic to pansystolic, band-shaped, soft and blowing:
  - PMI is on the right side of the chest usually at ICS 4.
- AF can occur with right-sided atrial dilatation but is reported less commonly.

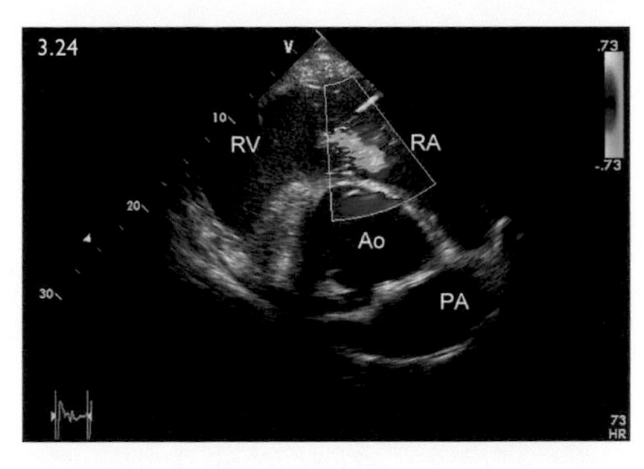


FIG. 3.24 Right AV (tricuspid) valvular regurgitation. Right-heart long-axis view taken from the right 4th ICS, highlighting the AV valve. This horse had a history of poor performance, AF and a grade 2/6 right-sided holosystolic murmur. Colour-flow Doppler echocardiography reveals high-velocity regurgitant flow in green. RV = right ventricle; RA = right atrium; Ao = aorta; PA = pulmonary artery.

- right-sided congestive heart failure may develop in severe cases.

## Differential diagnosis

- Endocarditis, functional murmurs and VSD.

## Diagnosis

- thoracic auscultation reveals a right-sided, systolic murmur.
- echocardiography is required for diagnosis and prognosis (Fig. 3.24):
  - 2D descriptions of the tricuspid valve and extent of regurgitation.
  - right-sided chamber dimensions and shape (especially relative to the left side).
  - pulmonary artery dimensions.
  - colour-flow and pulsed-wave Doppler aid in determining the extent of regurgitation:
    - evidence of trace/mild TR is common without an associated murmur.
- onset of signs of congestive heart failure is a poor prognostic indicator.

## Management

- no treatment indicated unless signs of congestive right-sided heart failure:
  - heart failure treatment is palliative and aimed at controlling clinical signs.
- related to primary severe respiratory disease, treat the underlying condition.
- TR severity may change overtime in response to detraining.

## Prognosis

- mild to moderate TR is excellent:
  - clinical signs, including performance effects, are uncommon.
  - progression, if any, is generally slow over years.
- serial echocardiograms provide the most accurate prognosis for use and lifespan.

# Aortic valve disease

## Definition/overview

- most common valve to have pathological change.
- most diagnosed in older horses (>12 years of age).

## Aetiology/pathophysiology

- most often affected by degenerative changes that are characterised by:
  - nodules, fenestrations and thickening of the cusps.
- valve prolapse can occur.
- valve fails to close completely and blood flows back into left ventricle in diastole.
- aortic insufficiency is generally well tolerated:
  - over time and depending on the size of the regurgitant jet.
  - left ventricular volume overload may occur:
    - increased risk of ventricular arrhythmias with overload.

## Clinical presentation

- diastolic murmur best heard from the left heart base:
  - frequently heard on the right as a slightly less intense diastolic murmur.
  - soft and typically crescendo–decrescendo.
  - musical or 'honking' murmurs are also described.
  - intensity of murmur does not correlate to severity of regurgitation.
- clinical signs are uncommon unless the disease is advanced.
- severe disease:

  - bounding peripheral arterial pulses are detected.
  - represent rapid diastolic drop-off in pressure.
- AF may develop with subsequent left-sided chamber enlargement.

## Differential diagnosis

- diastolic flow murmur is the top differential diagnosis.
- Pulmonic valvular regurgitation can present with a similar murmur.

## Diagnosis

- characteristic murmur suggests aortic valvular disease.
- echocardiography confirms the diagnosis (Fig. 3.25).
- exercising electrocardiogram (telemetry) is recommended for riding and driving horses with moderate or severe AR.
- trivial or minute AR:
  - commonly encountered with colour-flow Doppler echocardiography.
  - no clinical significance and is not associated with a murmur.

## Treatment

- Aortic valve disease is not treatable.
- left-chamber volume overload cases may benefit from administration of an ACE-I such as benazepril or quinapril:
  - slows progression to left-sided failure, prolonging a horse's performance career.
  - not well documented.
- signs of heart failure – treatment is palliative.

## Management

- serial echocardiograms are indicated (every 6–12 months) to monitor for progression.
- exercising ECG may become indicated as disease progresses or if there are any complaints of exercise intolerance.

## Prognosis

- good to excellent for life and performance with mild AR.
- poor when signs of cardiac failure are present.

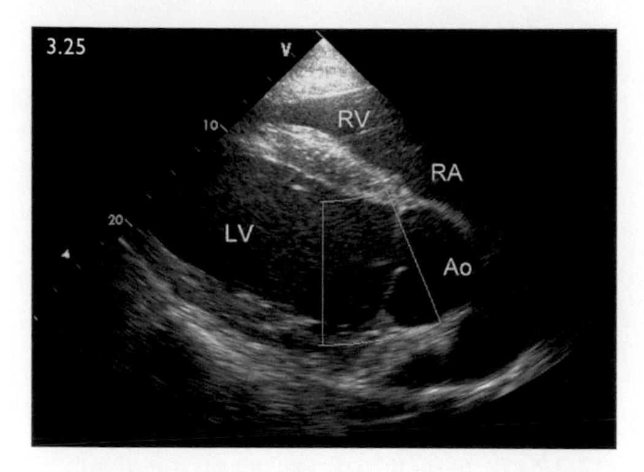


**FIG. 3.25** Aortic regurgitation. Right-heart long-axis view highlighting the aortic valve taken from the right 4th ICS. Colour-flow Doppler echocardiography reveals AR as a red flame-like flow present during diastole. No murmurs were associated with this finding. RV = right ventricle; RA = right atrium; LV = left ventricle; Ao = aorta.

# Bacterial endocarditis

## Definition/overview

- uncommon.
- acute or chronic disease associated with bacterial colonisation and development of vegetative lesions either on the valves or on the non-valvular endocardium.
- left-sided valves most affected:
  - aortic valve > left AV valve.
  - pulmonic valve and non-valvular endocardium uncommonly involved.

## Aetiology/pathophysiology

- endocardial injury may precede bacterial colonisation but is not a prerequisite:
  - injury from the repetitive impact of turbulent flow from regurgitant jets in valvular insufficiency or VSDs.
- subsequent damage to the valve following successful treatment of the infection:
  - permanent valvular dysfunction due to deformation and abnormal coaptation.
- *Actinobacillus* sp. and *Streptococcus* sp. are the most common isolates from blood cultures of infective endocarditis.
- source of infection is unknown:
  - septic jugular vein thrombophlebitis is a possible predisposing factor.
- no sex or breed predilections, but young adults appear to be affected more frequently.

## Clinical presentation

- commonly present with fever, which may be intermittent and only clinical sign.
- tachycardia, tachypnoea, weight loss, anorexia and depression are common.
- cardiac murmur may or may not be present:
  - fever, new murmur and ill-thrift should raise the clinical suspicion.
- clinical signs of cardiac failure may be the presenting complaint with chronic disease.
- septic foci in the myocardium lead to arrhythmias.
- clinical signs associated with septic emboli to other organs/structures may be the primary complaint:
  - renal infarction.
  - variable lameness and/or synovial structure distension is seen in some cases.

## Differential diagnosis

- parasitic endocarditis (uncommon, aortic valve), congenital heart disease, acquired valve insufficiency.
- abscessation, neoplasia, sepsis, polyarthritis, pneumonia and colitis.

## Diagnosis

- diagnostic approach as per any workup for a fever of unknown origin:
  - include a comprehensive physical examination and minimum data base.
- confirmed with:
  - echocardiographic identification of a lesion:
    - valvular deformity and vegetative lesions readily identified in most cases:
      - early disease lesions may appear as a small valvular nodule or mild thickening.
    - non-valvular endocardial lesions are more difficult to identify.
    - absence of a lesion on the first echocardiogram does not rule out endocarditis:
      - serial exams are warranted and careful daily auscultation.
  - positive blood culture:
    - samples from multiple sites using aseptic technique:
      - ideally during febrile episodes.
    - single blood cultures can be unrewarding.
    - collected prior to administration of antimicrobial agents.
- blood samples may reveal:
  - leucocytosis characterised by a mature neutrophilia and hyperfibrinogenaemia.
  - leucopaenia in acute onset cases.
  - non-regenerative anaemia typical of chronic disease.
  - cardiac troponin I increases if:
    - lesion extends to the myocardium.
    - septic emboli have settled in the myocardium.
- arrhythmias are possible and tachycardia is variable.

## Treatment

- aggressive treatment is essential.
- broad-spectrum intravenous antibiotic treatment is indicated:
  - initially penicillin G (22,000 IU/kg i/v every 6 hours) for Gram-positive coverage.
  - Gram-negative coverage either:
    - gentamicin (6.6 mg/kg i/v every 24 hours):
      - avoid in horses with renal azotaemia.
    - enrofloxacin (5–7.5 mg/kg i/v every 24 hours).
  - re-evaluate therapy based on culture and susceptibility results and clinical response to initial therapy after 5 days.
  - long-term antimicrobial treatment is required to sterilise the lesion:
    - intravenous course often followed by long-term oral antimicrobials.
    - duration is typically no less than 4 weeks.
    - therapy should extend:
      - at least 2 weeks beyond resolution of clinical signs.
      - normalisation of the leucogram and fibrinogen concentration.
      - improvement of lesion size and character on serial echocardiographic examinations.
- supportive therapy on a case-by-case basis:

- NSAIDs in pyrexic or depressed animals.
- gastroprotectants – may help improve appetite.
- i/v fluid therapy.
- digital cryotherapy for preventing sequelae such as laminitis.

### Prognosis

- poor to guarded for both survival and return to athletic performance.
- onset of heart failure is a grave prognostic indicator and euthanasia should be considered for humane reasons.

# MYOCARDIAL DISEASES

## Introduction

- cardiac dysfunction associated with myocardial disease, termed cardiomyopathy, is not well documented in horses.
- changes in the myocardium consistent with focal fibrosis are well documented in horses at necropsy.
- myocardial disease or injury may occur because of:
  - toxic insult.
  - infectious process.
  - neoplasia.
  - trauma, degeneration and inflammation.
  - nutritional cardiomyopathy related to vitamin E and selenium deficiency:
    - not well documented in the horse.
- occurrence of systolic or diastolic failure not documented in the horse.
- clinical signs of myocardial disease consist of:
  - tachycardia.
  - arrhythmia.
  - poor cardiac output.
  - acute myocardial disease can mimic colitis/colic.
  - focal lesions are more likely to present with arrhythmia.
  - extensive lesions more commonly lead to heart failure.

## Primary myocarditis

### Definition/overview

- inflammation of the cardiac muscle characterised on histopathology by:
  - myocardial cell death/damage and infiltration of inflammatory cells.
- working clinical diagnosis made based on cardiac abnormalities (abnormal rate, arrhythmia, impaired function) and increased cardiac troponin I.

### Aetiology/pathophysiology

- consequence of viral or bacterial respiratory disease:
  - inflammation/infiltration of inflammatory cells disrupts:
    - cell function, myocardial contraction, and conduction of electrical impulses.
- results in reduced cardiac output (CO) due to reduced stroke volume.

### Clinical presentation

- client may suspect colic due to signs such as depression, reduced appetite and weakness.
- recent respiratory disease, fever or anorexia may precede signs.
- arrhythmia with variable clinical signs.
- low cardiac output signs:
  - tachycardia with hypokinetic peripheral arterial pulses.
  - increased respiratory rate and effort.
  - possible signs of right-sided disease.
- exercise intolerance, congestive heart failure and collapse are possible.

### Differential diagnosis

- ionophore toxicity, cantharidin toxicity, vitamin D toxicity, severe colic or colitis and chronic respiratory disease.

### Diagnosis

- history of recent viral or bacterial disease and potential exposure to toxic agents.

- cardiac troponin I increased with active/ongoing myocardial inflammation.
- blood lactate concentration may be increased due to reduced cardiac output, pulmonary oedema, or both.
- electrocardiogram results are variable:
  - normal.
  - frequent ectopic beats, VT or other arrhythmias.
- echocardiography may demonstrate:
  - hypokinetic left ventricle.
  - regional abnormal wall motion.
  - cardiac chamber dilatation.
  - increased spontaneous luminal echo contrast.
- pulsed-wave tissue Doppler imaging may provide a more sensitive method of detecting subtle myocardial dysfunction.
- continuous electrocardiographic (telemetry) monitoring is indicated.

## Treatment

- underlying disease should be diagnosed and treated if possible.
- supportive therapy is indicated to help organ function and prevent hypoxaemia:
  - **intravenous fluid therapy contraindicated in horses with reduced contractility; may result in death.**
- NSAIDs or corticosteroids (if active infection ruled out) may be beneficial.
- strict stall rest to reduce myocardial oxygen demand on the heart:
  - rest period should extend well beyond resolution of clinical signs.

## Long-term management

- repeat echocardiogram and 24-hour telemetry at rest before horse re-enters training programme of any kind.
- exercise ECG should be performed prior to return to any forced exercise.
- exercise intolerance and arrhythmia may persist due to permanent myocardial damage.

## Prognosis

- guarded to good depending on severity.
- horses with increased cardiac troponin and arrhythmia following a viral infection have a more favourable prognosis than horses with overt cardiac dysfunction.

# Toxic myocarditis

## Definition/overview

- ionophore exposure is the most reported:
  - ionophore antibiotics:
    - used in cattle feed for growth promotion.
    - coccidiostats in poultry feed.
  - horses are exquisitely sensitive to ionophores.
  - $LD_{50}$ for salinomycin, monensin and lasalocid are 0.6 mg/kg, 2–3 mg/kg and 21.5 mg/kg, respectively.
- other causes of toxic myocarditis include:
  - cantharidin (blister beetle) toxicosis in alfalfa hay or alfalfa hay-derived cubes.
  - exposure to toxic plants containing cardiac glycosides:
    - oleander, milkweed, foxglove, azalea/rhododendron, mountain laurel, avocado leaves.

## Aetiology/pathophysiology

- mechanism of action of cardiotoxins varies.
- acute damage to myocardial cells leads to:
  - dysfunction and myodegeneration (loss of cardiomyocytes).
  - replacement with fibrosis.
- results in decreased cardiac function.
- arrhythmia due to variation in signal conduction, blockage and aberrant conduction.

## Clinical presentation

- clinical signs vary and are dependent on the nature of the exposure:
  - onset can be immediate or delayed for weeks.
  - delayed signs:
    - subacute/chronic low-level exposure.
    - horses recovered from an acute phase.
- colic may be suspected due to depression, reduced appetite, ataxia, weakness, extensive sweating and/or diarrhoea.
- death within 24–48 hours of exposure.
- arrhythmia with variable clinical signs.
- GI, hepatic, renal and skeletal muscle abnormalities may be present.

## Differential diagnosis

- viral myocarditis, vitamin D toxicity, severe colic or colitis.

## Diagnosis

- suspected when multiple animals are affected.
- recent feeding of a new shipment or source of feed including garden trimmings:
  - samples of feed should be saved for analysis.
  - definitive diagnosis on identification of a toxin in feed, pasture or surroundings.
- gastric contents analysed in horses that die acutely.
- urine collected for cantharidin detection.
- cardiac troponin I is increased with active/ongoing myocardial inflammation:
  - screening tool to identify horses with a sublethal exposure to ionophores.
- increased CK, AST, creatinine and bilirubin are all reported.
- hypocalcaemia and hypokalaemia may be present.
- electrocardiogram normal or likely to demonstrate frequent ectopic polymorphic complexes, VT or other arrhythmias.
- continuous electrocardiographic (telemetry) monitoring is indicated.
- echocardiography may demonstrate a whole range of changes including:
  - hypokinetic left ventricle.
  - heterogeneity of the myocardium.
  - cardiac chamber dilatation.

## Management

- no specific treatments are available.
- toxin confirmed in the feed source – remove and replace.
- nasogastric administration of activated charcoal (0.75–2 g/kg) or biosponge (1 lb powder/3 L of warm water) to adsorb toxins:
  - **mineral oil may potentiate absorption of some toxins and is not advised.**
- monitored intravenous fluid therapy indicated in acute cases:
  - enhances renal excretion of toxins.
  - supports circulation.
- therapy to support organ function and prevent laminitis.
- careful use of NSAIDs or corticosteroids depending on underlying aetiology.
- anti-arrhythmic therapy in life-threatening arrhythmias:
  - **caution as all anti-arrhythmics have pro-arrhythmic potential.**
- Digoxin is contraindicated in cases of ionophore toxicity.
- strict stall rest to reduce myocardial oxygen demand.
- period of rest well beyond resolution of clinical signs.

## Long-term management

- repeat echocardiogram and 24-hour telemetry at rest before horse re-enters training.
- exercise ECG performed prior to return to any forced exercise.
- exercise intolerance and arrhythmia may persist due to permanent myocardial damage.

## Prognosis

- good to grave depending on toxin ingested and severity of clinical signs.
- all horses with ionophore exposure, even those with sublethal exposure:
  - guarded for return to performance due to delayed onset of cardiac disease.
- survivors of acute cantharidin toxicity have a more favourable prognosis.

# Myocardial injury following acute haemorrhage or inflammatory disease

## Definition/overview

- injury or insult to the cardiac muscle due to ischaemic or inflammatory disease:
  - severe acute haemorrhage.
  - severe inflammatory or ischaemic GI lesion.
- persistent tachycardia and or ventricular arrhythmia despite improvement or resolution of the underlying disease.
- common especially in the 24–72 hours following the inciting event.
- cardiac troponin I increased and may continue to rise before decreasing rapidly if the underlying insult is removed.

- continuous telemetry documents the frequency and morphology of ectopic beats:
  - sustained VT can develop.
- monitor electrolytes, especially ionised magnesium concentrations.
- echocardiography may be normal or demonstrate hypokinetic left ventricle, cardiac chamber dilatation, etc.
- underlying disease typically has already been treated at the time of onset/ recognition.
- correct electrolyte disturbances.
- anti-arrhythmic therapy may be indicated.
- strict stall rest to reduce myocardial oxygen demand on the heart.
- rest beyond resolution of clinical signs (month or more) before rechecking.
- repeat 24-hour telemetry at rest before horse re-enters training programme of any kind.
- exercise ECG should be performed prior to return to any forced exercise.
- short-term prognosis is good to guarded:
  - depends on severity and ability to treat the underlying condition.
- exercise intolerance and arrhythmia may persist due to permanent myocardial damage.

# PERICARDIAL DISEASE

- congenital pericardial disease is very rare.
- acquired pericardial disease is uncommon.
- usually caused by inflammation of the pericardial sac and pericarditis:
  - classified as effusive, fibrinous or constrictive.
- benign pericardial effusion may present during disturbances in fluid homeostasis such as congestive heart failure or hypoproteinaemia.

## Pericarditis

### Definition/overview

- inflammation of the pericardial sac.

### Aetiology/pathophysiology

- often unknown and classified as idiopathic.
- bacterial pericarditis occurs sporadically or as a sequela of acute or chronic bacterial pleuropneumonia.
- fibrinoeffusive pericarditis associated with mare reproductive loss syndrome (MRLS):
  - first described in 2001 and linked to outbreaks of eastern tent caterpillar.
- *Actinobacillus* sp. is the most consistently cultured organism.
- *Enterococcus faecalis*, *E. coli*, *Streptococcus zooepidemicus*, *Mycoplasma*, *Corynebacteria* and *Rhodococcus* are reported as well.
- volume of fluid increases in the pericardial sac leading to increases in pressure:
  - predominantly affects the low-pressure, right side of the heart first:
    - signs of right-sided congestive heart failure.
    - compression of the right atrium impedes venous return, leading to venous congestion.
- clinical signs are more severe with sudden onset of effusion as the pericardial sac has no time to stretch.

### Clinical presentation

- tachycardia with a weak rapid pulse.
- muffled heart sounds.
- jugular vein distension with or without pulsation.
- ventral oedema.
- peripheral perfusion is decreased, extremities cool and the CRT prolonged.
- fever is common, but inconsistent.
- depression and colic-like signs are reported, but also inconsistent.

### Differential diagnosis

- right-sided congestive heart failure.
- colitis, pleuropneumonia and benign effusion secondary to hypoproteinaemia.

### Diagnosis

- inflammatory leucogram with hyperfibrinogenaemia is common.

- increased size and rounding of the cardiac silhouette on lateral thoracic radiographs:
  - occasionally, a gas cap is present.
- ECG often demonstrates decreased QRS amplitude (hallmark of pericardial effusion):
  - electrical alternans (variable complex size) may be present.
- echocardiography confirms the diagnosis and allows monitoring for progression and response to therapy:
  - valuable for guiding pericardiocentesis.
  - reveals character of fluid, presence of fibrin, air (anaerobic infection).
  - flattened appearance of right heart and reduced functional indices of left heart.
- pericardiocentesis is both diagnostic and therapeutic:
  - samples submitted for cytological evaluation and bacterial culture.

## Management

- Pericarditis with cardiac tamponade:
  - restriction of cardiac pump causing hypotension and low cardiac output.
  - severe life-threatening condition.
  - requires immediate aggressive management via pericardiocentesis (Table 3.5).
- mild effusion is not an emergency.
- stall rest indicated in all cases.
- antimicrobial therapy is often indicated:
  - broad-spectrum intravenous antimicrobial therapy when septic, fibrinous effusion.

**TABLE 3.5** Techniques for performing pericardiocentesis

| | |
|---|---|
| **Indications** | • Cardiac tamponade<br>• Guide therapy (culture, cytology) |
| **Contraindications** | • Small amount of fluid (risk of cardiac/coronary laceration)<br>• If echogenic pleural fluid present (risk of contamination of pericardial space) |
| **Adverse effects** | • Vessel laceration (thoracic or coronary) may lead to death<br>• Ventricular ectopy if epicardium contacted<br>• Extension of infection from pleural space |
| **Cautions** | • Pericardiocentesis without ultrasound guidance should be performed with great caution<br>• Left side preferred due to lower risk of coronary artery laceration<br>• Indwelling drains risk ascending infection and pneumopericardium |
| **Equipment** | • Ultrasound machine (optimal)<br>• 10–12-gauge intravenous catheter<br>• Alternatively, small-bore chest tube (if large volume) |
| **Location** | • 5th ICS on the left side (right side only if large volume of fluid present, due to increased risk of coronary artery laceration)<br>• Variable level, between costochondral junction and shoulder<br>• Avoid lateral thoracic vein |
| **Technique** | • Surgical preparation<br>• Local anaesthetic<br>• Scalpel incision – size dependent on catheter choice<br>• Gentle force needed to penetrate thoracic cavity<br>• One hand used to advance catheter<br>• Stabilise catheter with second hand near skin to control depth of advancement<br>• Advance catheter cautiously until slight pop detected or fluid aspirated<br>• Sudden loss of resistance as catheter enters thoracic cavity/pericardium necessitates controlled advancement to avoid laceration<br>• Allow fluid to drain<br>• Aspirate with caution – risk of myocardial contact<br>• Avoid pneumopericardium by use of one-way valve<br>• Indwelling drain possible |

  - anaerobic spectrum broadened by adding metronidazole:
    - gas in the effusion on echocardiography or radiography.
- Idiopathic cases with anechoic fluid and no evidence of fibrin:
  - idiopathic, non-infectious effusions.
  - do not require antimicrobial therapy.
  - corticosteroids may be of benefit in such cases.
- NSAIDs (flunixin meglumine) indicated for both anti-inflammatory and analgesic effects.
- pericardial drainage via pericardiocentesis or indwelling catheter is critical in cases with tamponade and septic effusions.

### Long-term management

- serial echocardiogram to monitor response to therapy:
  - repeated after period of stall rest and before starting any training programme.
- exercise ECG should be performed prior to return to any forced exercise.
- 6-monthly echocardiography performed for at least the first year following the diagnosis:
  - monitor for development of constrictive pericarditis.
  - especially in horses recovering from severe septic fibrinous pericarditis.

### Prognosis

- generally poor.
- early diagnosis with aggressive, appropriate therapy may be successful.
- young racehorses diagnosed with fibrinoeffusive pericarditis, when managed appropriately have had successful racing careers.

## Pericardial neoplasia

### Definition/overview

- extremely rare.
- secondary extension of intrathoracic neoplasia is more common than primary neoplasia.
- restriction of cardiac filling from:
  - extensive fluid accumulation in pericardial sac.
  - restriction by mass.
- clinical presentation is similar to pericarditis.
- main differentials are bacterial, viral or idiopathic pericarditis, right-sided congestive heart disease, pleural effusion and other intrathoracic neoplasia.
- diagnosis is similar to pericarditis:
  - echocardiography and cytological evaluation of fluid are key.
  - identification of neoplastic cells in pericardial effusion is diagnostic.
- pericardiocentesis may provide short-term relief.
- prognosis is grave and euthanasia is indicated.

## Heart failure

### Definition/overview

- overt cardiac failure is uncommon in horses.
- defined as inadequate pump function leading to circulatory failure.
- other conditions can result in circulatory failure (thoracic mass, severe hypovolaemia).

### Aetiology/pathophysiology

- some primary cardiac conditions can result in the development of heart failure:
  - bacterial endocarditis.
  - chordae tendinae rupture.
  - cardiomyopathy.
  - myocarditis.
  - congenital cardiac disease.
  - pericarditis.
  - severe mitral valve disease.
- time course from onset of primary disease to heart failure varies greatly.
- right-sided heart failure can occur:
  - secondary to chronic respiratory disease, such as chronic severe equine asthma.
  - pulmonary hypertension develops and causes pressure overload on the right heart:
    - due to chronic pulmonary vasoconstriction associated with alveolar hypoxia.
    - known as cor pulmonale.
- heart failure is a dynamic condition which is not necessarily present at all times:

- animals with cardiac disease but sufficient cardiac reserve may have no clinical signs of heart failure at rest.
- at increased heart rates, the same animal may exceed its cardiac reserve and demonstrate signs of heart failure.

- eccentric hypertrophy (muscular hypertrophy combined with chamber dilatation):
  - typical response to volume overload.
  - commonly observed in horses with heart failure.
- heart failure may be either acute or chronic:
  - acute failure:
    - initiating event is sudden in occurrence (chordae tendinae rupture).
    - signs of acute left-sided heart failure are due to acute respiratory distress from pulmonary oedema.
    - signs of acute right-sided heart failure are less obvious.
    - peripheral oedema forms over a few days.
  - chronic failure:
    - underlying lesion present for weeks to years.
    - gradual decompensation occurs.
    - clinical signs are more insidious in onset.
    - more rapid decompensation once compensatory mechanisms are overloaded.

## Clinical presentation

- acute left-sided heart failure often presents in dramatically:
  - fulminant pulmonary oedema.
  - horse is distressed and dyspnoeic.
  - copious volumes of frothy fluid pouring from both nostrils.
  - death may occur before intervention is possible.
- chronic left-sided failure usually presents with:
  - fatigue and weight loss.
  - history of mild/moderate chronic respiratory disease not responsive to typical treatment:
    - frothy nasal discharge (occasionally just a dried white discharge).
  - syncope episodes may develop.
  - pulmonary venous pressures increase and lead to right-sided overload.
  - reduced peripheral arterial pulse quality/intensity (hypokinetic pulse).
  - tachycardia with or without arrhythmia.
  - cardiac murmur:
    - left-sided systolic of variable intensity and commonly maximally over the heart base.
    - may not be present if due to myocardial disease.
  - clinical signs of right-sided heart failure are often evident.
- right-sided failure (Fig. 3.26) causes:
  - decreased cardiac output and decreased renal blood flow, salt and water retention.
  - increases body water content and increases systemic venous pressures leading to peripheral oedema and venous engorgement:
    - pectoral/sternum, distal limb and ventrum.
    - rapid jugular filling and distension with pulsation
  - pendulous appearance to the abdomen (ascites) is uncommon.
  - decreased peripheral perfusion may be a result of left- and or right-sided heart failure.
  - weight loss, lethargy, renal or hepatic signs, and diarrhoea may result.
  - atrial dilatation and associated AF is common in heart failure.
  - cardiac murmur is variable (depends on underlying aetiology).

## Differential diagnosis

- acute left-sided heart failure:
  - acute pneumonia, acute oesophageal obstruction (pulmonary oedema fluid from nostrils mistaken for saliva), anaphylaxis, pulmonary abscess rupture.
- chronic failure:
  - all causes of weight loss (poor dentition, neoplasia, malabsorption), other causes of peripheral oedema.
- acute right-sided heart failure:
  - overly aggressive fluid therapy, especially in neonates or horses with compromised renal function.

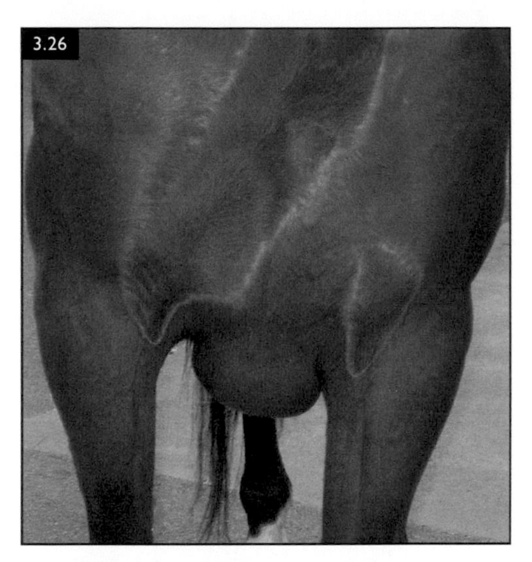

FIG. 3.26 Ventral oedema. Severe ventral oedema in a 7-year-old Standardbred mare in heart failure. The mare was tachycardic, had venous distension, AF, cardiac dilatation, reduced functional indices and severe left AV valve regurgitation. The ventral oedema resolved following 1 week of therapy with furosemide and digoxin.

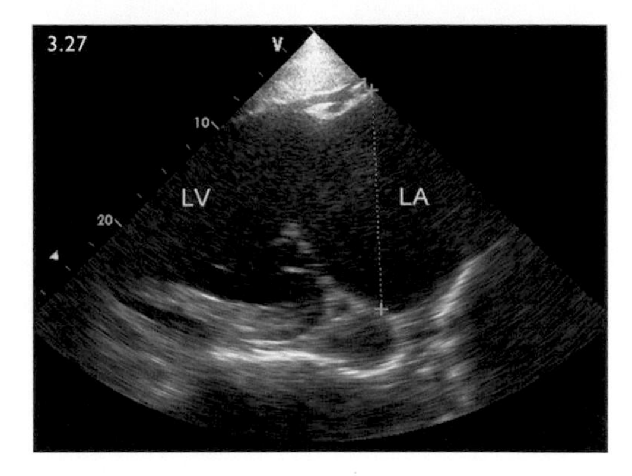


FIG. 3.27 Left atrial dilatation. Left-sided long-axis view taken from the left 5th ICS. Two-year-old Standardbred colt with a history of poor performance and a grade 5/6 left-sided holosystolic murmur. Left atrial measurements indicated left atrial dilatation with a diameter of 16.05 cm. The horse was in sinus rhythm, with no signs of heart failure. LV = left ventricle; LA = left atrium.

## Diagnosis

- clinical signs.
- chest radiography may show increased pulmonary interstitial pattern (non-specific).
- ECG:
  - tachycardia but magnitude can vary.
  - AF is common in horses with heart failure.
  - ventricular or supraventricular premature contractions.
- echocardiography:
  - confirm diagnosis and monitor progression/response to therapy (Fig. 3.27):
  - multiple cardiac chamber enlargement may be evident:
    - heart may not fit within the 30 cm of depth of the screen.
    - left ventricle often has a globoid, bowl-shaped appearance.
  - multiple valvular insufficiencies may exist.
  - pulmonary artery enlargement (diameter similar to or greater than aortic diameter) supports pulmonary hypertension.
  - chordae tendinae rupture will cause part of the mitral valve to appear to flail.

## Management

- restore cardiac function as soon as possible to abate the clinical signs of failure.
- stall rest to decrease cardiovascular demands.
- furosemide to treat life-threatening pulmonary oedema:
  - i/v administration (1–2 mg/kg q30 minutes):
    - until respiratory pattern is improved.
    - nasal pulmonary oedema fluid ceased.
  - then 0.5–2 mg/kg i/v or i/m every 8–12 hours.
- salt supplementation should be discontinued.
- ACE inhibitors:
  - benazepril, ramipril and quinapril.
  - decrease afterload but are expensive.
- torsemide (for long-term management).
- hydrochlorothiazide – diuretic used to reduce the level of potassium in the blood.
- potassium chloride:
  - replace potassium with chronic loop-diuretic use.

- hydralazine 0.5 mg/kg i/v q12 h – vasodilator to reduce afterload:
  - risk for hypotension and requires caution.
- digoxin is a positive inotrope and negative chronotrope:
  - improve cardiac output and decrease tachycardia.
  - narrow therapeutic index (1–2 ng/ml).
  - levels in the blood should be monitored regularly.

### Long-term management

- educate the owner to monitor for:
  - increase in resting heart rate.
  - rapid jugular refill time.
  - increased respiratory rate and effort.
  - nasal discharge characteristic of pulmonary oedema.
- serial echocardiogram every 3–6 months to monitor response to therapy.
- monitor serum electrolyte concentrations and renal function periodically.
- **horses with congestive heart failure must be retired:**
  - managed on pasture turnout successfully for months to years.
  - depending on the case and financial commitment.

### Prognosis

- generally poor and dependent on the underlying aetiology.
- chronic, slowly progressive heart failure due to valvular insufficiency is most likely to be managed for a limited period.
- right-sided heart failure due to primary respiratory disease is good:
  - if treatment of the respiratory condition is achieved.
- failure due to myocardial dysfunction or major ruptured chordae tendinae have a poor prognosis and euthanasia is often appropriate.

# MISCELLANEOUS CARDIOVASCULAR CONDITIONS

## Aortic root disease

### Definition/overview

- uncommon condition, most reported in aged stallions.
- rupture with or without aneurysm of the aortic root, resulting in collapse and death.
- aortocardiac fistula occurs when the aortic root ruptures into the interventricular septum or right side of the heart.
- Friesian horses appear to be predisposed to aortopulmonary fistula or aortic rupture and commonly present with signs of colic.

### Aetiology/pathophysiology

- unknown aetiology:
  - collagen disorder is suspected in Friesian horses.
- failure of the connective tissue of the aortic root and/or aortic wall due to necrosis or fibrosis.
- rupture may dissect from the right coronary sinus down the interventricular septum.
- junction of the right atrium, septum and aortic root will appear abnormal:
  - reminiscent of a VSD.
- haemopericardium or formation of pseudoaneurysms are other possible outcomes.

### Clinical presentation

- sudden death.
- acute onset heart failure (can mimic severe acute colic in some cases).
- tachycardia, discomfort and distress.
- features that suggest a primary cardiac problem:
  - bounding arterial pulse.
  - continuous murmur over the right ventricle.
  - diastolic component of the right-sided murmur is almost diagnostic.
- ventricular arrhythmia may be present.
- post the acute phase:

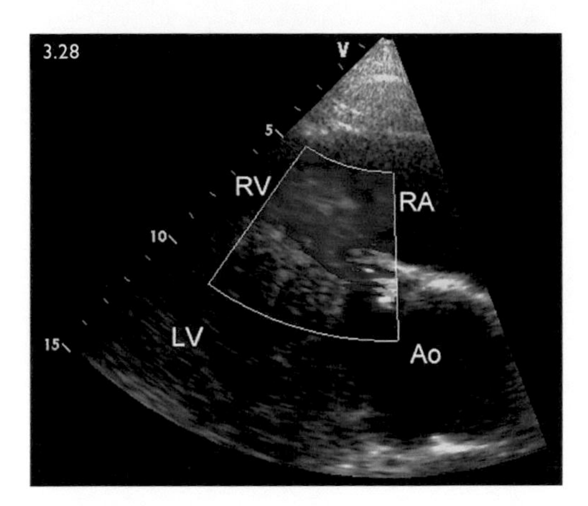


FIG. 3.28 Aortic root rupture in a 7-year-old Thoroughbred stallion with a history of sudden onset of colic signs and a new right-sided continuous murmur. Right-sided long-axis view, taken from the right 4th ICS, showing the right heart and aorta in magnification. Colour-flow Doppler reveals diastolic flow in red from the aorta into the right ventricle due to rupture of the aortic root, dissection into the interventricular septum and creation of an AV fistula. RV = right ventricle; RA = right atrium; Ao = aorta; LV = left ventricle.

- signs of right-sided heart failure develop in horses with aortocardiac fistula.
- Friesians with aortopulmonary fistula are commonly reported to demonstrate:
  - recurrent colic, increased body temperature, tachycardia, and respiratory signs.

## Differential diagnosis

- depends on the type and location of the rupture:
  - on echocardiography might include AR, VSD and myocarditis.
  - on sudden collapse or death, include pulmonary artery rupture, ventricular arrhythmia and non-cardiac causes.

## Diagnosis

- clinical examination with careful auscultation.
- cardiac troponin I may identify myocardial inflammation/injury.
- ECG:
  - sinus tachycardia or ventricular arrhythmia are common.
- echocardiography is required for diagnosis (Fig. 3.28):
  - 2D echocardiography may reveal a hypoechogenicity at the aortic root or interventricular septum.
  - colour-flow Doppler echocardiography to document abnormal flow from the aortic root across the septum.
  - rupture commonly occurs at the right coronary cusp of the aortic valve with rightward extension into the right ventricle.
  - with dissection into both ventricles, an acquired VSD might be observed.

## Management

- prognosis is poor.
- cases that survive should be treated symptomatically.
- recurrence or sudden acute death may occur.
- these horses and horses with sinus of Valsava aneurysm should be considered unsafe:
  - **not be ridden and handled only by adults informed of the condition.**

# Aortoiliac thrombosis

## Definition/overview

- uncommon condition.
- thrombosis of any part of the terminal aorta or aortic quadrification.
- ischaemic myopathy of the hindlimb muscles supplied by the terminal aorta or specific affected branch.

## Aetiology/pathophysiology

- unknown aetiology.
- *Strongylus vulgaris* implicated in some reports.
- sequelae of severe sepsis in foals.

## Clinical presentation

- variable signs.
- one or both hindlimbs may be affected.
- vague signs of mild lameness.
- lameness after or during exercise may become severe:
  - horse may stop and, in some cases, hold the leg up or be reluctant to bear weight.

- examination of affected limb may reveal:
  - slightly decreased temperature and peripheral pulse.
  - reduced filling of the saphenous vein.
- sweating and colic-like signs may be present.

## Differential diagnosis

- any cause of hindlimb lameness.
- colic.
- exercise-induced myopathy.

## Diagnosis

- palpation per rectum to evaluate the terminal aorta and branches for pulse quality:
  - aorta is more commonly affected.
  - pulse quality reduced.
- transrectal ultrasonography per rectum to identify thrombus (Fig. 3.29):
  - mixed to homogeneous echogenicity within vessel.
  - colour Doppler ultrasonography is useful to assess the blood flow.

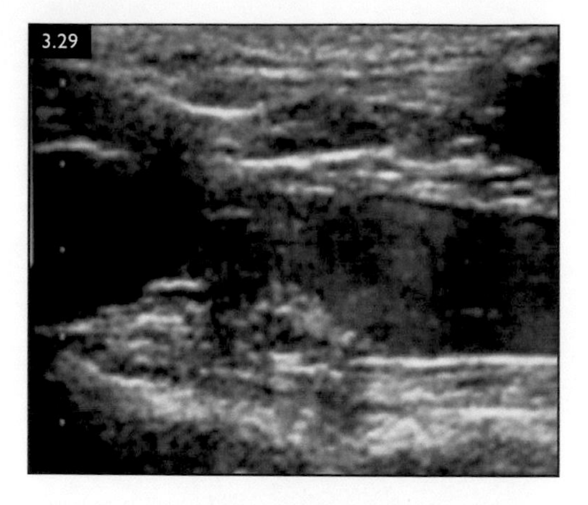


**FIG. 3.29** Transrectal ultrasonogram of a horse with aortoiliac thrombosis showing a large thrombus attached to the ventral wall of the terminal aorta, which is causing interruption of the blood flow downstream of the mass (hyperechoic area on right side of vessel).

## Management

- rest and anti-inflammatory medications.
- surgical thrombectomy can be attempted in cases where the femoral artery is primarily affected.
- anti-thrombotic agents (low molecular weight heparins or unfractionated heparins) and platelet aggregation inhibitors like clopidogrel are indicated.
- larvicidal deworming agents if *Strongylus* sp. are suspected.
- prognosis is guarded to poor.

# Thrombophlebitis

## Definition/overview

- common complication following venous catheterisation or when medications are administered via venipuncture.
- jugular vein is most affected due to its predominant use for venous access.
- thrombus size and presence or absence of infection dictate the clinical signs and treatment.

## Aetiology/pathophysiology

- local thrombogenic stimulus precedes the condition:
  - inflammation from systemic disease or local infection.
  - mechanical or chemical irritation.
- pro-thrombotic underlying disease:
  - severe colitis, pleuropneumonia, protein-losing enteropathy/nephropathy, etc.
  - can result in thrombus formation despite strict aseptic technique and careful catheter maintenance.
- injection of an irritating substance.
- poor aseptic technique during injection or catheter placement.
- pulmonary thromboembolism or endocarditis may be a sequela.

## Clinical presentation

- palpation of a firm, smooth, often cylindrical structure within the vessel lumen.
- venous filling may be impaired, slow or absent:
  - at the site of the thrombus.
  - cranial to the thrombus when the vein is held up distally.
- visible bulge or corded appearance to the affected site (Fig. 3.30).
- lack of definition of the jugular groove owing to adjacent or overlying soft-tissue involvement (oedema or cellulitis).

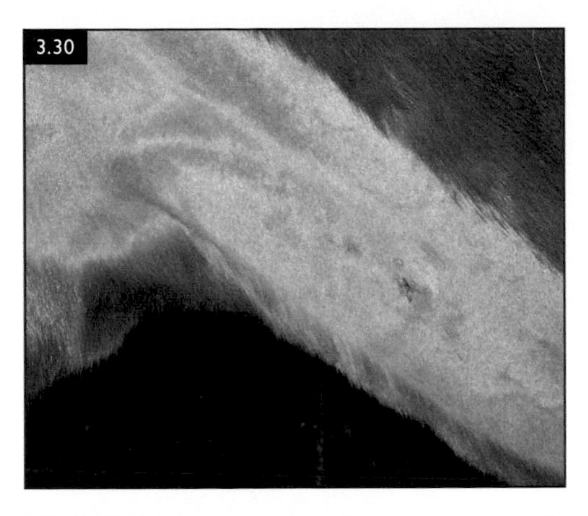

**FIG. 3.30** This horse had undergone recent colic surgery and long-term placement of a jugular vein catheter. It developed jugular vein thrombophlebitis, which clinically palpated as a hot, thready and fibrous vein proximal and distal to the skin puncture site, which was moist

- pain, heat and swelling at the site are hallmarks of septic thrombi:
  - fever (low grade to marked) is common with sepsis.
  - drainage from the skin puncture site is common.
- non-septic thrombi are typically not painful but can be associated with:
  - acute onset of oedema and superficial venous distension.
  - tissues that are typically drained by the affected vein:
    - unilateral facial swelling (especially over the masseter).
    - prominent superficial facial veins.
- complete occlusion of both jugular veins can result in severe pharyngeal oedema causing acute respiratory distress.
- small, partial thrombi are often unrecognised unless they interfere with intravenous catheter flow rates or are sources of fever.

## Differential diagnosis

- differentiation of septic thrombi from non-septic thrombi is important.
- perivascular reaction or mass.

## Diagnosis

- palpation and careful physical examination.

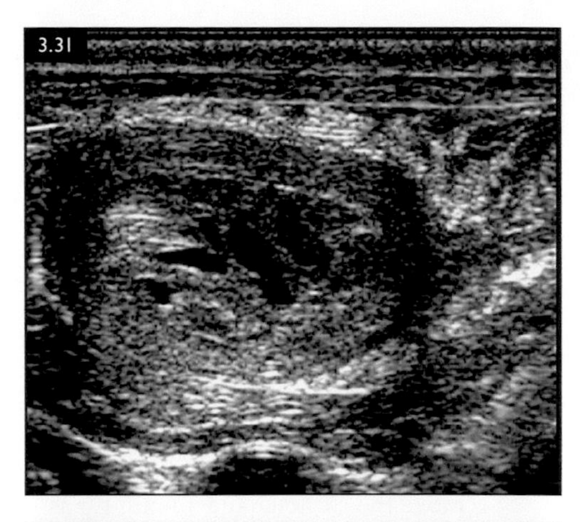

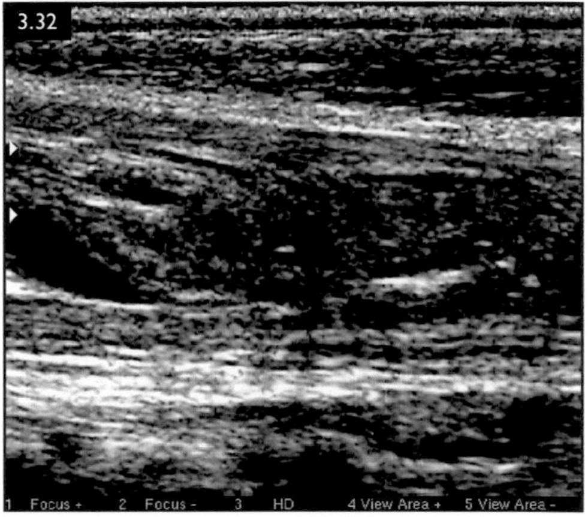


**FIGS 3.31, 3.32** Ultrasonograms of a thrombophlebitis of the jugular vein after catheter placement in a colic surgery case. Transverse (3.31) and longitudinal (3.32) views of the vein clearly showing intravascular thrombus occluding much of the vein.

- transcutaneous ultrasonography (Figs. 3.31, 3.32):
  - location and extent of the thrombus.
  - whether there is any blood flow around the affected portion of the vein.
  - appearance of the thrombus:
    - mixed echogenicity, especially hyperechoic foci characteristic of gas, are likely to be septic.
- complete blood count, fibrinogen and/or serum amyloid A concentrations.
- aerobic and anaerobic bacterial culture of any accessible infectious material:
  - catheter tip (distal end) aseptically cut off and submitted.

- material from the draining tract at the puncture site.
- common isolates include *Staphylococcus* sp., *Streptococcus zooepidemicus*, *Actinobacillus* sp. and *E. coli*.

## Management

- thrombus an incidental finding on an examination, no management required:
  - avoid venipuncture in that vein.
- small thrombi can resolve spontaneously without any specific treatment.
- catheter in place:
  - remove the catheter and submit tip as described.
  - alternative sites for venous access (catheter or venipuncture) used if necessary:
    - peripheral sites are preferred and include:
      - distally in the same vein.
      - lateral thoracic, cephalic or saphenous.
  - avoid catheterising or repeated venipuncture in the opposite jugular:
    - risk of bilateral jugular vein occlusion and severe pharyngeal oedema.
- acute thrombus and no evidence of pain or local/systemic inflammation:
  - topical management (hot compress, and diclofenac topical cream).
  - platelet aggregate inhibitor (clopidogrel).
- suspected or confirmed septic thrombi require:
  - systemic antimicrobial medication.
  - pain management.
  - topical therapy and platelet aggregate inhibitor.
- collateral circulation results in resolution of oedema in days to weeks.
- rare cases require surgical resection of the vein or marsupialisation of an abscessed vein.

## Prognosis

- excellent when a single vein is affected:
  - most cases recover with conservative medical therapy.
- collateral circulation leads to horses with complete venous occlusion having complete resolution of clinical signs.
- bilateral involvement is more likely to influence performance.
- venous abscessation is more difficult to treat and may require surgery.

CHAPTER 4

# Haemolymphatic System

## Diagnostic approach

### Blood collection

- collection is relatively easy:
  - usually the jugular vein.
- other options include:
  - saphenous, lateral thoracic and cephalic veins.
- only taken when the animal is calm.
- blood collected into tubes containing:
  - EDTA:
    - binds calcium to stop blood clotting.
    - ideal for haematology – purple-top blood tube.
- Citrate:
  - chelates calcium.
  - preferred for coagulation assays – blue-top blood tube.
- Heparin:
  - binds antithrombin III and not suitable for haematology.
  - blood-gas analysis and biochemical assays – green-top blood tube.
- Plain:
  - no anticoagulant – red top blood tube.
  - biochemical assays and serology.
- appropriate volume of blood per tube for correct anticoagulant/blood ratio for valid results.
- analyse samples quickly as each analyte has variable stability:
  - refrigerate at 4°C if a delay beyond 2 hours:
    - mix samples thoroughly before analysis.
    - analyse within 24 hours.
- perform blood smears immediately.
- haemostasis assays require very careful collection to prevent activation of the process:
  - discard initial 1–2 ml.
  - analyse within 4 hours.
  - or harvest plasma:
    - ship remaining sample on ice and freeze at –20° C until analysis.

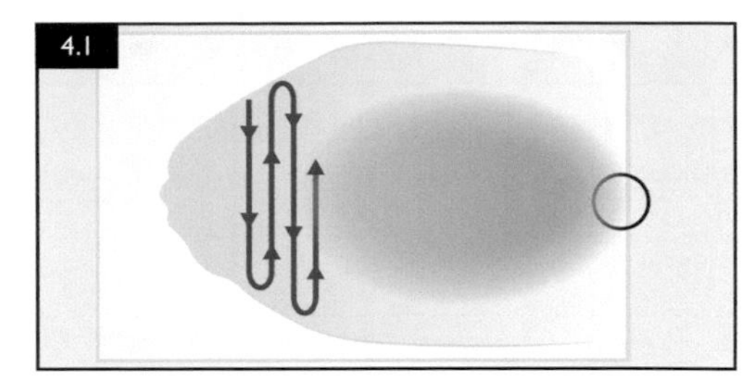


**FIG. 4.1** Wright-stained blood smear. The arrows indicate a systematic approach for performing a leucocyte differential in the monolayer of the slide.

### Blood cell enumeration

- automated counting of platelets, RBCs and leucocytes is available.
- other measured values include:
  - haemoglobin concentration.
  - mean corpuscular volume (MCV).
  - mean platelet volume (MPV).
- calculated values include:
  - haematocrit.
  - mean corpuscular haemoglobin (MCH).
  - mean corpuscular haemoglobin concentration (MCHC).
  - red cell distribution width (RDW).

### Blood smear evaluation

- feather-edged and monolayer smears are a necessary part of complete analysis.
- low magnification scan to evaluate:
  - overall distribution of cells on the smear.
  - platelet clumps or unusually large cells:
    - higher magnification to evaluate individual cell lines (Fig. 4.1).
- Platelets:
  - round to oval structures.
  - smallest blood cells and anucleate.
  - stain very palely with Romanowsky-type stains (Wright's, Diff-Quik).
  - may aggregate together.

DOI: 10.1201/9781003451921-4

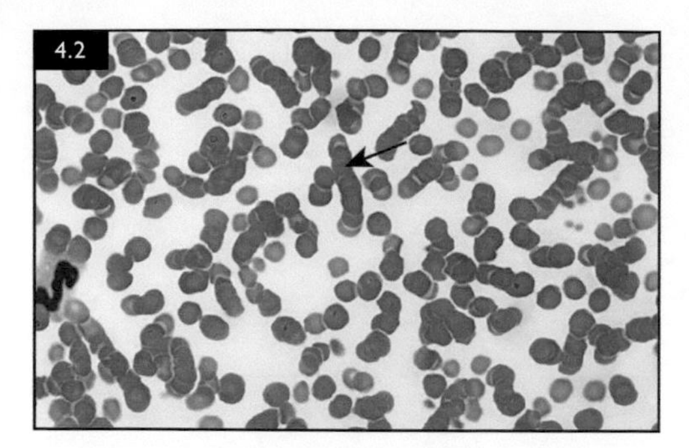


**FIG. 4.2** Normal equine blood smear. The arrow indicates RBC rouleaux formation (Wright's stain).

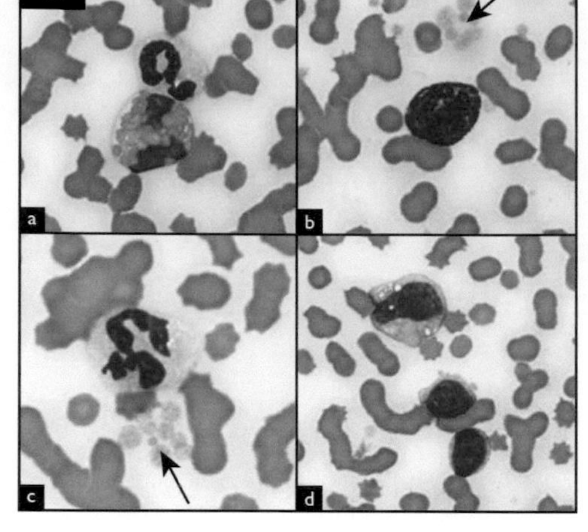


**FIG. 4.3** Blood smear demonstrating morphology of normal WBCs and platelets: (a) neutrophil with an eosinophil below; (b) basophil; (c) neutrophil; (d) monocyte with two lymphocytes below. The arrows indicate clumps of platelets (Wright's stain).

- RBCs:
  - round and stain orange to red with Romanowsky-type stains.
  - exhibit minimal central pallor.
  - uniform in size in health.
  - polychromatophilic RBCs/reticulocytes not readily observed in horse.
  - frequently exhibit rouleaux formation (Fig. 4.2).
- Leucocytes (WBC):
  - examined at ×400 to ×1000 magnification.
  - minimum of 100 cells used for differential (Fig. 4.3).
- Neutrophils (polymorphonuclear cells [PMNs]):
  - most abundant leucocyte.
  - segmented nuclei with neutral staining cytoplasm:
    - occasional fine pink granules.
    - toxic changes include:
      - cytoplasmic basophilia and vacuolation.
    - bone marrow cellular development interference in inflammatory diseases:
      - Döhle body formation and nuclear degeneration.
- Lymphocytes (L):
  - second most common leucocyte.
  - small round cells with round nuclei with scant, slightly basophilic cytoplasm.
- Monocytes (M):
  - infrequently observed but largest appearing leucocyte.
  - abundant grey-blue cytoplasm with small discrete vacuoles.
  - nuclei of any shape except round, not segmented and pale staining.
- Eosinophils (E):
  - uncommonly observed but very distinctive.
  - segmented nuclei.
  - numerous large round pink-orange granules filling the cytoplasm.
- Basophils (B):
  - rare with segmented nucleus and few to many small purple cytoplasmic granules.

## Collection and evaluation of bone marrow

- collected from the wing of the ilium or sternebrae (Fig. 4.4):
  - can be collected and evaluated concurrently.
- aspiration for cytological evaluation.
- core tissue biopsy removed and fixed for histopathological evaluation.
- interpreted by pathologists:
  - cellularity.
  - synchronous maturation.
  - proportions of developing erythrocytic, granulocytic and megakaryocytic cells.

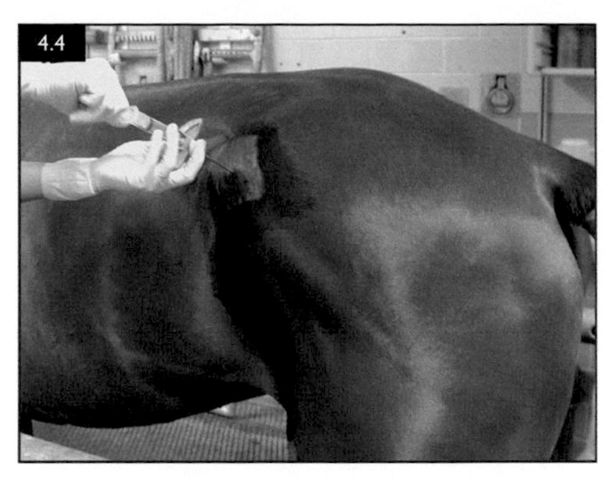

FIG. 4.4 Collection of bone marrow from the wing of the ilium using an 11-gauge, 4-inch Jamshidi needle and a heparinised 12-ml syringe.

## Evaluation of coagulation (Fig. 4.5)

- citrated plasma.
- prothrombin time (PT) and activated partial thromboplastin time (APTT) most performed:
  - normal PT indicates adequate extrinsic and common pathways of coagulation cascade.
  - normal APTT indicates the presence of adequate intrinsic and common pathways.
- other assays include fibrinogen quantitation, thrombin clotting time and indicators of fibrinolysis (fibrin degradation products [FDPs] and D-dimers).

## Immunohaematology

- **Crossmatch:**
  - determines the compatibility of donor and recipient cells.
  - performed for horses requiring a blood transfusion.
  - **major crossmatch:**
    - determines whether recipient has naturally occurring serum antibodies to antigens on the donor's RBCs.
    - antibody binding is indicated by agglutination (Fig. 4.6), haemolysis or a Coombs test.
  - **minor crossmatch:**
    - assesses whether recipient's erythrocytes form complexes with donor's serum.

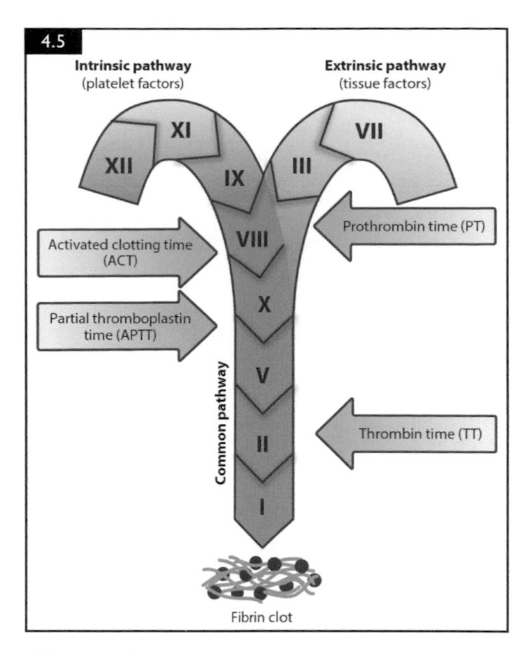


FIG. 4.5 Diagram of the coagulation pathways, clotting factors and routine tests used to assess the system.

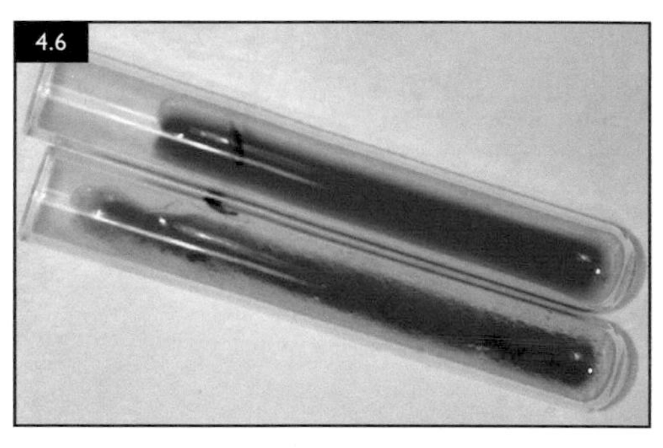

FIG. 4.6 Major crossmatch. The upper test tube is the negative control (no agglutination) and the lower test tube is a strongly positive result (agglutination present), indicating an incompatible crossmatch.

    - not often performed.
- **Blood typing:**
  - assessment of antigens present/absent on surface of an individual's RBCs or lymphocytes.
  - several RBC groups (or systems) including A, C, D, K, P, Q and U.
  - used for predicting the occurrence of neonatal isoerythrolysis (NI), analysing pedigree and identifying animals.

- **Coombs test:**
  - confirm the presence of antibody on the surface of RBCs (Fig. 4.7).
  - positive test in animals with:
    - immune-mediated haemolytic anaemia (IMHA).
    - NI and EIA virus infection.
  - performed on EDTA anticoagulated blood.

## Protein determination

- **Total plasma protein:**
  - estimated by refractometry.
  - reasonable in a clear specimen:
    - excess turbidity or lipaemia interferes with refractive index, producing false elevated results.
  - measured with the biuret colourimetric assay on automated analysers:
    - individual proteins separated by serum protein electrophoresis for detailed analysis.

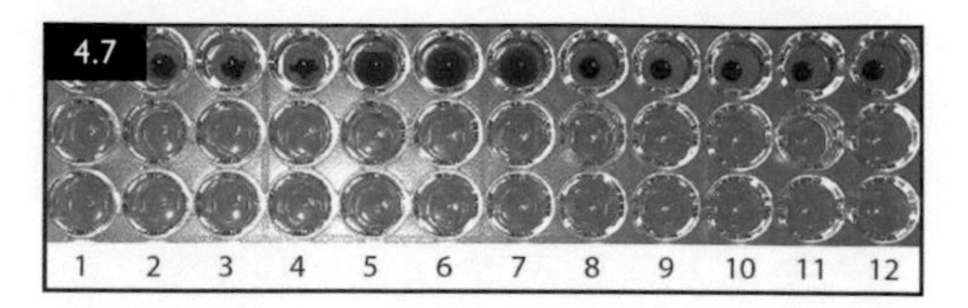


**FIG. 4.7** Positive Coombs test with 1:64 titre (negative <1:16). Test performed in a 96-well plate with progressive dilutions of the sample. Well number 1 (far left well) is the negative control, well numbers 2–7 are positive (lattice formation, best observed in wells 5–7) and well numbers 8–12 are negative (RBCs fall to the bottom of the well to form a pellet).

# DISORDERS OF THE HAEMOLYMPHATIC SYSTEM

## Anaemia

- decrease in oxygen-carrying haemoglobin content of blood.
- characterised by ability of bone marrow to respond to the deficit by expanding production of erythrocytes (RBC) (Table 4.1):
  - regeneration indicated by bone marrow erythrocytic hyperplasia:
    - increased anisocytosis, macrocytosis and a larger RDW (Fig. 4.8):
      - **these features may occur in other circumstances.**
    - evaluation of serial haemograms still the best way to establish a response.
    - bone marrow aspirates or core biopsies also helpful (Fig. 4.9).

| **TABLE 4.1** Parameters used to determine regenerative anaemia in the horse |
|---|
| • Serial haemograms |
| • Increased mean corpuscular volume (MCV) |
| • Increased RBC distribution width (RDW) |
| • Decreased mean corpuscular haemoglobin concentration (MCHC) |
| • Bone marrow evaluation |

## Traumatic and acute haemorrhage

### Definition/overview

- trauma anywhere with high concentrations of small blood vessels or a single large blood vessel resulting in overt blood loss:

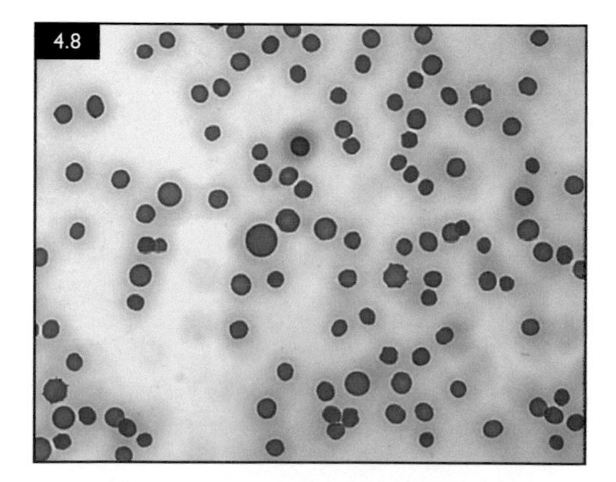


**FIG. 4.8** Blood smear from a horse with regenerative anaemia showing marked anisocytosis. Haematocrit 0.09 l/l (9%) (reference interval: 0.28–0.44 l/l [28–44%]); RDW 38.5 (reference interval: 18–21) (Wright's stain).

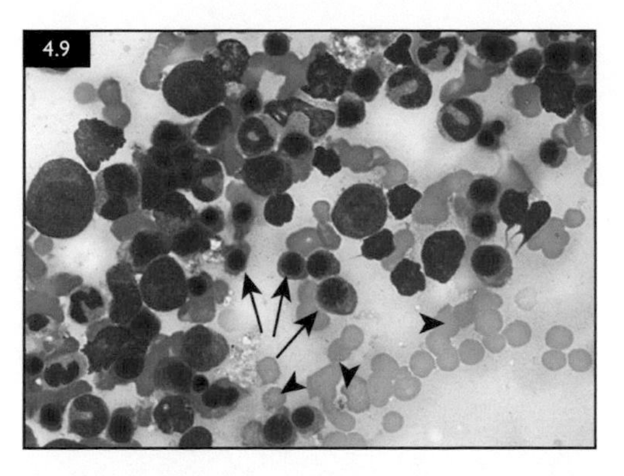


FIG. 4.9 Bone marrow aspirate from the horse in Fig. 4.8. The sample confirms that the anaemia is regenerative, as indicated by the presence of marked erythrocytic hyperplasia. Arrows indicate developing nucleated RBCs and arrowheads indicate polychromatophilic cells (Wright's stain).

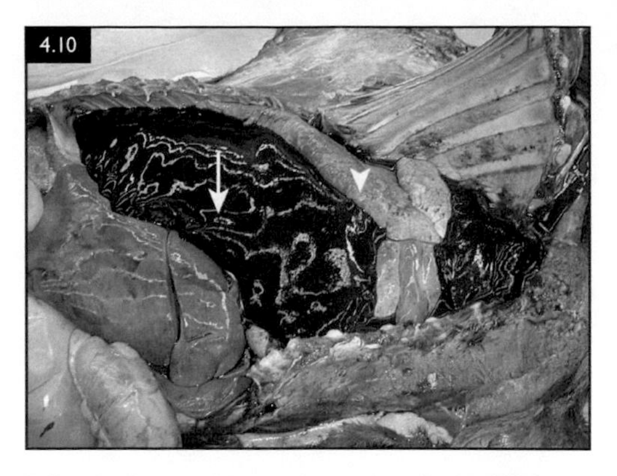


FIG. 4.10 Severe haemothorax in a foal secondary to rib fractures. Note the size of the blood clot (arrow) compared to the lung (arrowhead).

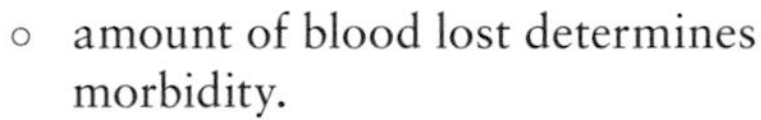

- amount of blood lost determines morbidity.
- lacerations of major arteries can result in death within minutes.

## Aetiology/pathophysiology

- rapid loss of large quantities of blood can result in hypovolaemic shock and death.
- anaemia develops as extravascular fluids move into the vascular space to replace lost volume.
- erythrocytic hyperplasia in the bone marrow replaces the erythrocyte deficit over days to weeks if the haemorrhage does not persist.

## Clinical presentation

- signs of traumatic skin lesions and external haemorrhage:
  - other signs of trauma including lameness, bruising and swelling.
- internal haemorrhage into cavities, such as the thorax, abdomen or uterus:
  - often less obvious (Fig. 4.10).
- other signs depend on the volume of blood lost:
  - pale mucous membranes.
  - tachycardia and tachypnoea.
  - weakness and lethargy.

## Differential diagnosis

- other reasons for bleeding:
  - coagulopathy.
  - disseminated intravascular coagulation (DIC).
  - neoplasia.
  - immune-mediated thrombocytopenia (IMTP).

## Diagnosis

- history and clinical signs.
- difficult to estimate amount of blood loss:
  - owner estimates are often excessive.
  - **immediate CBC does not reflect loss:**
    - takes hours for haematocrit to decline.
    - several days for any increase if the haemorrhage has ceased.
    - haematocrit, MCV and MCHC increase as young RBCs released.
    - serial haemograms useful to follow clinical progress.
- excessive blood loss with minor trauma should lead to investigation of haemostasis.
- additional diagnostic tools:
  - ultrasonography and radiography (Fig. 4.11).
  - abdominocentesis and thoracocentesis.
  - rectal palpation.

## Management

- cessation of haemorrhage is essential:
  - apply direct pressure if still bleeding.
- surgical intervention:

  - internal haemorrhage, severe trauma or uncontrolled arterial bleeding.
- replacement of lost blood volume:
  - balanced electrolyte fluid therapy and blood products.
  - blood transfusion (Table 4.2) considered with:
    - severe haemorrhage.
    - clinical signs of anaemia:
      - tachycardia, tachypnoea, pale mucous membranes, weakness.
- rest.

### Prognosis

- good if blood loss is not life-threatening or can be stopped prior to the development of hypovolaemic shock.

## Chronic haemorrhage

### Definition/overview

- blood loss that occurs slowly over several weeks to months.
- presence of bleeding might not be apparent until clinical signs of anaemia develop.
- identification of the source of bleeding may be difficult.

### Aetiology/pathophysiology

- many variable causes:
  - gastric SCC.
  - ulceration from NSAID therapy.
  - blood-sucking parasites.
  - coagulopathies.
- appropriate bone marrow response initially with erythrocytic hyperplasia.
- iron stores become depleted if continues and non-regenerative anaemia develops.

### Clinical presentation

- weakness and lethargy.
- mucosal pallor if severe anaemia.
- clinical signs of specific causes:
  - tarry faeces or haematochezia.
  - skin lesions if external parasites.
  - petechial, mucosal or body cavity haemorrhage in coagulopathies.

### Differential diagnosis

- other causes of non-regenerative anaemia:
  - anaemia of chronic disease.
  - primary bone marrow disorders.

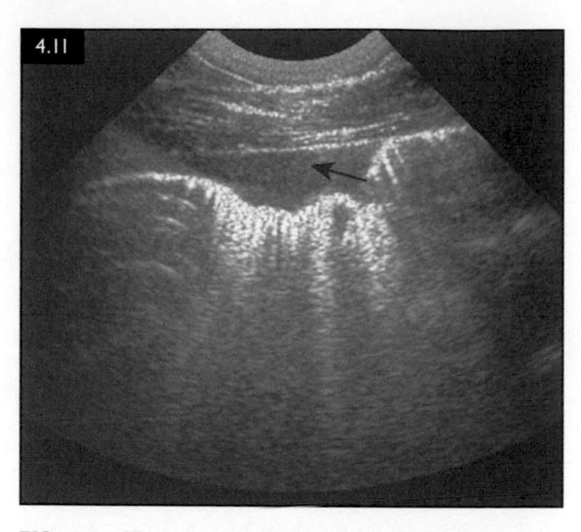


**FIG. 4.11** Transabdominal ultrasound image of a foal with haemoabdomen secondary to trauma. Note the echogenicity of the peritoneal fluid (arrow), which is suggestive of haemoabdomen.

### Diagnosis

- history and clinical signs of chronic haemorrhage.
- blood samples for CBC:
  - non-regenerative or poorly regenerative microcytic, hypochromic anaemia.
  - RBCs increased amount of central pallor:
    - insufficient haemoglobin production.
  - fragmented RBCs and thrombocytosis.
  - decreased iron stores (serum iron and ferritin).
- gastroscopy for tumours:
  - gastric ulcers rarely cause anaemia in the horse but occasionally in foals.
- faecal examination (occult blood and parasites).
- examine skin for ectoparasites.

### Management

- treat the underlying cause of bleeding.
- oral or parenteral iron supplementation.
- blood transfusion only if clinical signs of anaemia are present.

### Prognosis

- poor if the cause is untreatable, such as a tumour.
- good if the cause can be established and eliminated.

**TABLE 4.2** Blood transfusion protocol

| |
|---|
| • Select an appropriate donor. A clinically normal adult horse should be chosen. The horse should be negative for EIA virus, have never received a blood or plasma transfusion, never foaled and have a normal packed cell volume (PCV) |
| • Crossmatching is ideal, particularly if the animal has had a prior transfusion. A major crossmatch identifies incompatibility of donor RBCs with recipient serum. A minor crossmatch evaluates the inverse |
| • Blood should be collected into sterile containers with anticoagulant (acid–citrate–dextrose or citrate–phosphate–dextrose). The anticoagulant/blood ratio should be 1:9 |
| • Blood should be collected using sterile technique. A healthy horse can donate up to 20% of its blood volume (approximately 8 l for a 500-kg horse) every 30 days. Blood should be used immediately, if possible, but whole blood can be stored refrigerated for up to 3 weeks |
| • An intravenous catheter should be placed in the recipient. Blood must be given via a transfusion filter set to remove any clots |
| • Baseline heart rate, respiratory rate and temperature should be obtained. Blood should be administered at a rate of 0.1 ml/kg over the first 15 minutes, then increased to 20 ml/kg/h if no adverse reactions are observed. Adverse reactions include tachypnoea, tachycardia, restlessness, urticaria, muscle fasciculation and collapse |
| • If adverse reactions are encountered, the transfusion should be ceased and flunixin meglumine (1.1 mg/kg i/v) given. If anaphylaxis is encountered, adrenaline (epinephrine) (0.01–0.02 mg/kg of 1:1000 i/v) should be administered, along with aggressive intravenous fluid therapy. Corticosteroids (prednisolone sodium succinate, 4.5 mg/kg i/v) are often administered concurrently. If the reaction was mild, transfusion can be recommenced 15–30 minutes after flunixin administration. If adverse reactions redevelop or the reaction was severe, the blood should be discarded, and another source obtained |
| • Transfused RBCs have a short lifespan (4–6 days), so the beneficial effects of blood transfusion will be transient. Icterus and an increase in free bilirubin will be expected within a few days of transfusion |

# Inherited haemostasis disorders

## Definition/overview

- rare in the horse.
- prekallikrein deficiency in miniature and Belgium horses.
- factor VIII deficiency (haemophilia A), factor IX deficiency (haemophilia B), factor XI deficiency and von Willebrand disease.

## Aetiology/pathophysiology

- inheritance of a specific genetic defect leading to abnormality of protein production:
  - prekallikrein deficiency:
    - autosomal recessive.
  - haemophilia A:
    - male sex-linked recessive.

## Clinical presentation

- affected animals usually do not haemorrhage spontaneously:
  - usually noted after trauma, surgery or venipuncture.
- prekallikrein deficiency – occasionally prolonged bleeding after castration:
  - frequently able to clot normally.
- haemophilia A patients do not usually exhibit spontaneous bleeding unless factor VIII activity is <5%.

## Differential diagnosis

- DIC.
- anticoagulant poisoning.
- liver disease.

## Diagnosis

- APTT can be prolonged.
- demonstration of deficiency or absence of specific factor activity.

### Management

- blood transfusion if severe blood loss with clinical signs, but not practical in the long term.

### Prognosis

- successful treatment is not possible but clinical signs are often absent or minor.
- poor if recurrent spontaneous bleeding.

## Anticoagulant toxicity

### Definition/overview

- administration of warfarin can lead to haemorrhage if:
  - diet contains less vitamin K.
  - concurrent administration of highly protein-bound drugs.
- ingestion of mouldy sweet clover (*Melilotus* spp.) in hay or anticoagulant rodenticide.

### Aetiology/pathophysiology

- interference with activation of the vitamin K-dependent coagulation proteins:
  - protein-bound drugs or hypoalbuminaemia increases free active toxin.
- vitamin K required for activation of procoagulant factors II, VII, IX and X and anticoagulant protein factors C and S.
- affected animals cannot form clots and bleed.

### Clinical presentation

- within 3–8 weeks of ingestion of mouldy sweet clover and 3–5 days of ingestion of anticoagulant rodenticides.
- multiple-site bleeding, often from the nose, GI and urinary tracts.
- bleeding into body cavities and joint spaces.
- subcutaneous haematomas may occur with relatively mild trauma.

### Differential diagnosis

- other causes of multiple-site haemorrhage:
  - disseminated intravascular coagulopathy.
  - inherited haemostatic defect.

### Diagnosis

- combination of clinical signs and history of exposure to the toxin:
  - evaluate diet for mouldy sweet clover.
  - presence of rodenticides in environment.
- prolonged PT and APTT:
  - factor VII has the shortest half-life.
  - PT may initially be the only abnormal test.
- evidence of anaemia, hypoproteinaemia and liver enzyme response on blood samples.
- response to vitamin K therapy.

### Management

- remove the animal from the source of toxin.
- stall rest.
- volume replacement:
  - intravenous fluids or blood products.
  - fresh plasma for clotting factors in ongoing haemorrhage.
- vitamin K administration:
  - vitamin $K_1$ (1.0–1.5 mg/kg s/c or i/m q4–12 h) for 3 days:
  - until PT has returned to reference interval:
    - improvement often observed within 24 hours.
  - alfalfa is rich in vitamin $K_1$.
- **Vitamin $K_3$ is not effective, can be toxic and should not be administered.**

### Prognosis

- good if recognised early, proper therapy is instituted, and haemorrhage is not life-threatening.

## Neonatal isoerythrolysis

- See Book 2, pages 176–178.

## Immune-mediated haemolytic anaemia (IMHA)

### Definition/overview

- destruction of erythrocytes by the immune system:
  - **Primary IMHA** when the process is directed against a self-antigen.
  - **Secondary IMHA:**

FIG. 4.12 Grossly visible agglutination of RBCs. (Photo courtesy RM Jacobs.)

- ♦ subsequent to infectious, inflammatory, neoplastic or drug-related stimuli.
- ♦ more common in horses.

## Aetiology/pathophysiology

- sequela to lymphoma, EIA, other viral and bacterial infections.
- following administration of drugs including:
  - penicillins, cephalosporins and sulphonamides.
- formation of antigen–antibody complexes on the surface of the RBCs recognised as foreign by the immune system and removed from circulation:
  - loss of tolerance of a self-antigen.
  - unmasking of an existing antigen.
  - presence of a new antigenic molecule.
- haemolysis can occur extravascularly:
  - macrophages in the spleen, liver and/or bone marrow:
    - ♦ increased unconjugated bilirubin resulting in icterus.
- haemolysis intravascularly:
  - haemoglobinaemia and haemoglobinuria.

## Clinical presentation

- weakness.
- pale and/or icteric mucous membranes.

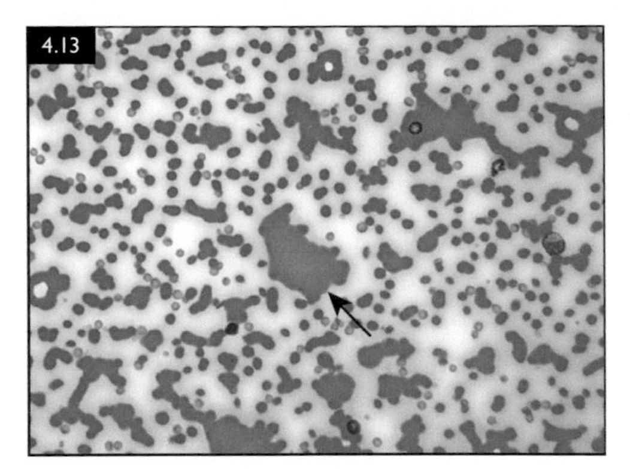

FIG. 4.13 Blood smear with an arrow indicating RBCs in a large grape-like cluster indicative of microagglutination (Wright's stain).

- tachycardia and tachypnoea where anaemia is severe and rapid onset.
- pigmenturia.

## Differential diagnosis

- other causes of haemolytic anaemia:
  - Heinz body-induced haemolysis.
  - parasitic haemolysis.

## Diagnosis

- suggestive clinical signs and history.
- laboratory evidence of haemolysis:
  - macrocytic, hypochromic with anisocytosis anaemia.
  - agglutination might be present grossly and on smears (Figs. 4.12, 4.13).
  - RBC ghosts might be observed on the blood smear.
  - direct antiglobulin test (Coomb's) often positive.
  - increased unconjugated bilirubin.
  - increased urea and creatinine if pigment-associated renal failure.
- haemoglobinuria is often present.
- other tests to identify underlying cause.

## Management

- treat underlying disease.
- discontinue ongoing drug therapy if possible or change.
- stabilise PCV within 24–48 hours.
- blood transfusion in severe cases is required (see *Table 4.2*).
- intravenous fluid therapy is useful in reducing haemoglobin-associated renal damage.

- restricted exercise until PCV returns to normal.
- immunosuppressive therapy:
  - minimise further antibody formation.
  - corticosteroids (dexamethasone 0.05–0.2 mg/kg i/v or i/m q12–24 h):
    - tapering based on response to treatment.
- azathioprine (3 mg/kg p/o q24 h) in refractory cases.

### Prognosis

- good if the underlying problem can be identified and addressed.
- poor to grave in lymphoma and unidentifiable causes.

## Red maple leaf toxicosis

### Definition/overview

- consumption of dried leaves of the red maple tree (*Acer rubrum*).
- Heinz body haemolytic anaemia.

### Aetiology/pathophysiology

- dried leaves only which drop onto pasture in autumn:
  - quantity required to cause damage is variable between horses and seasons.
- undetermined toxic compound causing oxidative damage to RBCs.
- primarily intravascular haemolysis and Heinz body formation.
- methaemoglobinaemia can develop.

### Clinical presentation

- variable from acute death to gradual development of disease.
- weakness, lethargy, anorexia, pale mucous membranes and icterus.
- cyanosis may not develop if anaemia is severe.
- haemoglobinuria.

### Differential diagnosis

- other causes of Heinz body haemolytic anaemia:
  - onion ingestion, phenothiazine toxicosis and lymphoma.

### Diagnosis

- history of exposure to dried red maple leaves – **examine the pasture.**

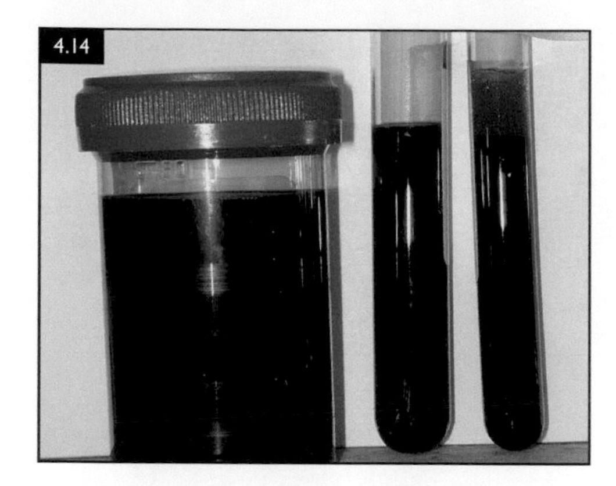

**FIG. 4.14** Left: dark brown-black urine due to methaemoglobinuria. Right: two blood tubes with similar discolouration due to methaemoglobinaemia from a horse with red maple toxicosis. (Photo courtesy L Arroyo.)

- clinical signs of acute onset of anaemia.
- haemolytic anaemia with Heinz bodies in the blood.
- haemoglobinaemia and haemoglobinuria (Fig. 4.14).

### Management

- no specific treatment.
- prevent further exposure to leaves immediately and in the future.
- intravenous fluid therapy in dehydrated animals or haemoglobin renal failure.
- blood transfusion depending on clinical signs and PCV.
- restrict exercise and monitor closely for complications, including laminitis.
- ascorbic acid (30–50 mg/kg i/v q12 h) is useful as an antioxidant.

### Prognosis

- guarded to poor:
  - early detection and treatment improve the chances of recovery.
- renal damage, tissue hypoxia and DIC are serious complications.

## Other causes of Heinz body haemolytic anaemia

### Definition/overview

- other oxidative compounds leading to haemolysis:
  - onions, garlic and phenothiazine.

## Aetiology/pathophysiology

- oxidation denatures haemoglobin, with subsequent Heinz body formation.
- damaged cells are removed by macrophages in the spleen or lysed intravascularly.

## Clinical presentation

- pale mucous membranes.
- depression and weakness in severe cases.
- haemoglobinuria.

## Differential diagnosis

- Red maple toxicosis.

## Diagnosis

- history of onion or garlic ingestion, or administration of phenothiazine.
- haemolytic anaemia with Heinz bodies on the blood smear (Fig. 4.15).

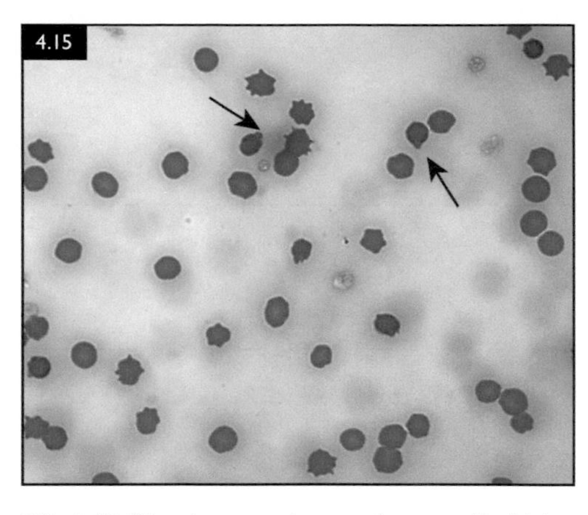


**FIG. 4.15** Blood smear from a horse with Heinz body haemolytic anaemia. The arrows indicate Heinz bodies (Wright's stain).

## Management

- stop source of toxic substances.
- similar to red maple leaf toxicosis.

## Prognosis

- depends on severity of anaemia and presence of secondary hypoxic organ damage.

# Equine infectious anaemia (EIA)

## Definition/overview

- worldwide viral infection in all equids.

## Aetiology/pathophysiology (Fig. 4.16)

- Equine infectious anaemia virus (EIAV):
  - can exhibit latency and intermittent recrudescence.
- transmitted by:
  - insect vectors:
    - tabanid, deer and stable flies, and mosquitoes.
  - fomites:
    - blood-contaminated needles, syringes, surgical instruments, blood products.
- enters cells of the mononuclear–phagocytic system and endothelial cells.
- disease characterised by three distinct stages: acute, chronic and inapparent.
- immune-mediated erythrocyte destruction and suppressed production of cells in the bone marrow leads to haematological changes.

## Clinical presentation

- **Acute form:**
  - fever, depression, ventral oedema and mucosal petechiation.
  - less common is life-threatening epistaxis.
  - mild cases are not easily detected.
  - within 1–4 weeks after infection and persists for approximately one week.
- **Subacute and chronic forms:**
  - anorexia, ventral oedema, weight loss and intermittent pyrexia.
  - haemolytic anaemia.
  - associated with recurrent episodes of viral replication.
  - normal between episodes.
  - within 1 year, the severity of episodes wanes and the carrier stage develops.
- **Inapparent carrier form:**
  - clinically normal but reservoir of infection.

## Differential diagnosis

- other causes of anaemia and thrombocytopenia:
  - equine viral arteritis (EVA).
  - immune-mediated disease.

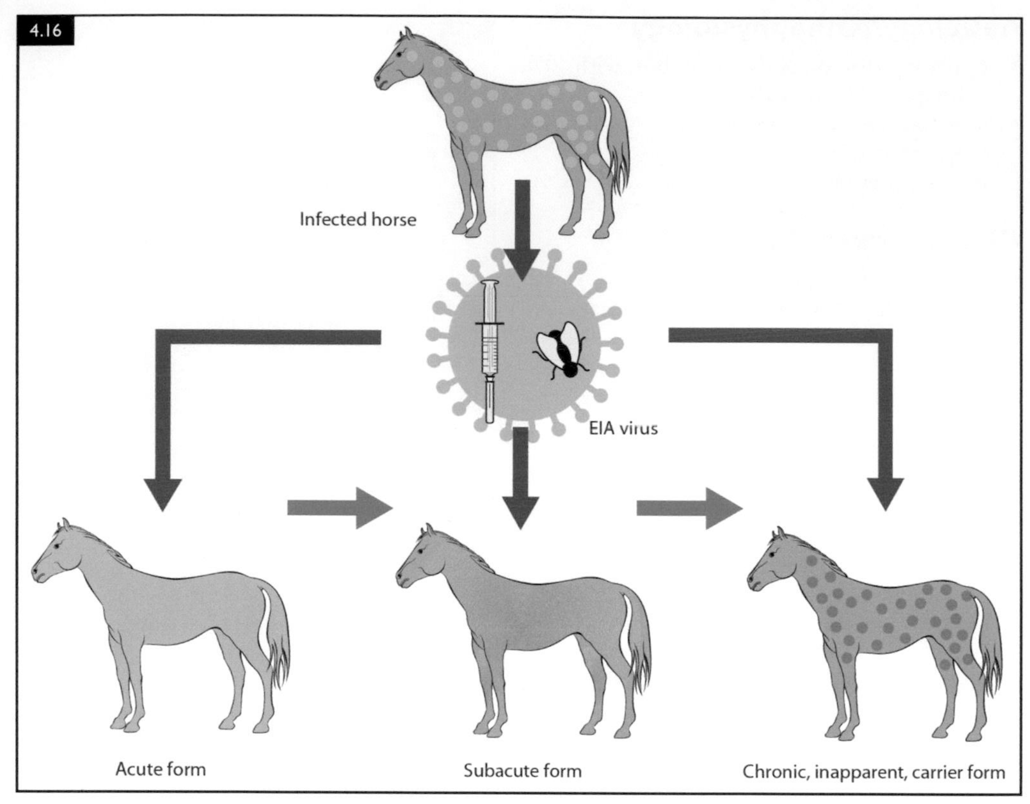


**FIG. 4.16** Pathogenesis of equine infectious anaemia (EIA). Insect vectors transmit EIA virus from an infected horse to susceptible horses. Iatrogenic infection through re-use of needles or other cross-contamination of blood can also occur. Acute, subacute and chronic (carrier) forms may ensue.

  - ehrlichiosis.
  - purpura haemorrhagica.
- chronic cases require differentiation from lymphocytic neoplasia and other persistent inflammatory diseases.

## Diagnosis

- combination of fever, haemolytic anaemia and thrombocytopenia:
  - anaemia worst in subacute/chronic stages.
- leucopenia, lymphocytosis and monocytosis.
- definitive diagnosis on detection of:
  - EIAV-specific antibody p26 in agar-gel immunodiffusion assay (Coggins test):
    - positive within 45 days of acute infection.

## Management

- **no specific treatment.**
- supportive therapy can help clinical recovery.
- **notifiable disease in most countries.**
- fly control measures are an important management tool in endemic areas.
- iatrogenic transmission prevented by cleaning and sterilisation of all materials.

## Prognosis

- poor with no specific antiviral treatment and regulatory demands.
- death is unusual but the carrier stage is the ultimate outcome.

# Babesiosis (piroplasmosis)

## Definition/overview

- infection with intraerythrocytic protozoal organisms of the *Babesia* and *Theileria* genera.
- tropical and subtropical regions where *Dermacentor*, *Hyalomma* and *Rhipicephalus* ticks are found.

### Aetiology/pathophysiology

- *Theileria equi* (formerly *Babesia equi*) and *Babesia caballi*:
  - *T. equi* tends to be more pathogenic and additionally infects lymphocytes.
- intraerythrocytic infection leads to intravascular haemolysis.

### Clinical presentation

- severity of signs depends on the immunity of the animal.
- **Naïve horse:**
  - 1–4 weeks post-infection.
  - fever, depression, lethargy, anaemia, icterus and petechial haemorrhages.
  - haemoglobinuria may be present with *T. equi* infection.
  - death may occur within 48 hours.
- **Chronic carrier state following acute infection:**
  - clinical disease is uncommon, but infection persists.
  - mild infections may cause exercise intolerance.

### Differential diagnosis

- other causes of haemolytic anaemia:
  - IMHA, EIA and Heinz body haemolytic anaemia.

### Diagnosis

- horse with fever, haemolytic anaemia and haemoglobinuria.
- confirmed by observation of characteristic protozoa in RBCs.
- serological and molecular tests are available for both organisms:
  - complement fixation test commonly used.

### Management

- *B. caballi*: imidocarb (2.2 mg/kg i/m q24 h for 2 doses).
- *T. equi*: imidocarb (4 mg/kg i/m q72 h for 4–6 doses):
  - infection more difficult to treat and rarely eliminated.
  - injections divided into at least four different sites.
- imidocarb may prevent development of natural immunity:
  - monitoring for signs of anticholinesterase effects of imidocarb:
    - ◆ colic, hypersalivation and diarrhoea.
    - ◆ donkeys are very sensitive; only use at lower doses or not at all.
- animals are prone to re-infection in endemic areas.
- supportive therapy including fluids and blood products.
- stable rest.
- notifiable disease in many countries:
  - persistently infected animals acting as reservoirs of infection are common.

### Prognosis

- favourable for *B. caballi* infection if early diagnosis and treatment.
- guarded for *T. equi* infection as treatment is less effective.
- regulatory concerns may limit the movement of infected animals.
- positive horses must be euthanased in some countries.

## Anaemia of inflammatory or chronic disease (AID)

### Definition/overview

- mild/moderate decrease in RBCs in response to an inflammatory condition.
- caused by inflammatory cytokines.
- usually develops insidiously but acute onset within 3–10 days has been observed.

### Aetiology/pathophysiology (Fig. 4.17)

- inflammatory process releases cytokines which suppress RBC production:
  - abscess.
  - systemic infection or inflammation.
  - immune-mediated disease.
  - neoplasia.
- increased sequestration of iron in macrophages of the bone marrow and liver.
- suppression of erythropoiesis.
- decreased survival time of RBCs.

### Clinical presentation

- often identified during investigation of the primary underlying disease.
- clinical signs of underlying inflammatory disease are pre-eminent.

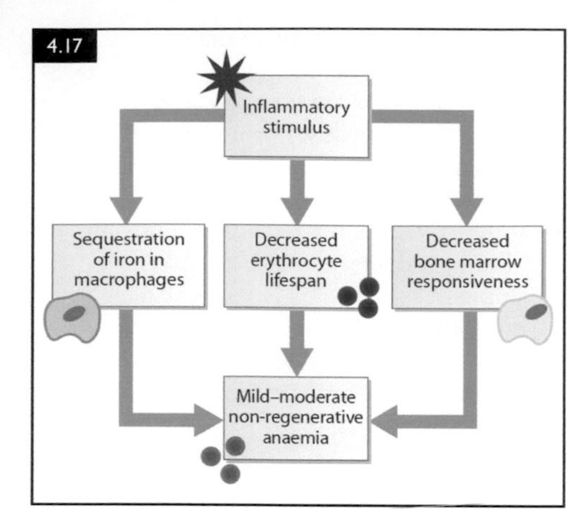


**FIG. 4.17** Pathogenesis of anaemia of inflammatory or chronic disease.

## Differential diagnosis

- iron-deficiency anaemia.

## Diagnosis

- mild/moderate, normocytic, normochromic anaemia.
- presence of an underlying inflammatory disease.
- decreased serum iron and normal to slightly low total iron-binding capacity (TIBC):
  - normal to increased ferritin and bone marrow iron stores.

## Management

- treatment and resolution of the underlying inflammatory lesion.
- **specific treatment for anaemia usually not required.**
- iron supplementation is not effective.

## Prognosis

- depends on the underlying disease and its successful treatment.

# Iron-deficiency anaemia

## Definition/overview

- iron is important for haemoglobin formation:
  - inadequate stores, particularly in young animals, can lead to anaemia.
- uncommon in adults and usually associated with chronic external blood loss.

## Aetiology/pathophysiology

- chronic external blood loss:
  - parasitism
  - bleeding GI ulcers.
  - neoplasia
  - coagulopathy.
- poor iron intake in neonates on a milk diet and without access to soil.
- inadequate iron stores limits erythropoiesis.

## Clinical presentation

- anaemia may be inapparent until severe.
- common initial signs include:
  - lethargy, exercise intolerance and pale mucous membranes.

## Differential diagnosis

- anaemia of chronic disease.

## Diagnosis

- microcytic, hypochromic anaemia.
- decreased serum iron and normal to increased TIBC.
- sparse bone marrow iron stores.
- accumulation of erythrocytes in the later stages of development (Figs. 4.18, 4.19).

## Management

- identify and treat the inciting cause.
- iron supplementation:

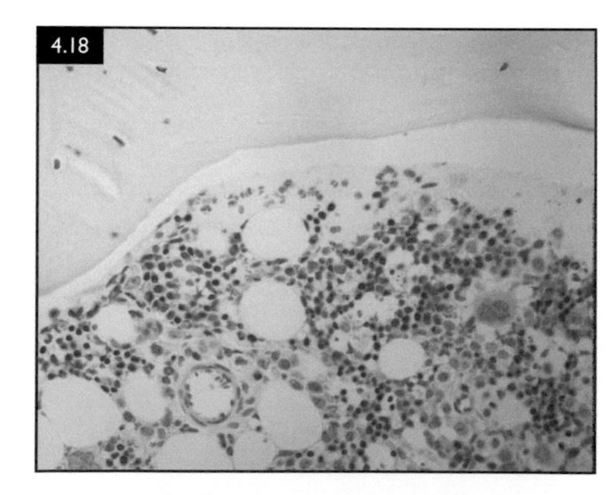

**FIG. 4.18** Bone marrow core biopsy from a horse with iron-deficiency anaemia. The lack of blue staining indicates negative staining for iron (Perl iron stain).

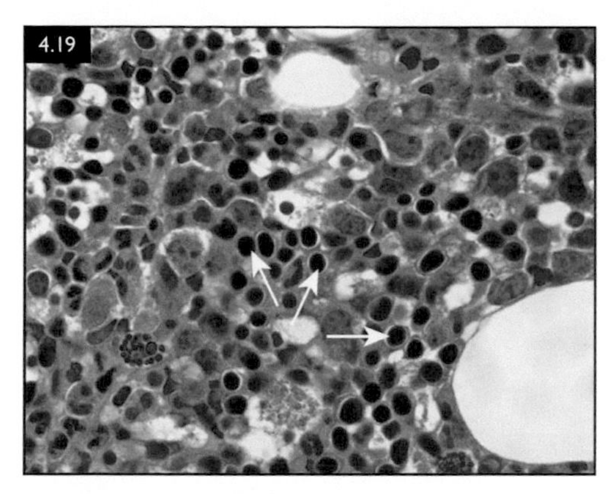


FIG. 4.19 Bone marrow core biopsy from a horse with iron-deficiency anaemia. There are increased numbers of late developing erythrocytic cells (arrows) and no stainable iron. Polychromatophils are not present due to the lack of iron required for complete maturation (H&E stain).

  - oral iron supplementation preferred (ferrous sulphate 1.0–4.0 g/450 kg p/o q24 h).
  - i/v iron supplementation given slowly (iron cacodylate 1 g/adult horse).
  - **iron dextran should not be administered:**
    - high incidence of adverse reactions and death.
  - **iron should not be administered to foals in the first 2 days of life.**
  - weeks of supplementation may be required in severely iron depleted animals.
- monitor haematocrit 1–3 times weekly during initial treatment.
- serum iron and TIBC should be evaluated every 2 weeks:
  - iron supplementation ceased when serum iron, TIBC and PCV within reference intervals.

### Prognosis

- good if the reason for the iron depletion can be identified and eliminated.

## Aplastic anaemia and pure red cell aplasia

### Definition/overview

- **Aplastic anaemia (AA)**
  - bone marrow stem cell disorder characterised by:
    - decreased production of all blood cell types.
    - replacement of normal haematopoietic tissue with adipose tissue.
- **Pure red cell aplasia (PRCA):**
  - severe non-regenerative anaemia due to depletion of developing erythrocytic precursors in bone marrow.
  - leucocytes and platelets unaffected.
- rare diseases in the horse.

### Aetiology/pathophysiology

- unknown cause but consider:
  - immune-mediated and idiosyncratic drug reactions.

### Clinical presentation

- AA:
  - loss of platelets:
    - petechial haemorrhages and epistaxis.
  - loss of neutrophils:
    - intermittent fever and weight loss.
- PRCA:
  - moderate to severe anaemia.
  - pallor, weakness and lethargy.

### Differential diagnosis

- other causes of pancytopenia or non-regenerative anaemia.
- iron deficiency.
- anaemia of chronic disease.

### Diagnosis

- pancytopenia (AA) or non-regenerative anaemia (PRCA) on repeated CBCs.
- examination of bone marrow demonstrating (Figs. 4.20, 4.21):
  - marked hypoplasia to aplasia and replacement with adipose tissue (AA).
  - severely decreased/absent developing erythrocytic precursors (PRCA).

### Management

- treat any primary process if identified.
- supportive care including broad-spectrum antibiotics and blood transfusions.

### Prognosis

- guarded to poor generally but some cases respond to immunosuppressive therapy.

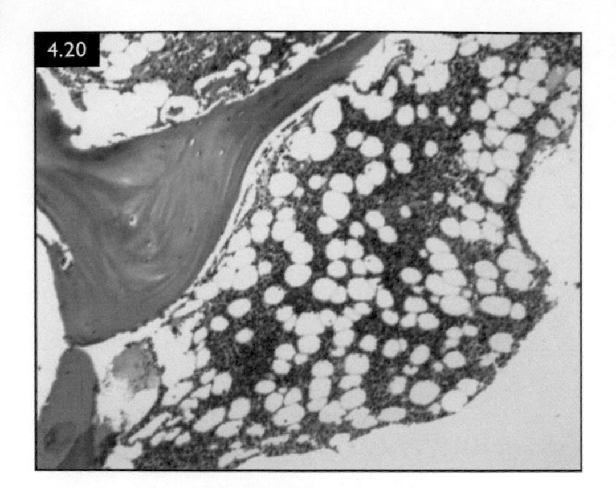

**FIG. 4.20** Bone marrow core biopsy with a normal cellularity (approximately 50% cellular and 50% fat) (H&E stain).

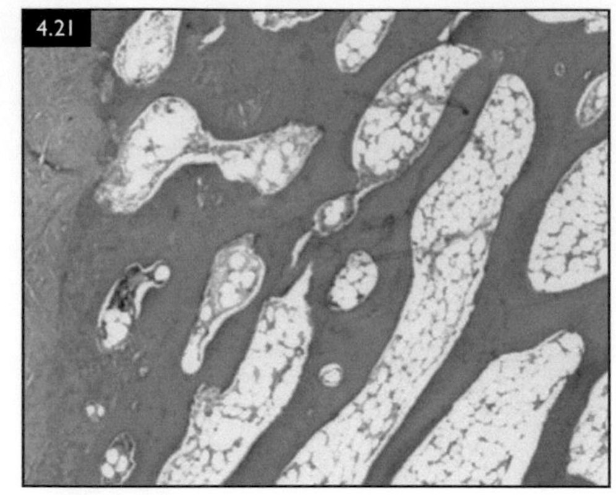

**FIG. 4.21** Bone marrow core biopsy from a horse with aplastic anaemia. There is complete absence of all haematopoietic cells, with only fat remaining (H&E stain).

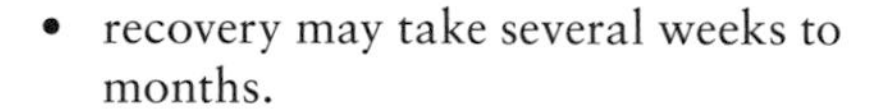

- recovery may take several weeks to months.

## Immune-mediated thrombocytopenia (IMTP)

### Definition/overview

- inappropriate destruction of mature platelets:
  - presence of, or exposure to, a perceived foreign antigen on the surface of the cell.
  - subsequent removal of these cells by the mononuclear phagocytic system.

### Aetiology/pathophysiology

- autoimmune where the perceived foreign antigen is a self-antigen.
- secondary due to the ability of some drugs to act as haptens.
- infectious agent exposes an antigen or has a cross-reactive antigen.
- thrombocytopenia develops due to removal of platelets by the mononuclear phagocytic system in the spleen and liver.
- haemostatic abnormalities develop depending on the severity of thrombocytopenia:
  - no spontaneous haemorrhage > $30 \times 10^9$/l platelets.
  - clinical signs < $10 \times 10^9$/l platelets.

### Clinical presentation

- bleeding from multiple mucosal surfaces:
  - petechial or ecchymotic haemorrhages.
- epistaxis, melaena and hyphaema may be present.
- excessive bleeding following trauma, surgery or venipuncture.
- horses are normal otherwise unless there is underlying secondary disease.

### Differential diagnosis

- other causes of petechial haemorrhages from thrombocytopenia:
  - disseminated intravascular coagulopathy.
  - infectious diseases.

### Diagnosis

- clinical signs in conjunction with severe thrombocytopenia.
- macroplatelets indicate early release of younger platelets (Fig. 4.22).

### Management

- treat any underlying cause.
- stop ongoing drug therapy if possible or modify type and dose.
- immunosuppressive drugs to reduce antigen-antibody complexes:
  - often for a minimum of 3 weeks.
  - dexamethasone: 0.05–0.2 mg/kg i/v or i/m q12–24 h:
    - tapered gradually based on response to treatment.
  - azathioprine: 3 mg/kg p/o q24 h:

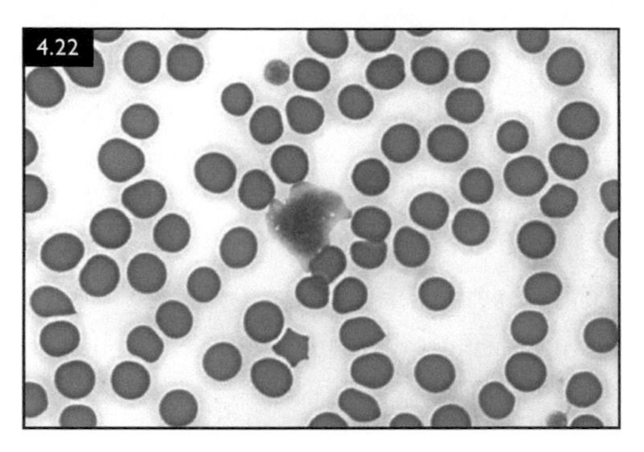


**FIG. 4.22** Blood smear from a horse recovering from thrombocytopenia. Note the large platelet in the centre of the field with a normal-sized platelet above (Wright's stain).

    - ♦ refractory cases or where complications develop (laminitis).
- recurrence after treatment has stopped has been reported.

### Prognosis

- depends on existing predisposing causes:
  - poor for lymphoma.
  - secondary to drug administration more favourable.

## Disseminated intravascular coagulopathy (DIC)

### Definition/overview

- complication of systemic inflammatory diseases:
  - endotoxaemia, Gram-negative sepsis and metastatic neoplasia.
- activation of clotting process and consumption of clotting factors:
  - concurrent formation of thrombi and tendency to bleed.

### Aetiology/pathophysiology

- accompanies many serious diseases:
  - GI accidents and endotoxaemia.
  - neoplasia, severe burns, liver disease, IMHA and snake bites.
- widespread or localised endothelial injury (e.g. vasculitis) results in:
  - initiation of both platelet plug formation and thrombin generation.
  - increased platelet consumption.
  - deposition of fibrin strands in vascular spaces.
- rapid consumption of pro- and anticoagulant factors and fibrinolysis.

### Clinical presentation

- clinical signs of the primary disease process:
  - colic, colitis, pleuropneumonia, systemic sepsis, neoplasia, etc.
- mucosal surface bleeding and petechial or ecchymotic haemorrhages (Fig. 4.23).
- if thrombi develop, clinical signs related to dysfunction of the affected organ will occur (e.g. dyspnoea with pulmonary thrombi).

### Differential diagnosis

- other causes of mucosal surface bleeding:
  - IMTP.
  - warfarin toxicosis.
  - inherited or acquired platelet function defects.

### Diagnosis

- supportive laboratory results in an animal with serious inflammatory illness (Table 4.3).

### Management

- treatment of the underlying disease is essential.
- specific DIC treatments include:
  - heparin to minimise further clot formation.
  - plasma transfusion to replace consumed clotting factors.

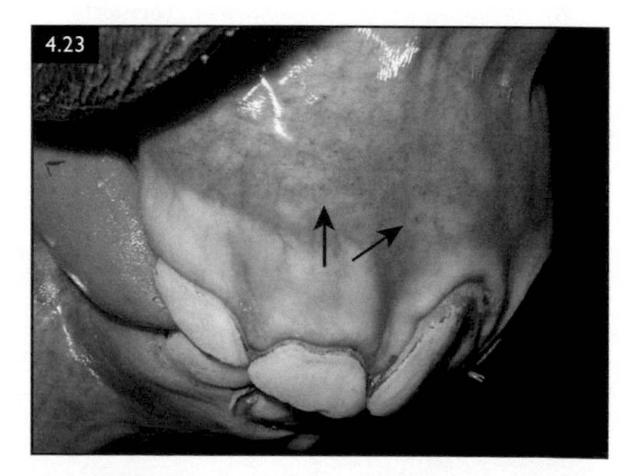


**FIG. 4.23** Small petechial haemorrhages (arrows) on the oral mucous membranes of a horse with thrombocytopenia. (Photo courtesy J Scott Weese.)

| **TABLE 4.3** Laboratory abnormalities supporting diagnosis of DIC |
|---|
| • Thrombocytopenia |
| • Prolonged prothrombin time (PT) |
| • Prolonged activated partial thromboplastin time (APTT) |
| • Fragmented RBCs (schistocytes) |
| • Increased fibrin degradation products (FDPs) or D-dimers |
| • Decreased antithrombin |
| • Decreased specific factor activity |
| • Hypofibrinogenaemia |
| • Prolonged thrombin clotting time (TCT) |

### Prognosis

- serious complication of a severe disease, both of which can be difficult to treat:
  - guarded to grave prognosis.

## Equine anaplasmosis (equine granulocytic ehrlichiosis)

### Definition/overview

- seasonal rickettsial disease characterised by haematological, GI and neurological signs.
- reported in the USA, Canada, northern Europe and Brazil.

### Aetiology/pathophysiology

- *Anaplasma phagocytophilum* (formerly *Ehrlichia equi*).
- transmitted by tick (*Ixodes* spp.) bites.
- tropism for granulocytes, predominantly neutrophils.
- combination of increased demand, immune-mediated destruction and altered bone marrow microenvironment:
  - decreases WBCs, platelets and occasionally RBCs.

### Clinical presentation

- clinical signs develop within 14 days of exposure and can be vague/non-specific.
- **over 4 years old:**
  - fever, depression and anorexia.
  - icterus and limb oedema.
  - ataxia which may lead to trauma.
- **under 4 years old:**
  - milder signs and often only fever is present in horses aged < 1 year.
  - typically peak by 3–5 days and persist for 14–16 days.

### Differential diagnosis

- other causes of vasculitis, icterus and ataxia.
- other infectious agents such as *Theileria equi and* EIAV

### Diagnosis

- clinical signs in a geographical area where the disease occurs:
  - during appropriate season for tick-borne disease.
- leucopenia and thrombocytopenia.
- hyperbilirubinaemia is common.
- characteristic rickettsial morulae within granulocytes on a blood/peritoneal fluid smear (Figs. 4.24, 4.25):
  - present in middle of febrile period in 20–75% of neutrophils.
- serology will reveal a fourfold rise in antibodies after several weeks.
- molecular assays are available and very specific.

### Management

- Oxytetracycline: 7 mg/kg i/v q24 h for 5–7 days:
  - rapid response with improvement in 12–24 hours.
  - relapse is uncommon and usually within 30 days.

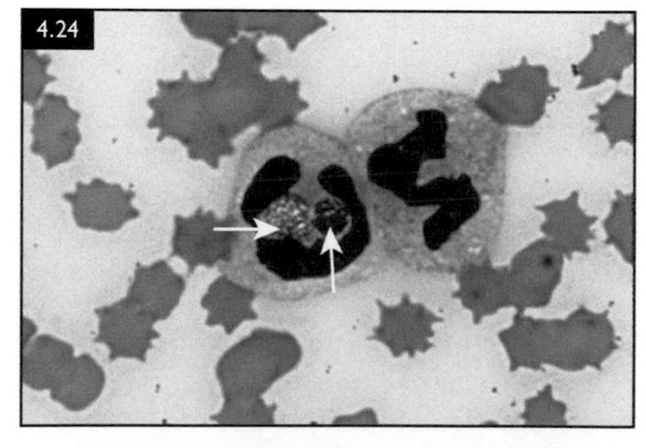


**FIG. 4.24** Blood smear from a horse with equine anaplasmosis. The arrows indicate the infectious agent, *Anaplasma phagocytophilum*, present in a neutrophil (Wright's stain).

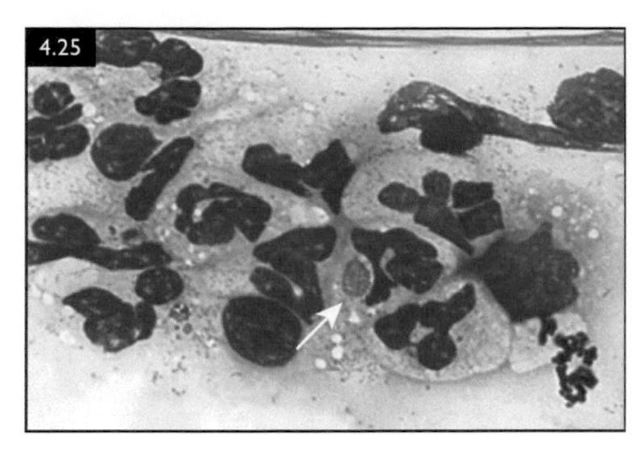

FIG. 4.25 Peritoneal fluid from a horse with equine anaplasmosis. The arrow indicates a morula of *Anaplasma phagocytophilum* in a neutrophil (Wright's stain).

- Doxycycline: 7 mg/kg i/v q24 h for 3–7 days.
- supportive care is important and includes stall rest:
  - NSAIDs are useful in pyrexic cases.
  - fluid therapy with balanced electrolytes in dehydrated/anorexic horses.
  - non-ataxic horses should be walked to control limb oedema.
  - cold hosing of limbs and limb bandages can be helpful.
  - tick repellents in infested areas.

### Prognosis

- good as most animals recover within a few days after treatment.
- self-limiting infection in untreated cases with resolution of signs in 2–3 weeks.
- severe ataxia worsens the prognosis due to the risk of trauma.

## Purpura haemorrhagica

### Definition/overview

- serious sequela to recent respiratory infection:
  - most frequently following *Streptococcus equi* infection (strangles).

### Aetiology/pathophysiology

- usually develops 2–4 weeks after infection with:
  - *S. equi*, *S. nonepidemic*, EHV-1, equine influenza virus.

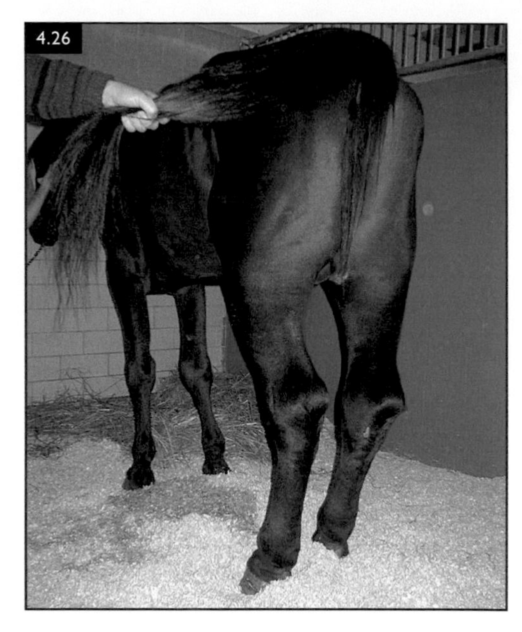

FIG. 4.26 Severe limb oedema in a horse with purpura haemorrhagica.

- hypersensitivity to infectious antigens is suspected as cause.
- resulting vasculitis.

### Clinical presentation

- variable clinical signs including:
  - head, ventral body and limb oedema (Fig. 4.26).
  - petechiation, ecchymosis.
  - fever and anorexia.
  - rapid weight loss.
  - pharyngeal oedema and inflammation, leading to respiratory noise and dysphagia.
- less commonly:
  - GI tract mucosal inflammation with pain and ileus.
  - renal failure.

### Differential diagnosis

- other causes of petechial haemorrhage and vasculitis:
  - EVA
  - EIA
  - anaplasmosis.

### Diagnosis

- suspected with history of recent *S. equi* infection/vaccination, or other upper respiratory tract infection.
- clinical signs.

- confirmed on histopathology via immunofluorescence of:
  - globulin deposition along vessel walls in skin or mucous membrane biopsy.
- CBC changes may include:
  - neutrophilia and hyperfibrinogenaemia, but rarely thrombocytopenia.
- serological titres for the inciting cause.

## Management

- early and aggressive therapy is required.
- corticosteroids:
  - dexamethasone 0.05–0.2 mg/kg i/m or i/v q24 h especially early on.
  - prednisolone 0.5–1.0 mg/kg p/o q12–24 h.
  - taper off gradually over a 14–21-day period.
- antimicrobials help prevent bacteraemia/sepsis and eliminate antigenic stimulus:
  - sodium penicillin, 22,000 IU/kg i/v q6 h.
  - procaine penicillin, 20,000 IU/kg i/m q12 h.
  - ceftiofur sodium, 2.2 mg/kg i/v or i/m q12 h.
- supportive care:
  - intravenous fluid therapy may be necessary in severe cases.
  - soft feed in dysphagic animals.
  - controlling limb oedema:
    - frequent walking.
    - cold water hosing.
    - application of support wraps.
  - analgesics and NSAIDs help decrease pain and inflammation.

## Prognosis

- depends on aggressive early therapy and absence of secondary organ damage and sepsis.
- mortality rate of 30% reported.
- poor with skin sloughing and poor initial response to treatment.
- recovery period can be prolonged.

# Lead toxicity

- See neurology section, page 44.

# Polycythaemia

## Definition/overview

- increase in RBC count, haemoglobin concentration or haematocrit.

## Aetiology/pathophysiology (Fig. 4.27)

- dehydration results in a lowered plasma volume:
  - relative increase in RBC and protein concentrations.
- relative polycythaemia develops with excitement:
  - epinephrine-induced splenic contraction results in transient increase in RBC concentration only.
- absolute polycythaemia with:
  - neoplastic proliferation of mature RBCs:
    - erythropoietin independent (polycythaemia vera).

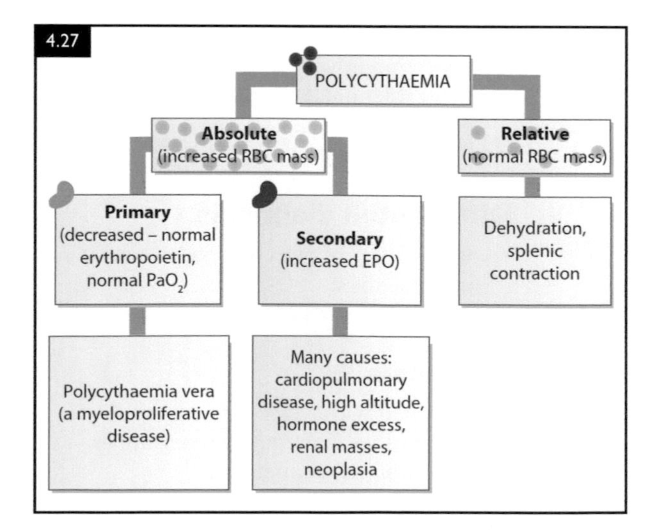


**FIG. 4.27** Causes of polycythaemia in the horse. Polycythaemia is classified as either relative (commonly a result of dehydration) or absolute (either primary, a myeloproliferative disease, or secondary, because of increased erythropoietin concentrations).

- secondary to conditions causing hypoxaemia:
  - cardiopulmonary disease.
  - high altitude.
  - erythropoietin-producing renal tumours.
- hyperviscosity, poor perfusion, decreased oxygenation and haemostatic disturbances occur with a PCV of > 60%.

### Clinical presentation

- dark-red mucous membranes.
- severe haematocrit increases can be associated with neurological and ocular signs.

### Diagnosis

- clinical signs and history.
- laboratory evidence of:
  - increased RBC concentration, haemoglobin and/or haematocrit.
  - serial measurements during rehydration therapy can identify relative polycythaemia.
- determine hydration status.
- measurement of erythropoietin concentration and arterial partial pressure of oxygen ($PaO_2$) can help define absolute polycythaemia causes:
  - polycythaemia vera: normal to low erythropoietin and normal $PaO_2$.
  - secondary absolute polycythaemia: increased erythropoietin and low $PaO_2$.

### Management

- treatment of dehydration with appropriate fluid therapy.
- polycythaemia vera more difficult to treat:
  - periodic blood removal (phlebotomy) with concurrent administration of a balanced electrolyte solution.
  - 10 ml/kg of blood every 2–3 days until PCV ≤ 50%
- secondary absolute polycythaemia caused by hypoxia:
  - compensatory mechanism – address primary cause.
  - phlebotomy only if PCV > 60% and clinical signs.

### Prognosis

- good for relative polycythaemia if fluid therapy is administered.
- guarded for secondary causes.

## Myeloproliferative disease (myeloid leukaemia)

### Definition/overview

- tumours of developing non-lymphocytic haematopoietic cells including:
  - derived from erythrocytes, granulocytic cells, monocytic cells and platelets.
  - collectively referred to as leukaemias and rare in horses.

### Aetiology/pathophysiology

- neoplastic transformation of a specific cell line.
- usually unknown cause.
- increased autonomous production of these cells in bone marrow:
  - infiltrates other tissues (spleen, liver and lymph nodes).
  - released into blood.
- replacement of normal haematopoietic cells resulting in cytopoenias in other cell lines due to lack of production.

### Clinical presentation

- vague signs of disease:
- poor performance, lethargy and weakness:
  - inappetence and weight loss.
- more specific signs include:
  - haematological involvement:
    - pale mucous membranes and petechial haemorrhages.
  - neutropenia may lead to secondary infections and intermittent fever.

### Differential diagnosis

- circulating atypical cells:
  - myeloproliferative and lymphoproliferative disorders.

### Diagnosis

- observation of atypical cells or inappropriate accumulation of mature cells on blood smears or in bone marrow (Fig. 4.28).
- peripheral blood cytopenia, with accompanying dysplastic features in developing haematopoietic cells.

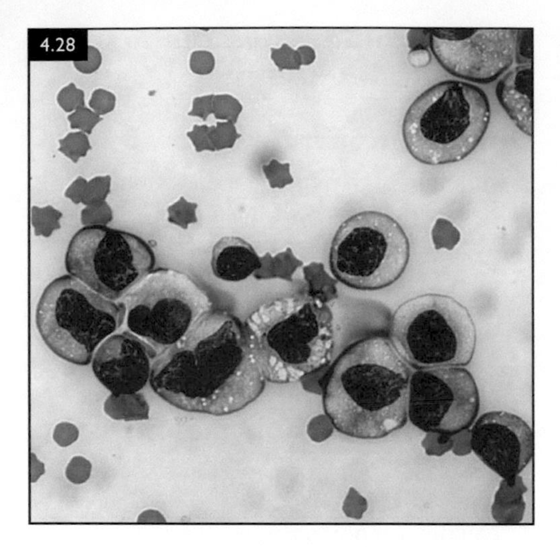


**FIG. 4.28** Blood smear from a horse with acute myeloid leukaemia. A monotypic population of large round cells with round to indented nuclei and multiple prominent nucleoli is pictured (Wright's stain). (Photo courtesy RM Jacobs.)

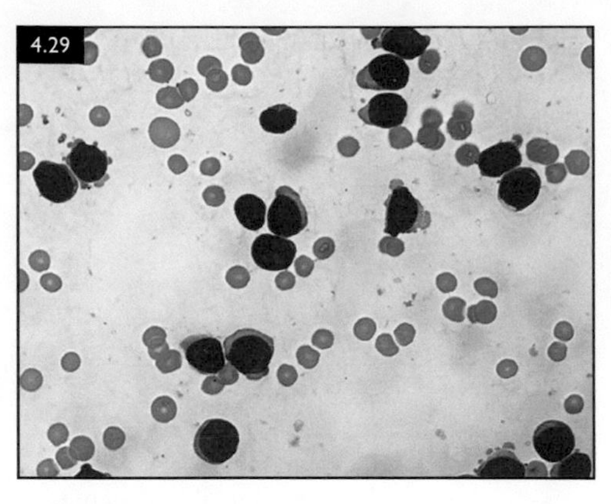


**FIG. 4.29** Bone marrow aspirate from a horse with lymphocytic leukaemia. There is a monotypic population of neoplastic lymphocytes present. Normal haematopoietic tissue is absent due to complete effacement by neoplastic cells (Wright's stain).

### Management/prognosis

- no treatment described.
- poor to grave prognosis.

## Lymphoproliferative disease (lymphocytic leukaemia)

### Definition/overview

- neoplastic proliferation of immature (acute lymphocytic leukaemia [ALL]) or mature (chronic lymphocytic leukaemia [CLL]) lymphocytes.
- rare tumours.

### Aetiology/pathophysiology

- unknown.

### Clinical presentation

- clinical signs often vague and non-specific:
  - poor appetite and weight loss.
  - weakness and lethargy.
  - oedema
  - colic.
  - petechial haemorrhage.
  - fever and lymphadenopathy.
  - **all more common in young horses.**
- infiltration of other organs frequently occurs:
  - lymph nodes, liver, spleen, lungs, kidneys and GI tract.

### Differential diagnosis

- haematopoietic tumours especially lymphoma.

### Diagnosis

- atypical cells on blood smears and bone marrow with ALL (Fig. 4.29).
- increased numbers of small lymphocytes with CLL.
- variable cytopenia of other cell lines:
  - erythrocytes, neutrophils and platelets, especially with ALL.
- hypercellularity and increased proportion of blast cells (> 20%) on bone marrow examination for ALL.

### Management

- no treatment is effective.
- euthanasia is common after diagnosis.

### Prognosis

- poor to grave.

## LYMPH NODE DISEASES

## Lymphoma

### Definition/overview

- infiltration of solid tissues with neoplastic lymphocytes.
- several forms of lymphoma occur:
  - generalised or multicentric, cutaneous, mediastinal and alimentary.
- **not uncommon in young horses and even foals.**

### Aetiology/pathophysiology

- unregulated proliferation of a cell line at a certain stage of development:
  - increased proportion of these cells in haemolymphatic tissues.
- replacement of normal cells creating distortion of normal tissue architecture.
- rarely, lymphoma becomes leukaemic (neoplastic cells in circulating blood).
- most affected tissues include lymph nodes, spleen and liver.

### Clinical presentation

- most common are weight loss, lethargy and enlarged lymph nodes (Fig. 4.30).
- other signs depending on tumour location:
  - dyspnoea, colic, neurological and ocular signs.
- splenic enlargement or internal masses possible on rectal palpation.
- cutaneous form has a few to many firm nodules scattered over the body.
- intestinal form has been associated with the development of IMHA.

### Differential diagnosis

- carefully differentiated from lymphocytic hyperplasia.

### Diagnosis

- predominance of neoplastic lymphocytes altering the normal architecture of a solid tissue on fine-needle aspiration or excisional biopsy (Fig. 4.31).
- characterisation of cell type (i.e. T- or B-cell) by identification of surface markers.

### Management

- cutaneous lymphoma may respond to corticosteroid therapy:
  - may relapse after treatment is stopped.

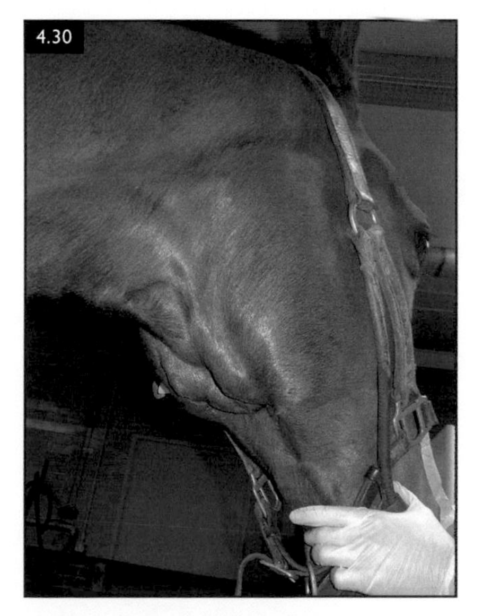

**FIG. 4.30** Enlarged submandibular lymph nodes in a horse with lymphoma.

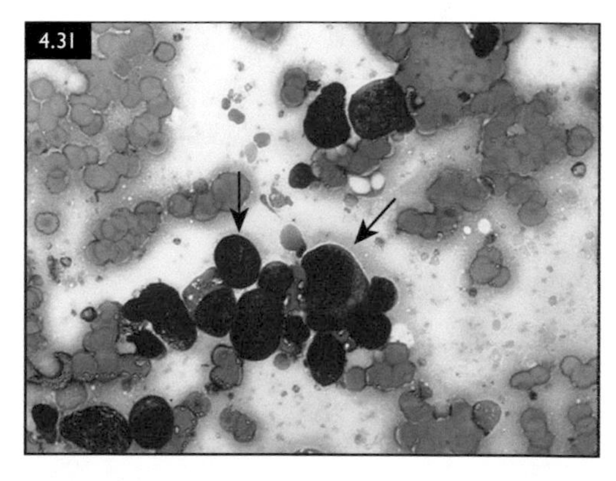

**FIG. 4.31** Lymph node FNA from a horse with lymphoma. Heterogeneous population of small, medium and large lymphocytes; however, the arrows indicate medium to large lymphocytes with atypical morphology (multiple prominent nucleoli) (Wright's stain).

- some cases can survive for varying periods without treatment.
- other forms of lymphoma do not respond well to treatment.
- specific treatments for lymphoma of different body systems are covered elsewhere.

### Prognosis

- poor to grave overall.
- chemotherapeutic options are very expensive and unrewarding.
- cutaneous lymphoma has a better prognosis.

## Lymphadenitis

### Definition/overview

- inflammation of lymph nodes or lymphoid tissue in other organs such as the spleen, intestinal tract and thymus.
- infiltration of neutrophils, eosinophils, macrophages, lymphocytes, plasma cells or a combination of these cell types, depending on the inciting stimulus.
- part of a primary systemic disease.

### Aetiology/pathophysiology

- initiated by various microbial organisms (bacteria, viruses, parasites, fungi), immune-mediated diseases and neoplasia.
- production of cytokines resulting in a chemotactic gradient that attracts inflammatory cells to the tissue.

### Clinical presentation

- enlarged palpable lymph nodes which may be warm and painful.
- fever.
- other clinical signs related to any primary pathology.

### Differential diagnosis.

- lymphocytic neoplasia and hyperplasia.

### Diagnosis

- inflammation on aspiration or excisional biopsies of affected tissues.

### Management

- treatment of the primary disease process:
  - antimicrobials.
  - anti-inflammatory drugs.

### Prognosis

- depends on the ability to eliminate the inciting inflammatory stimulus.

## Lymphocytic hyperplasia

### Definition/overview

- sometimes referred to as 'reactive' lymph node.
- any lymphocytic tissue increased in size due to the presence of increased numbers of plasma cells and/or immature lymphocytes.

### Aetiology/pathophysiology

- many causes including:
  - infectious, inflammatory, immune-mediated and neoplastic.
- development of a lymphocyte-mediated immune response:
  - T-cell- and/or B-cell-mediated.

### Clinical presentation

- palpable enlarged lymph nodes.
- signs of systemic disease such as fever, lethargy and anorexia.

### Differential diagnosis

- Lymphoma.
- Lymphadenitis.

### Diagnosis

- increased numbers of immature lymphocytes and/or plasma cells on cytology or histology samples.

### Management

- treatment of the underlying disease process.

### Prognosis

- depends on the ability to successfully treat the underlying disease process.

# HYPERPROTEINAEMIA

## Haemoconcentration

### Definition/overview

- loss of plasma volume due to loss of body fluids and dehydration.
- apparent or relative increase in measured protein indices.

### Aetiology/pathophysiology

- decrease in plasma volume and not an increase in the production of proteins.
- fluid loss with intestinal disease, renal disease, salivation and sweating.
- water restriction is an uncommon cause of haemoconcentration.

### Clinical presentation

- depends on degree of dehydration:
  - prolonged tenting of skin.
  - sunken eyeballs.
  - tacky mucous membranes.
  - weight loss.

### Differential diagnosis

- pathological increases in protein such as hyperglobulinaemia.

### Diagnosis

- clinical signs supportive of dehydration.
- increased biochemical or refractometric measurement of total protein concentration.

### Management

- restoration of body volume via fluid therapy:
  - oral fluid therapy in mild haemoconcentration.
  - i/v balanced electrolyte fluid therapy in moderate to severe cases.
- management of the underlying disease and stopping ongoing losses.

### Prognosis

- depends on successfully treating the underlying disease.

## Hyperglobulinaemia

### Definition/overview

- relative or absolute increase in the globulin fraction of total protein measurement.
- Globulins are a diverse group of proteins including:
  - acute-phase proteins, transport proteins and immunoglobulins.
  - divided into alpha, beta and gamma fractions based on electrophoresis.

### Aetiology/pathophysiology

- **relative** hyperglobulinaemia usually due to dehydration:
  - decrease in plasma volume.
- **absolute** hyperglobulinaemia:
  - increased production of globulin proteins.
  - one (monoclonal gammopathy):
    - lymphocytic tumours, especially plasma cell tumours (Fig. 4.32).
  - more than one (polyclonal gammopathy):
    - chronic inflammatory disease.

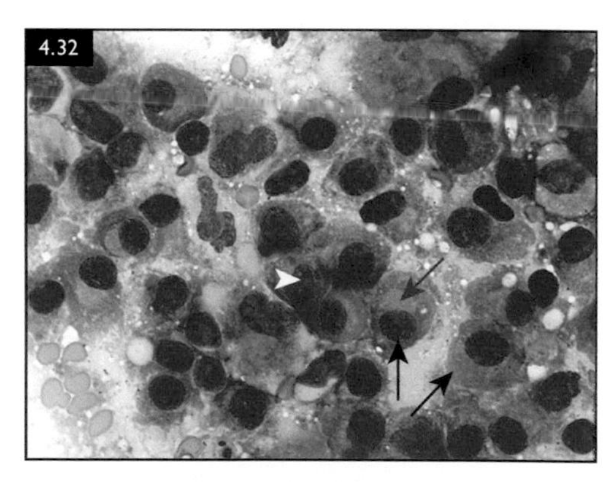


**FIG. 4.32** FNA from a cutaneous mass in a horse with multiple myeloma and a monoclonal gammopathy. The black arrows indicate the monotypic population of plasma cells. The red arrow indicates the clear Golgi zone present in several of these cells. The arrowhead indicates a single prominent nucleolus present in an atypical plasma cell (Wright's stain).

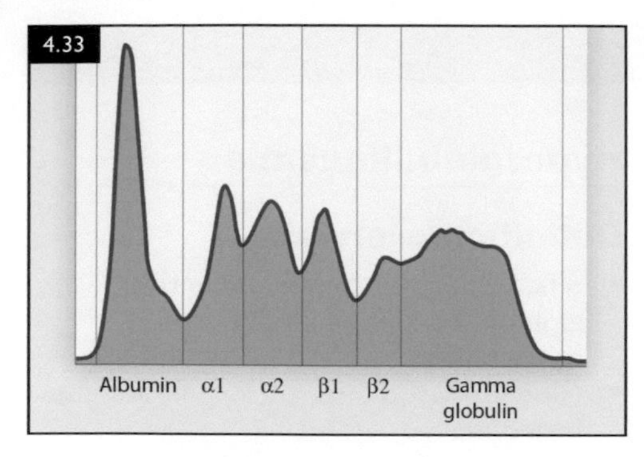


**FIG. 4.33** Serum protein electrophoresis. Polyclonal gammopathy (increase in alpha, beta and gamma globulins). Observed with chronic inflammation or infection.

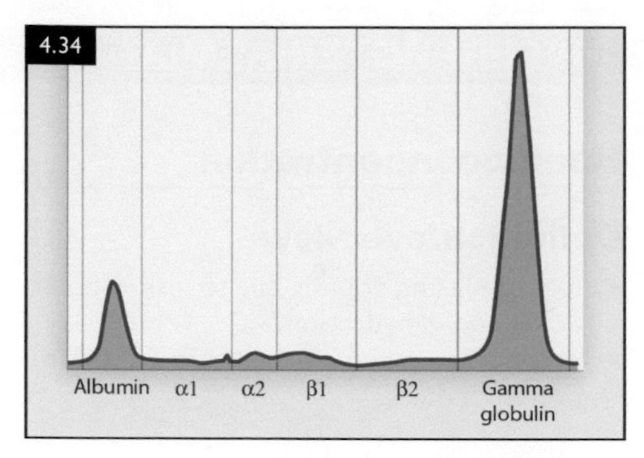


**FIG. 4.34** Serum protein electrophoresis. Monoclonal gammopathy from a horse with B-cell lymphoma. The sharp, narrow-based peak represents the clonal production of a single gamma globulin from the tumour cells.

## Clinical presentation

- vary with the cause of the hyperglobulinaemia.
- can cause sludging of blood flow to vital organs and clinical dysfunction if severe.

## Differential diagnosis

- other causes of hyperglobulinaemia.

## Diagnosis

- biochemical measurements of total protein and albumin concentration:
  - relative hyperglobulinaemia accompanied by:
    - hyperalbuminaemia and dehydration.
- protein electrophoresis to determine monoclonal or polyclonal (Figs. 4.33, 4.34).

## Management

- address the underlying disorder.

## Prognosis

- good for relative cases if the fluid can be replaced.
- depends on the cause if absolute – causes such as neoplasia are difficult to treat.

# SPLENIC DISEASES

## Introduction

- spleen is an important organ:
  - filters the blood.
  - acts as a secondary lymphoid organ.
  - serves as a reservoir of RBCs, WBCs and platelets.
  - location for extramedullary haematopoiesis.
- size of the spleen is highly variable in normal animals.
- spleen has a relatively homogeneous character on ultrasound (Fig. 4.35).
- splenic disease is uncommon.

## Splenic rupture

### Definition/overview

- uncommon condition.
- life-threatening with risk of severe intra-abdominal haemorrhage.

### Aetiology/pathophysiology

- subsequent to trauma to left abdominal wall (Fig. 4.36).
- rupture of a splenic haematoma, tumour or abscess.
- minor bleeding contained within the splenic capsule.

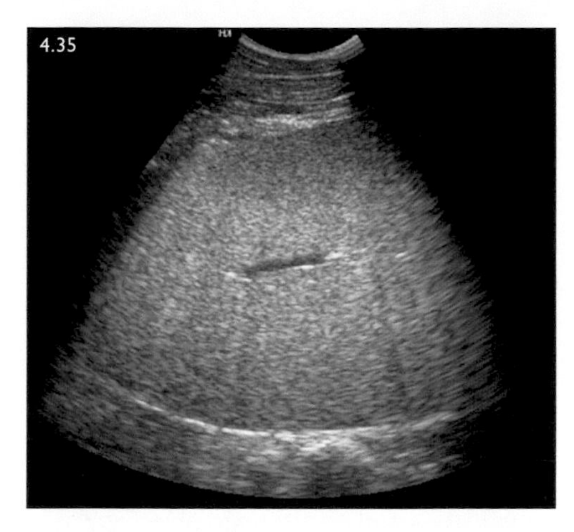

FIG. 4.35 Ultrasonographic appearance of a normal spleen.

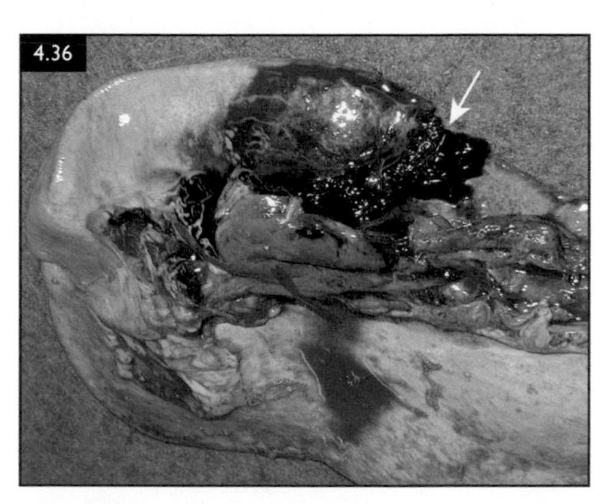

FIG. 4.36 Splenic laceration in a foal (arrow), caused by a kick.

- rupture and damage to large vessels leads to:
  - major haemorrhage into the abdominal cavity.
  - haemorrhagic shock and potentially death.

## Clinical presentation

- depends on the amount of haemorrhage:
  - mild bleedings often inapparent or mild colic.
  - severe bleeding results in hypovolaemic shock:
    - tachycardia
    - tachypnoea.
    - weakness.
    - pale mucous membranes and cold extremities.
    - abdominal distension.

## Differential diagnosis

- colic of GI tract origin.
- peritonitis
- sepsis
- poisonings.

## Diagnosis

- clinical signs can be non-specific in milder cases.
- abdominal distension.
- evidence of clinical signs of hypovolaemic/haemorrhagic shock.
- ultrasonographic examination of the abdomen for free blood and splenic damage.
- abdominocentesis under ultrasonography guidance confirms the presence of blood.
- haematology is not useful in confirming acute haemorrhage.

## Management

- depends on the cause and severity of haemorrhage.
- fluid therapy is essential if shock is present:
  - hypertonic (5–7%) saline may be useful initially.
  - followed by isotonic balanced electrolyte fluids.
- blood transfusion is particularly useful in severe shock or if PCV < 0.2 l/l is present.
- surgery if severe, uncontrolled bleeding:
  - **emergency preoperative stabilisation is vital.**
- clot stabilisation:
  - aminocaproic acid is used with little evidence supporting efficacy.

## Prognosis

- good if trauma-associated splenic rupture and bleeding stops.
- poor if splenic neoplasia.

# Splenomegaly

## Definition/overview

- abnormally large spleen.
- often unclear clinical significance.

## Aetiology/pathophysiology

- obstruction of venous return:
  - nephrosplenic entrapment of the large colon.
  - right heart failure.

- acute splenitis.
- haemolytic anaemia.
- purpura haemorrhagica.
- infiltrative disease (i.e. neoplasia).
- grass sickness.
- splenic infarction.

### Clinical presentation

- highly variable depending on the cause:
  - no clinical signs.
  - colic
  - anorexia
  - icterus.
  - depression
  - weight loss
  - pyrexia.
  - ventral oedema
  - tachycardia.

### Diagnosis

- depending on presentation, evaluate for:
  - localised or systemic infection, neoplasia, anaemia and colic.
- haematology results vary with cause.
- ultrasonography of the abdomen and, in particular, the spleen.
- rectal palpation to assess size, location and texture.

### Management

- address the inciting cause.
- successful treatment of primary splenomegaly with splenectomy is reported.

### Prognosis

- good with nephrosplenic entrapment of the large colon.
- guarded with other causes.
- poor with splenic neoplasia.

## Splenic neoplasia

### Definition/overview

- primary or secondary.
- uncommon.

### Aetiology/pathophysiology

- lymphoma (most common) especially in 2–8 year olds.
- melanoma
- haemangiosarcoma.

### Clinical presentation

- weight loss, intermittent colic, depression and anorexia initially.
- clinical signs of a primary tumour if a secondary splenic tumour.

### Differential diagnosis

- splenic abnormality on palpation or ultrasound:
  - splenic abscess.
  - splenic haematoma.

### Diagnosis

- hypoproteinaemia is common but non-specific.
- AID or IMHA anaemia may be present.
- irregular or enlarged spleen on rectal palpation and ultrasonography (Fig. 4.37).
- abdominocentesis often unhelpful.
- FNA or biopsy under ultrasound or laparoscopic guidance:
  - risk of haemorrhage with biopsy.
- thorough clinical workup looking for primary or secondary neoplasia.

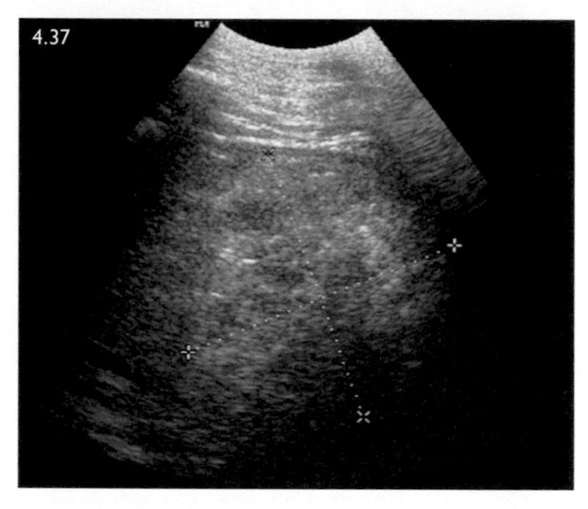


**FIG. 4.37** Transabdominal ultrasound image of a splenic lymphoma. Note the large splenic mass delineated by the calliper marks.

### Management

- limited options as disease often advanced on identification.
- splenectomy only helpful if no secondary lesions present.
- chemotherapeutic agents not adequately evaluated and cost prohibitive.

### Prognosis

- grave.

# Index

## A



## B



## C

## M



## N



## O



## P

## Q



## R



## S



## T



## U



## V

## W